SPRINGER TRACTS
IN MODERN PHYSICS

Ergebnisse
der exakten Natur-
wissenschaften

Volume 55

Springer-Verlag Berlin Heidelberg GmbH 1970

Manuscripts for publication should be addressed to:

G. Höhler, Institut für Theoretische Kernphysik der Universität, 75 Karlsruhe, Kaiserstraße 12

Proofs and all correspondence concerning papers in the process of publication should be addressed to:

E. A. Niekisch, Kernforschungsanlage Jülich, Institut für Technische Physik, 517 Jülich, Postfach 365

ISBN 978-3-662-15600-1   ISBN 978-3-540-36397-2 (eBook)
DOI 10.1007/978-3-540-36397-2

© by Springer-Verlag Berlin Heidelberg 1970
Originally published by Springer-Verlag Berlin Heidelberg New York in 1970
Softcover reprint of the hardcover 1st edition 1970

Library of Congress Catalog Card Number 25-9130.

Low Energy Hadron Interactions

Invited Papers presented at the
Ruhestein-Meeting, May 1970

Compilation of Coupling Constants
and Low Energy Parameters

Contents

Low Energy
Pion-Pion Scattering
D. MORGAN and J. PIŠÚT   1

Analytic Extrapolations and
the Determination of
Pion-Pion Phase Shifts
J. PIŠÚT   43

Coulomb Corrections in the
Analysis of πN Experimental
Scattering Data
G. C. OADES   61

Kaon-Nucleon Interactions
below 1 GeV/c
B. R. MARTIN   73

The $\Lambda K N$ Coupling and
Extrapolation below the
$\bar{K} N$ Threshold
A. D. MARTIN   141

Nucleon-Nucleon Interactions
below 1 GeV/c
G. KRAMER   152

Coupling Constants from PCAC
H. PILKUHN   168

Regge Residues
C. MICHAEL   173

Vector Mesons in
Electromagnetic Interactions
M. GOURDIN   191

Coupling Parameters of
Pseudoscalar Meson
Photoproduction on Nucleons
W. PFEIL and D. SCHWELA   213

Compilation of Coupling Constants and
Low Energy Parameters
G. EBEL, D. JULIUS, G. KRAMER,
B. R. MARTIN, A. MÜLLENSIEFEN, G. OADES,
H. PILKUHN, J. PIŠÚT, M. ROOS,
G. SCHIERHOLZ, W. SCHMIDT, F. STEINER,
J. J. DE SWART   239

Low Energy Pion-Pion Scattering

D. MORGAN and J. PIŠÚT

Contents

§ 1. Introduction 1
§ 2. Note on Predictions for $\pi\pi$ Scattering from Current Algebra and the Veneziano
    Model 2
§ 3. Information from Experiments on $\pi N \to \pi\pi N$ 5
    3.1 Methods of Extracting $\pi\pi$ Data from the Study of Reactions $\pi N \to \pi\pi N$ at a
        few GeV 5
    3.2 General Empirical Framework 8
§ 4. Forward Dispersion Relation (FDR) 14
§ 5. Pion-Nucleon Dispersion Relations 29
§ 6. Information from $K e_4$ Decays and $K_L^0 \to 2\pi$ Decays 35
§ 7. Comments and Conclusions 37
References 40

§ 1. Introduction

The extraction of $\pi\pi$ elastic phase shifts from available experimental information differs from the other low energy processes to be discussed at this meeting in that there is no direct data. As a substitute, one has to rely on more or less indirect evidence in which one exploits the following special features:

(i) *The π is the lightest hadron* therefore dipion systems often feature among reaction products.

(ii) *One Pion Exchange (OPE) is pervasive* (perhaps anomalously so, e. g. in charged pion photo-production) — leading to the possibility of extrapolation to the pion-pole (Chew Low). This is where the majority of the existing data comes from — i. e. from the analysis of peripheral dipon production in $\pi N \to \pi\pi N$ and also in $\pi N \to \pi\pi\Delta$, $\pi d \to N N \pi\pi$.

(iii) *The structure of $\pi\pi$ elastic scattering is especially simple*. In fact up to 1 GeV, $\pi\pi$ scattering appears to be describable by the five phase shifts $\delta_0^0, \delta_0^2, \delta_1^1, \delta_2^0, \delta_2^2$ (notation δ_J^I). This hinges on the facts that one has no spin, Bose Statistics, short-range forces so that few partial waves are excited, $\pi\pi \not\to 3\pi$ because of G parity and $\pi\pi \to 4\pi$ does not set in until high energies. This simplicity goes some way to compensate for the indirectness of the evidence.

(iv) $\pi\pi$ *scattering is a crossing-symmetric reaction* in a relevant way. The separation of the physical regions on the Mandelstam plot is small compared to characteristic energy ranges of the dynamics. The simplicity of the $\pi\pi$ system of course also makes the results of the utmost theoretical importance and there are well-known predictions from chiral and dual models as will be discussed.

One can classify reactions yielding information on the $\pi\pi$ system into those in which just two pions appear, e. g. Ke_4 decay, $e^+e^-\to\pi^+\pi^-$, $K^0_{S,L}\to 2\pi$, $\pi N\to\pi N$ (extrapolated to the t-channel) and $\pi N\to\pi\pi N$ with the OPE term successfully isolated; and those in which three or more pions appear or two pions appear in company with another hadron, e. g. η, K^0_L, $K^\pm$, $N\bar{N}\to 3\pi$ and $\pi N\to\pi\pi N$ with the OPE term not isolated. Analyses of the latter reactions are strongly model dependent.

We shall discuss briefly in the following section, a summary of the predictions for $\pi\pi$ scattering from Current Algebra and the Veneziano model, (Section II), experimental information from $\pi N\to\pi\pi N$ and related processes (Section III), the application of forward dispersion relations (Section IV), information from the analysis of πN elastic scattering (Section V), and, in bare outline, the current status of information from Ke_4 and $K^0_L\to 2\pi$ decays (Section VI). In the conclusion, we attempt a summary of the present understanding of low energy $\pi\pi$ phenomena and look to future possibilities.

Important topics which have been omitted are the rigorous constraints from analyticity, positivity and crossing due to *Martin, Wanders, Roskies* and co-workers and also the whole area of partial wave dispersion relations.

Since, as was stressed at the beginning, $\pi\pi$ evidence is indirect, it follows that all inferences from data to elastic $\pi\pi$ scattering proceed via some form of extrapolation either explicitly or implicitly. The implicit methods habitually understate the errors or fail to state the assumptions by which errors are reduced. The general question of extrapolation methods in relation to problems such as $\pi\pi$ scattering is discussed by one of us in a companion chapter in the present volume*. The hazards of extrapolation which are exposed there should be kept in mind in the following sections.

§ 2. Note on Predictions for $\pi\pi$ Scattering from Current Algebra and the Veneziano Model

We attempt here no deep commentary on these widely ramifying subjects but merely set down a few formulas and the the actual numerical predictions to recall the machinery to the reader's mind.

---

* *J. Pišút*, p. 43.

(i) *Weinberg's Current Algebra Predictions (Chirality)*

The steps are as follows (*Weinberg*, 1966).

a) Expand the $\pi\pi$ amplitude for general off-shell 4 momenta up to quadratic terms to give

$$A^0 = 3A + B + C \text{ etc.} \tag{2.1}$$

with

$$A = \alpha + \beta(u+t) + \gamma s \quad \text{(no other dependence).} \tag{2.2}$$

Here the A^Is are the I spin eigenstates and A, B, and C are the functions which permute under crossing (notation of *Chew* and *Mandelstam*, 1960). The above expansion alone implies

$$2a_0 - 5a_2 = 18 m_\pi^2 a_1 . \tag{2.3}$$

b) Adler Zero: The PCAC requirement (*Adler*, 1965) that the amplitudes vanish for one of the pion 4-momenta going to zero implies

$$\alpha + m_\pi^2 (2\beta + \gamma) = 0 \tag{2.4}$$

from the requirement that the quantity A above should vanish for $s = t = u = m_\pi^2$

c) The requirement that the $I = 2$ part of the σ-commutator should vanish implies that the coefficient β should vanish. One now has scattering lengths uniquely determined except for an overall multiplicative factor

$$a_0 = 7\gamma m_\pi^2, \quad a_2 = -2\gamma m_\pi^2, \quad a_1 = 4/3\gamma . \tag{2.5}$$

In particular,

$$a_0/a_2 = -7/2 .$$

d) Finally the scale is fixed from the expression for the commutator of the axial charge and axial current

$$\{A_a^0(y), A_b^\mu(x)\}_{x_0 = y_0} = 2 i g \varepsilon_{abc} V_c^\mu \delta^3 (x - y) . \tag{2.7}$$

On inserting numbers, the following alternative predictions result:

$2a_0 - 5a_2$	a_0	a_2	a_1	PCAC constant from	
$0.54 \, m_\pi^{-1}$	0.157	-0.045	0.030	π decay rate.	(2.8)
$0.69 \, m_\pi^{-1}$	0.201	-0.058	0.038	Goldberger-Treiman Relation	

(ii) *Lovelace-Veneziano Model (Duality)*

Lovelace in his Veneziano model for the $\pi\pi$ system (*Lovelace*, 1968) begins by assuming an expression for the exotic $I_s = 2$ channel (the

1*

$I_s = 0$ and 1 amplitudes follow from crossing)

$$A^2(s, t) = -\beta \frac{\Gamma(1-\alpha(t))\,\Gamma(1-\alpha(u))}{\Gamma(1-\alpha(t)-\alpha(u))}. \tag{2.9}$$

Here the function $\alpha(u)$ has the usual linear form $\alpha(x) = \alpha_0 + \alpha' x$. *Lovelace* points out how the Adler zero referred to above is readily achieved with reasonable trajectory intercepts and no additional terms.

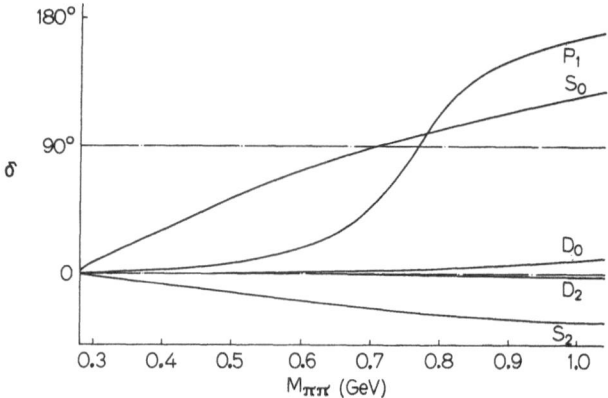

Fig. 1. Phases predicted by *Lovelace*'s single channel unitarization of his $\pi\pi$ Veneziano model (*Lovelace, 1969, Wagner, 1969*)

Normalising *Lovelace*'s Ansatz to the Weinberg form to which it is essentially equivalent within the Mandelstam triangle, *Osborn* (1969) showed that the Lovelace formula implied

$$f_{\varrho\pi\pi}^2 \Big/ \left(\frac{m_\varrho^2}{F_{\pi^2}}\right) = 0.331 \tag{2.10}$$

to be compared with the KSFR★ prediction of $1/2$ and that from the linear approximation of $1/\pi$. Inserting the pion decay rate value for the PCAC constant he obtained $\Gamma_\varrho = 91$ MeV while for the Goldberger-Treiman value he obtained

$$\Gamma_\varrho = 115 \text{ MeV}, \quad \Gamma_s = 520 \text{ MeV}, \quad \Gamma_{f_0} = 90 \text{ MeV} \tag{2.11}$$

The Veneziano formula above is manifestly non-unitary. *Lovelace* (1969) has proposed a unitarisation recipe wherby the partial wave amplitude from the Veneziano formula is interpreted as the main contribution to the K matrix (the factor $\varrho(s)$ in the unitarity relation in

★ *Kawarabayashi, K., Suzuki, M.:* Phys. Rev. Letters **16**, 255 (1966); *Riazuddin* and *Fayazuddin*, Phys. Rev. **147**, 1071 (1966).

addition to possessing right hand cuts from unitarity has also left hand cuts; these are arbitrary, although not unreasonable for the purpose at hand). This has been done both for a one-channel form, for $\pi\pi$ scattering considered in isolation, and also in a 2-channel framework with the $K\bar{K}$ channel added. The prediction for the $\pi\pi$ phase shifts in the one-channel unitarised model is shown in Fig. 1.

The 2-channel model yields scattering lengths.

$$a_0 = 0.29 , \qquad a_2 = -0.063 . \tag{2.12}$$

This model predicts a very asymmetric ε with the δ_0^0 phase rising rapidly above the resonance to give a zero and second resonance in the δ_0^0 phase. A phase shift δ_0^0 of this general form is shown in Fig. 12 (the dashed line).

§ 3. Information from Experiments on $\pi N \rightarrow \pi\pi N$

3.1. Methods of Extracting $\pi\pi$ Data from the Study of Reactions $\pi N \rightarrow \pi\pi N$ at a few GeV

This is an extensive topic and the detailed discussion lies outside the scope of these lectures (see *Argonne, 1969; Gutay et al., 1969*). Here we merely present a sketch of the procedures and problems before going on to examine the results.

Extraction of two body scattering information from data on three hadron final states is in general very complicated and ambiguous. However, for the reactions $\pi N \rightarrow \pi\pi N$ in the GeV region a simplication arises from their peripheral nature. In particular, all practical methods of inferring the $\pi\pi$ scattering hinge on the fact that there is a strong contribution from one pion exchange (OPE). Differences lie in how the background to this is handled.

We may indicate how the principal methods of analysis work by means of a picture (Fig. 2). With this in mind, one can bring out immediately certain general problems:

(i) The OPE contribution vanishes at momenta transfer $t = 0$. The question arises, can one assume the same for the background? If so, in implementing the CLG extrapolation method, one can extrapolate in $\dfrac{(t - \mu^2)^2}{t} \dfrac{d\sigma}{dt}$ ("pseudo peripheral" assumption or "evasive extrapolation"). This improves the "leverage" of the data. On the other hand, there may be contributions to the background which do not vanish at $t = 0$. The question then arises how should they be treated.

(ii) The factorisation model assumes the production amplitudes to be of the form $A_{\pi\pi}^l M_{l\mu}^{\lambda\lambda}$ where $A_{\pi\pi}^l = \sin\delta \, e^{i\delta}$ and the $M_{l\mu}^{\lambda\lambda}$ are *real* functions of the total energy s, and the momentum transfer t. In practice, they

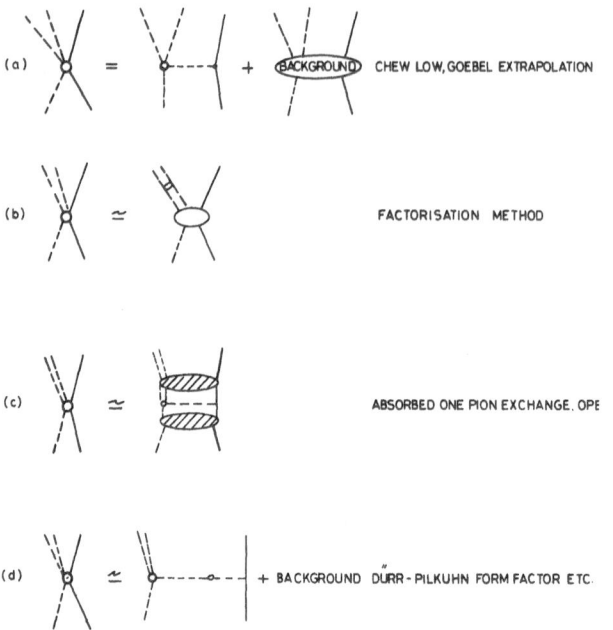

Fig. 2. Schematic of alternative approaches to extracting $\pi\pi$ phase-shifts from peripheral data

are assumed independent of $m_{\pi\pi}$. It is crucial that the only relative phase entering should be the $\pi\pi$ final state phase δ. However, it follows from the Regge pole relation between phase and energy dependence that any appreciable contribution to the background from, for example, ω exchange will spoil the phase assumption. This consideration applies directly to methods (b) (c) and (d) of Fig. 2 and poses the threat of an out of phase background when applying method (a).

(iii) As will be discussed below, the principal $\pi\pi$ partial waves below 1 GeV are the $I = 1P$ wave, the $I = 2S$ wave and the $I = 0S$ wave (phase shifts $\delta_1^1, \delta_0^2, \delta_0^0$.) A fair consensus has been reached as to the parameters for the former two, although there is still discussion as to details. The form of δ_0^0 has proved hard to isolate. The reason for this is that one depends on studying the decay angular distribution of the dipion system for distinguishing the partial waves. The S wave produces an isotropic distribution, longitudinal rho's (the ones that can be produced by OPE) produce a $\cos^2\theta_{\pi\pi}$ distribution, whilst transverse rho's produce a $\sin^2\theta_{\pi\pi}$ distribution (we refer to polarisation in the Jackson frame). Thus any attempt to isolate the S wave by examination of the isotropic term tends to be contaminated with de-polarised rho production. The $\cos\theta_{\pi\pi}$ term or, equivalently, the forward-backward asymmetry (FBA), is not subject to this difficulty although, as will be noted below, there are other hazards in its use. In any case, its determination leads to a two-way ambiguity as to the value of δ_0^0.

Before going on to discuss in detail what has been done and the results which have been achieved, let us amplify the point (i) above concerning the nature of the background to OPE at small t. There have been several recent discussions – *Kane* and *Ross* (1969), *Frogatt* and *Morgan* (1969), *Cho* and *Sakurai* (1969), *Williams* (1970), *Scharenguivel et al.* (1970b), *Kane* (1970). The problem has bearing both on the relation of rho production to charged pion photo-production via vector-dominance and also on the extraction of $\pi\pi$ information. Only the latter concerns us here. The possibilities are most succinctly stated in terms of t channel helicity amplitudes* because elementary OPE contributes to just one of them $-f_{++}^{(0)}$ (notation $f_{\lambda\lambda'}^{(\nu)}$ with λ, λ' and ν denoting the t-channel helicity state of respectively the nucleon, antinucleon and the dipion system; $\lambda = \pm$ means $\lambda = \pm\frac{1}{2}, |\nu = \pm\rangle = |\lambda_\varrho = +\rangle \pm |\lambda_\varrho = -\rangle, |\nu = 0\rangle = |\lambda_\varrho = 0\rangle$). In order to have a definite picture of the possibilities, let us, following *Froggatt* and *Morgan* (1969), set down expressions for the t channel helicity amplitudes at small t. Firstly one has the two amplitudes for production of dipion helicity zero (either S wave dipion production or longitudinal rho production

$$f_{++}^{(0)} = A_{\pi\pi}\left(\frac{\sqrt{-t}}{t-\mu^2} + \frac{\Gamma_0}{\sqrt{-t}} + O(-t)^{\frac{1}{2}}\right), \tag{3.1}$$

$$f_{+-}^{(0)} = A_{\pi\pi}\left(\frac{\Gamma_0}{\sqrt{-t}}(t/t_{\min}-1)^{\frac{1}{2}} + O(-t)^{\frac{1}{2}}\right).$$

One further has amplitudes for transverse rho production

$$f_{++}^{(-)} = A_{\pi\pi}\left(\frac{2\Gamma_1}{\sqrt{-t}}(t/t_{\min}-1)^{\frac{1}{2}} + O(-t)^{\frac{1}{2}}\right), \tag{3.3}$$

$$f_{+-}^{(-)} = A_{\pi\pi}\left(-\frac{2\Gamma_1}{\sqrt{-t}} + O(-t)^{\frac{1}{2}}\right) \tag{3.4}$$

with two further amplitudes for the $f_{\lambda\lambda'}^{(+)}$ amplitudes. In the above, $A_{\pi\pi} = \sin\delta\, e^{i\delta}$ for on shell $\pi\pi$ scattering and the normalisation has been chosen to make the formulae look simple. The cross-section for dipion production integrated over decay angles is given by the sum of squares of the six helicity amplitudes; the four shown above and the two others not written. The OPE contribution is the term $\sqrt{-t}/t-\mu^2$ in $f_{++}^{(0)}$, the terms in Γ_0 and Γ_1 are possible singular contributions to

---

* Aside from this, much of the discussion especially on Vector Dominance and Absorption is best stated in terms of s-channel helicity states.

the background. If they are absent, then the extrapolated intensity
vanishes at $t = 0$, as it would if only the OPE term were present. In this
case, it is legitimate to extrapolate in $\dfrac{(t-\mu^2)^2}{(-t)}\dfrac{d\sigma}{dt}$ to determine the
residue at the pion pole. *A priori*, one has no reason to expect Γ_0 and
Γ_1 to be zero. Models in fact suggest the contrary. The indications are
particularly strong for the Γ_1 coefficient, where models (e. g. *Cho* and
Sakurai, 1969; Williams, 1970) and fits to the density matrices (*Scharenguivel et al., 1970*) give a rather unaminous opinion $\Gamma_1/\sqrt{-t_{\min}} \sim 0.14$
(*Froggatt* and *Morgan, 1970*).

What does all this say for extrapolation? Clearly it is better not to
assume any special behaviour at $t = 0$, for example to extrapolate in
$(t-\mu^2)^2 \dfrac{d\sigma}{dt}$ without assuming a zero at $t = 0$. The likelihood of there
being non-zero $\Gamma's$ makes it dangerous to extrapolate ratios of coefficients
in the decay distribution, for example the forward backward asymmetry
(FBA), since one has the prospect of a numerator and denominator
each slightly modified from zero. (*Williams, 1970; Kane, 1970*). The
acceptance as reliable of on-shell $\pi\pi$ FBA inferred from extrapolation
(*Scharenguivel et al., 1969a*) involves an assumption that such effects
do not introduce appreciable systematic errors. The whole situation
should become clarified when high statistics data extending to small
t becomes available. Hypotheses on the background can then be confronted with all the intensity and density matrix information.

3.2 General Empirical Framework

The data for $M_{\pi\pi} < 1$ GeV is commonly analysed under the following
assumptions:

(i) The inelasticity is very small below 1 GeV.

(ii) All waves above D waves are negligible.

(iii) Therefore our problem concerns the phases $\delta_0^0, \delta_0^2, \delta_1^1, \delta_2^0, \delta_1^2$
(notation δ_J^I).

(iv) The following resonances exist in the $\pi\pi$ system below 1700 MeV
firstly, $\varrho(765)$, $f(1260)$, $f'(1514)$ and $\varrho_N(1660)$ or g are all well established;
probably also $\eta_{0^+}(1060)$ or S^* seen as an enhancement in $S^* \to K_S^0 K_S^0$
and possibly $\eta_V(1080)$ $(\to \pi^+\pi^-, I = 0, J > 2?)$; (for details and references
on the above see *Particle Data Group* 1970); finally, it is now accepted
that there exists an $I = J = 0$ resonance below 1 GeV, variously named
$\sigma, \eta_{0^+}(700)$, or, as here, ε. This last will be our major topic.

(v) It seems that the D waves are calculable from derivative forward
dispersion relations and the results are fairly independent on the as-

sumptions $-\delta_2^2$ negligible, δ_0^2 dominated by f_0. Experimental confirmation is needed.

(vi) The phase δ_1^1 is dominated by the rho with parameters something like the "Best values" listed in the current Particle Data Tables (*Particle Data Group*, 1970).

$$M_\varrho = 765 \pm 10 \text{ MeV}, \quad \Gamma_\varrho = 125 \pm 20 \text{ MeV}.$$

A number of values are listed in the Tables:

	M(MeV)	Γ(MeV)	
ϱ^0	774 ± 5	111 ± 5	from $e^+e^- \to \pi^+\pi^-$
ϱ^0	768 ± 10	140 ± 14	
ϱ^0		105 ± 15	from $\pi N \to \pi\pi N$, $\pi\pi$ phase shift by C. L.
ϱ^-	755 ± 5	110 ± 9	extrapolation; with energy independent width;
ϱ^0	768 ± 2	132 ± 13	energy dependent width.
ϱ^-	764 ± 2	147 ± 14	$\pi N \to \pi\pi N$ fits in physical region.

We show the graphs for two examples. Firstly the results on $\sigma_{\pi^-\pi^0}$. from C. L. extrapolation applied to data on $\pi^- p \to \pi^- \pi^0 p$ at 2.77 GeV/c. (*Baton, Laurens*, and *Reignier*, 1967). (Fig. 3). Secondly results on ϱ_0, from an analysis of the combined Orsay and Novosibirsk $e^+e^- \to \pi^+\pi^-$ data (Fig. 4 and 5) (see review by *Haissinski*, 1969).

(vii) δ_0^2 is determined from studies of FBA in $\pi^-\pi^0$ production and also from studies of final states involving $\pi^+\pi^+$ and $\pi^-\pi^-$. All analyses

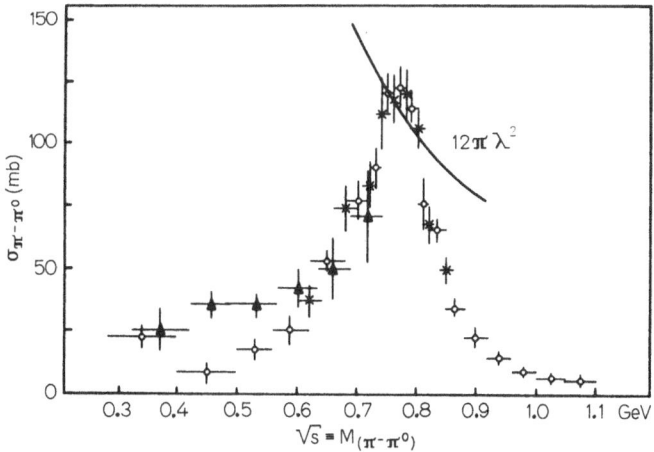

Fig. 3. Total $\pi^-\pi^0$ cross-section $\sigma_{\pi-\pi^0}$ by Chew-Low Extrapolation (*Baton, Laurens*, and *Reignier*, 1967)

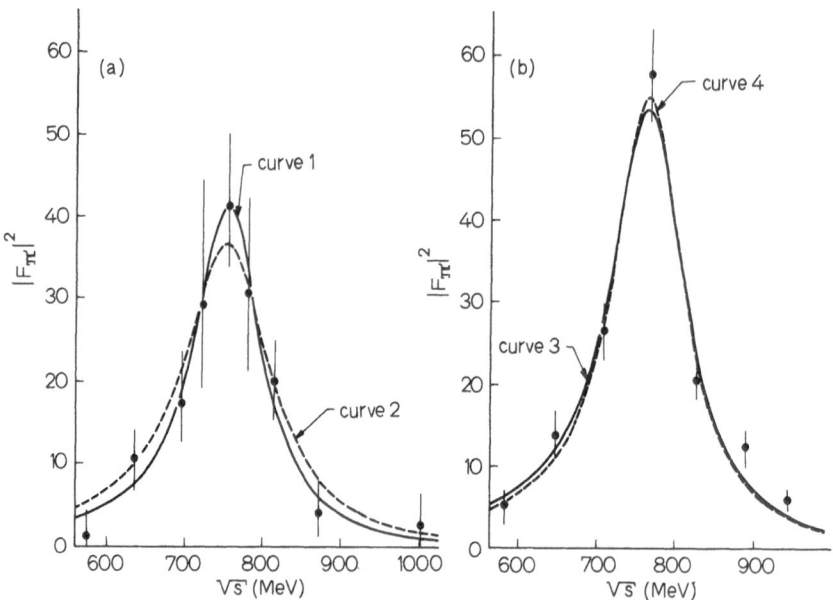

Fig. 4a and b. The pion form factor inferred from data on $e^+ e^- \rightarrow \pi^+ \pi^-$ (from *Haissinski*, 1969). Full Lines are for fits to a Breit-Wigner form — dashed line for fit to a form proposed by *Gounaris* and *Sakurai* (1968). a) Novosibirsk measurements (*Auslander et al.*, 1969); b) Orsay Measurements (*Augustin et al.*, 1969a)

agree that δ_0^2 should be small and negative but differ somewhat as to details (Fig. 6). (Data from *Walker et al.*, (1967); *Baton, Laurens*, and *Reignier*, 1967; *Katz et al.*, 1969; *Colton, Malamud*, and *Schlein*, 1969; and *Scharenguivel et al.*, 1969; the dotted curves show theoretical curves from *Morgan* and *Shaw*, 1970.)

One could summarise the present state of knowledge in the statement

$$\delta_0^2(m_\varrho^2) \sim -15 \pm 5° .$$

The principal problem concerns the form of δ_0^0. Experimental evidence comes from

(i) Forward Backward Asymmetry (FBA) in $\pi^+ \pi^-$ production.

(ii) Experiments on $\pi^0 \pi^0$ production.

In the FBA method, one studies the interference between $A_s^{+-} = (2/3) A_0^0 + (1/3) A_0^2$ and the P wave A_1^1 (neglecting D waves). There results the well-known up-down ambiguity in the determination of δ_0^0 (assuming δ_1^1, δ_0^2 known). A simplified explanation of this is that one essentially measures $\sin(2\delta_0^0 - \delta_1^1)$ and, if this quantity be denoted by $\cos \varepsilon$, δ_0^0 is determined as $\delta_0^0 = \frac{1}{2}\delta_1^1 + \pi/4 \pm \varepsilon$. This is not the exact form of the ambiguity, although it is a good approximation in the vicinity of the rho peak.

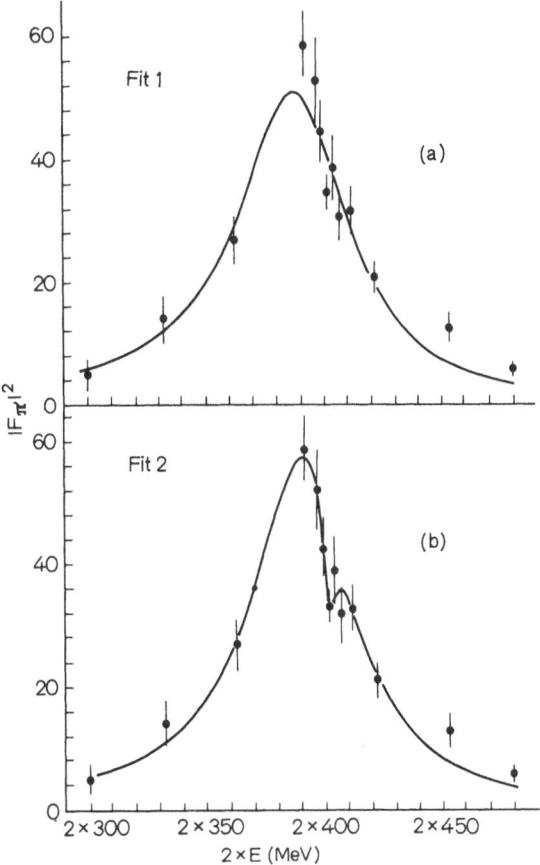

Fig. 5. The pion form factor with emphasis on the Omega mass region. Data on $e^+ e^- \to \pi^+ \pi^-$ (*Augustin et al.*, 1969b), showing fits with and without allowance for rho-omega interference (*Haissinski*, 1969)

An analysis based on the FBA method has been published by *Scharenguivel et al.*, (1969a). (Fig. 7 — points with triangles). Inputs to this analysis are the forms of the *P* wave phase shift δ_1^1 and the $I = 2S$ wave δ_0^2, for which the phase of *Baton*, *Laurens*, and *Reignier* (1967) ($\delta_0(m_\varrho^2) \sim -10°$) were used. Also shown for comparison are the results of *Malamud* and *Schlein* (1969) (Fig. 7 — points with circles.)

Fig. 8 shows an approximate reconstruction by one of us (D. M.) of the one-shell forward backward asymmetry α from the Purdue groups published values for α in the physical region (*Loos et al.*, 1970). The accompanying theoretical curves are discussed in § 4.

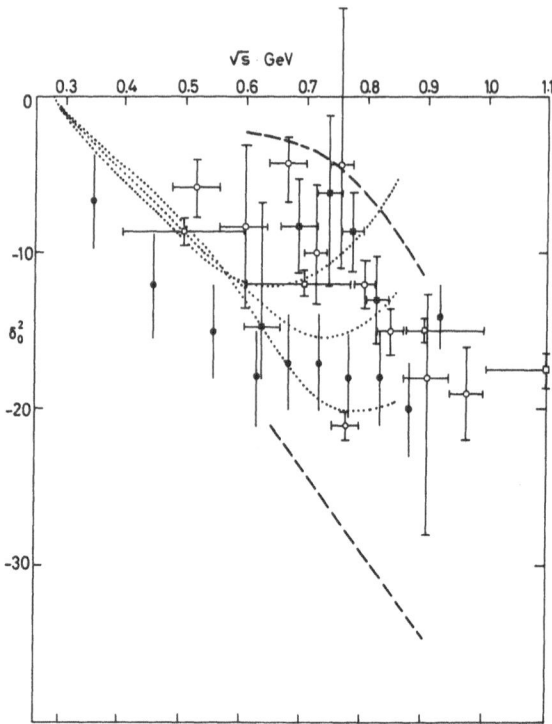

Fig. 6. Alternative determinations of the $I = 2S$ wave phase shift. Legend: ⟡— ■— *Baton, Laurens* and *Reignier* (1967), ● *Walker et al.* (1967), ⟁ *Katz et al.* (1969), ◇ *Colton et al.* (1969), ----- Limits on δ_0^2 due to *Scharenguivel* (1969b), ······ "Theoretical" curves from *Morgan* and *Shaw* (1970) for $\delta_0^2(m_\varrho^2) = -10°, -15°, -20°$ ($a_2 = -0.05$)

Determination of a_0/a_2

There have recently been two determinations of this quantity. Firstly, *Gutay, Meiere,* and *Scharenguivel* (1969a) have studied the relation between Δ^2 and s at the point Δ_0^2 where the forward-backward asymmetry vanishes in peripheral $\pi^+\pi^-$ production. Assuming an exact linear form in s, t, u for the real part of the amplitude, and imposing the Adler Consistency condition (§ 2) leads them to the relation

$$\Delta_0^2(s) = \frac{(2a_0 - 5a_2)}{(10a_0 + 23a_2)}(s - \mu^2). \tag{3.5}$$

They find that the experimental values of $\Delta_0^2(s)$ for various values of the dipion mass $\sqrt{s}$ do approximately conform to the above relation. The fit yields a value $a_0/a_2 = -3.2 \pm 0.1$. The admission of quadratic terms would increase the errors to something like $a_0/a_2 = -3.2 \pm 1.0$.

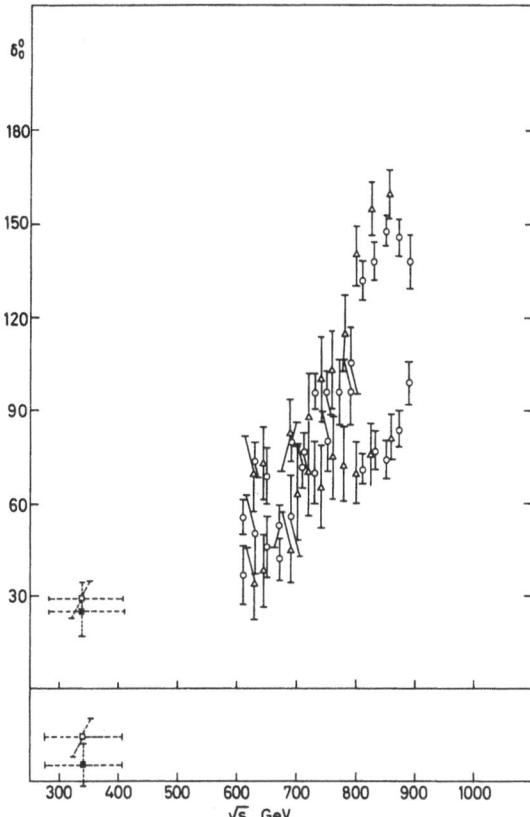

Fig. 7. Alternative determinations of the $I = 0S$ wave phase shift. Legend: λ *Scharenguivel et al.* (1969a); ϕ *Malamud* and *Schlein* (1969) ⊢··□··┤ ⊢··■··┤ Estimates from analysis of Ke_4 decay, with and without "enhancement" factor (*Ely et al.,* 1969)

    Cline, Braun and *Scherer* (1969a, 1969b) have deduced the same ratio from a study of the charge branching rations $R_1 = \sigma(\pi^0\pi^0)/\sigma(\pi^+\pi^+)$ and $R_2 = \sigma(\pi^0\pi^0)/\sigma(\pi^+\pi^-)$ near threshold. Both ratios depend essentially just on the value of a_0/a_2. (They apply corrections for P and D-waves.) They found a consistent fit to both branching ratios arriving at the value $a_0/a_2 = -3.2 \pm 1.1$ in agreement with the quite independent result of *Gutay, Meiere,* and *Scharenguivel* referred to above. The data considered by *Cline, Braun,* and *Scherer* will be further discussed in §4 (see also Fig. 14).

    Evidence favouring the Down branch for δ_0^0 above the rho and against the Down branch below the rho has come from the study of

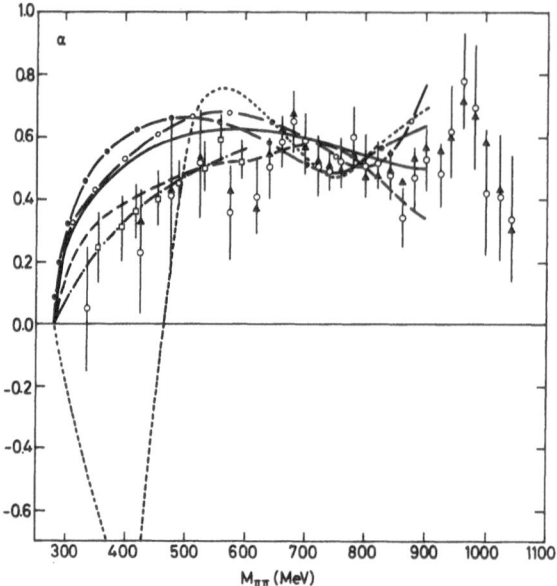

Fig. 8. Forward Backward Asymmetry for $\pi^+\pi^-$ scattering. Approximate reconstruction by one of us (D. M.) of the on-shell forward backward asymmetry α from the Purdue Group's published values for α in the physical region (*Loos et al.*, 1970). Legend: $\oint$ Linear Fit to Data for $|t| \leqq 12\ m_\pi^2$, $\blacktriangle$ Quadratic Fit to Data out to $|t| \sim 40\ m_\pi^2$, $\boxminus$ Determination by *Scharenguivel et al.* (1970a), $-\cdot-\cdot$ Fit of these authors. Also shown are theoretical curves based on *Morgan* and *Shaw* (1970) (SM) with $\delta_0^2(m_\varrho{}^2) = -20°$, no D waves, SM's form for $\delta_1^1(\Gamma_\varrho = 120$ MeV) and alternative SM solutions for δ_0^0 labelled thus: $-\!\!\bigcirc\!\!-\!\!\bigcirc\!\!-$ BDII, $-\!\!\bullet\!\!-\!\!\bullet\!\!-$ BU, $-\!\!-\!\!-\!\!-$ BDI, $-\,-\,-\,-$ UD, $-\,-\,-\,-\,-$ DUI

$\pi_0\pi_0$ reactions. This will be referred to in the discussion of phenomenology based on Forward Dispersion Relations.

The above is a very brief and selective discussion of the evidence from experiments in peripheral dipion production (with a mention of data from e^+e^- experiments). For further details, and especially to get a feeling for the uncertainty in this field, the reader is referred to *Argonne*, 1969 and also to the relevant portions, especially those concerned with the rho, of Particle Data group, 1970.

§ 4. Forward Dispersion Relation (FDR)

The forward dispersion relations, or more generally fixed t dispersion relations, afford a means of generating amplitudes incorporating the requirements of analyticity. The special features of $\pi\pi$ scattering already noted — Bose statistics, no spin, few partial waves and absence of significant inelasticity below 1 GeV and the fact that $\pi\pi$ scattering is usefully a crossing symmetric reaction — makes the FDR approach very suitable. The principal alternative is the use of partial wave dispersion relations (PWDR). Advantages of the FDR in the context of $\pi\pi$ scattering are

(i) the right hand cut is expressible in terms of σ_{tot};

(ii) the left hand cut is simply related to the right hand cut by crossing;

(iii) although unitarity is not formally a simple constraint, the simplicities noted above in the sub-GeV region, imply that unitarity can none the less be treated;

(iv) the behaviour at infinity of the amplitudes is rather directly related to general notions on high energy phenomena. This contrasts with the situation for the PWDR.

The form of the available experimental information on the $\pi\pi$ system has dictated the use of the FDRS and it has mainly consisted of phenomenology rather than dynamics. Since our best information relates to conspicuous resonances — notably the rho — which dominate the absorptive parts in the dispersion integrals, most applications have been concerned with calculating the amplitudes in the low energy region — in other words with making an extrapolation. The dispersion relations are as liable as any other technique to the hazards of extrapolation. Calculations gain their apparent precision from a concealed simplicity assumption which can only be vindicated by experiment. Results are crucially influenced by the assumed input (presence and absence of bumps and rapid local variations) much more than the technical details of how the dispersion relation are treated.

Notation

The scattering amplitudes $F^I(s, t, u)$ for the isospin eigenstates are normalised to give the partial wave expansion

$$F^I(s, t, u) = \sum_{l}' (2l + 1) P_l(\cos\theta_s) F_l^I(s) \tag{4.1}$$

with

$$F_l^I(s) = \frac{W}{2q} e^{i\delta_l^I} \sin\delta_l^I \tag{4.2}$$

where $s = 4m_\pi^2(v + 1)$, $v = q^2/m_\pi^2$, $W = 2m_\pi\sqrt{v + 1}$ and $\sum'$ denotes the summation over all states with $l + I$ even (generalised Bose statistics). In order to emphasize su crossing symmetry in fixed t dispersion relations, it is convenient to use the variable $z = \dfrac{s - u}{4m_\pi^2}$. For the case $t = 0$ (forward scattering), $z = 2v + 1$, $s = 2m_\pi^2(z + 1)$ and $u = 2m_\pi^2(z - 1)$. The cross-section formulae following from the above normalisation are

$$\frac{d\sigma_{el}^I}{d\Omega} = \frac{16\hbar^2}{W^2} |F^I(s, \cos\theta_s)|^2 = \frac{4(\hbar^2/m_\pi^2)}{v + 1} |F^I(s, \cos\theta_s)|^2, \quad (4.3)$$

$$\sigma_{tot}^I = \frac{1b\hbar^2}{qW} \operatorname{Im} F^I(s, t = 0) = \frac{8\pi(\hbar^2/m_\pi^2)}{\sqrt{v(v + 1)}} \operatorname{Im} F^I(s, t = 0) \quad (4.4)$$

$(\hbar^2 = 0.389 \, (\text{GeV}/c)^2 \text{ mb.}).$

Note the extra factor of two in the expression for σ_{tot}^I and of four in the expression for $\frac{d\sigma_{el}^I}{d\Omega}$ $\left(\sigma_{el}^I = 2\pi \int\limits_0^1 d(\cos\theta) \frac{d\sigma_{el}^I}{d\Omega}\right)$: These arise because the formalism treats the scattering of I-spin eigen states as for the scattering of identical particles. Scattering amplitudes for physical charge states are related to the above isospin amplitudes by the relations

$$F(++, ++) = F^2,$$
$$F(+-, +-) = \tfrac{1}{3}F^0 + \tfrac{1}{2}F^1 + \tfrac{1}{6}F^2,$$
$$F(+0, +0) = \tfrac{1}{2}(F^1 + F^2), \quad (4.5)$$
$$F(00, 00) = \tfrac{1}{3}(F^0 + 2F^2),$$
$$F(00, +-) = \tfrac{1}{3}\sqrt{2}(F^2 - F^0)$$

The formalism guarantees that the appropriate formulae for non-identical particle scattering are automatically recovered (One must then use

$$\sigma_{el} = 2\pi \int\limits_{-1}^1 d(\cos\theta) \frac{d\sigma_{el}}{d\Omega}\right).$$

The isospin crossing matrices are

$$C_{II'}^{(su)} = \begin{bmatrix} \tfrac{1}{3} & -1 & \tfrac{5}{3} \\ -\tfrac{1}{3} & \tfrac{1}{2} & \tfrac{5}{6} \\ \tfrac{1}{3} & \tfrac{1}{2} & \tfrac{1}{6} \end{bmatrix}; \quad C_{II'}^{(st)} = \begin{bmatrix} \tfrac{1}{3} & 1 & \tfrac{5}{3} \\ \tfrac{1}{3} & \tfrac{1}{2} & -\tfrac{5}{6} \\ \tfrac{1}{3} & -\tfrac{1}{2} & \tfrac{1}{6} \end{bmatrix}. \quad (4.6)$$

Eigenstates of t-channel I spin will be denoted by T^I so that $F^I = \sum\limits_{I'} C_{II'}^{(st)} T^{I'}$

Fixed t dispersion relations take a simple form in terms of the T^I. For fixed t, one can form two independent crossing even amplitudes ($s \leftrightarrow u$ crossing), a convenient choice being

$$F_1^{(+)} \equiv F^{00} = \tfrac{1}{3}F^0 + \tfrac{2}{3}F^2 = \tfrac{1}{3}T^0 + \tfrac{2}{3}T^2, \quad (4.7)$$
$$F_2^{(+)} \equiv F^{+0} = \tfrac{1}{2}F^1 + \tfrac{1}{2}F^2 = \tfrac{1}{3}T^0 - \tfrac{1}{3}T^2. \quad (4.8)$$

One also has the crossing odd amplitude

$$F^{(-)} \equiv F^{+-} - F^{++} = \tfrac{1}{3}F^0 + \tfrac{1}{2}F^1 - \tfrac{5}{6}F^2 = T^1. \quad (4.9)$$

In terms of the conventional Regge picture with poles P, P', ϱ etc. the asymptotic behaviour of the T^I in the forward direction takes the form

$$T^0 \to i C_P(s/s_0) + C_{P'}\left[i - \operatorname{ctg}\frac{\pi\alpha_{P'}}{2}\right]\left(\frac{s}{s_0}\right)^{\alpha_{P'}}$$

$$T^1 \to C_\varrho\left[i + \operatorname{tg}\frac{\pi\alpha_\varrho}{2}\right]\left(\frac{s}{s_0}\right)^{\alpha_\varrho} \qquad\qquad \left.\vphantom{\begin{array}{c}a\\a\\a\end{array}}\right\} \quad s \to \infty . \qquad (4.10)$$

$$T^2 \to C_2\left[i - \operatorname{ctg}\frac{\pi\alpha_2}{2}\right]\left(\frac{s}{s_0}\right)^{\alpha_2}$$

Here α_P has been set equal to unity, $\alpha_\varrho \approx 0.5$, and α_2 denotes an effective trajectory with $\alpha_2 < 0$.

Dispersion Relations

When written without subtractions, the dispersion relations take the form

$$F(z) = \frac{1}{\pi}\int_1^\infty \frac{\operatorname{Im}F(z'+i\varepsilon)}{z'-z}\,dz' + \frac{1}{\pi}\int_1^\infty \frac{\operatorname{Im}F(-z'+i\varepsilon)}{z'+z}\,dz' \quad (4.11)$$

where

$$\operatorname{Im}F^I(-z'-i\varepsilon) = \sum_{I'} C_{II'}^{(su)} \operatorname{Im}F^{I'}(z+i\varepsilon) . \qquad (4.12)$$

Specializing to the thresholds for the s and u channels ($z = \pm 1$) one obtains sum rules for the scattering lengths

$$F(\pm 1) = \frac{1}{\pi}\int_1^\infty \frac{\operatorname{Im}F(z+i\varepsilon)}{z\mp 1}\,dz + \frac{1}{\pi}\int_1^\infty \frac{\operatorname{Im}F(-z-i\varepsilon)}{z\pm 1}\,dz . \quad (4.13)$$

For pure $I = 2$ exchange, this becomes

$$\frac{1}{6}(2a_0 + a_2) = \frac{1}{\pi}\int_1^\infty dz \cdot \frac{2z}{z^2 - 1}\operatorname{Im}\left[\tfrac{1}{3}F^0 + F^1 + \tfrac{5}{3}\cdot F^2\right] . \quad (4.14)$$

This sum-rule has rather slow convergence and is therefore not very useful. For pure $I = 1$ exchange, $T^1(t)$ is an odd function, so that one can divide through by an additional factor of z to obtain

$$\frac{1}{6}[2a_0 - 5a_2] = \frac{1}{\pi}\int_1^\infty \frac{2\,dz}{z^2 - 1}\operatorname{Im}\left[\tfrac{1}{3}F^0 + \tfrac{1}{2}F^1 - \tfrac{5}{6}F^2\right] . \quad (4.15)$$

A sum rule for a_1, the P wave scattering length in m_π^{-3}, is obtained by taking $F(z) = \dfrac{F^{(1)}(z)}{z-1}$ to give

$$3a_1 = \frac{2}{\pi}\int_1^\infty \frac{\operatorname{Im}F^1(z+i\varepsilon)}{(z-1)^2}\,dz + \frac{2}{\pi}\int_1^\infty \frac{dz}{(z+1)^2}\operatorname{Im}\left[\tfrac{1}{3}F^0 - \tfrac{1}{2}F^1 - \tfrac{5}{6}F^2\right] . (4.16)$$

$\left(\text{The factor two comes from } \underset{z \pm 1}{\text{L}} \dfrac{z-1}{q^2} = \dfrac{2}{m_\pi^2}.\right)$ A sum rule with less
dependence on high energy absorptive parts arises on subtracting
(4.16) from (4.15) to give:

$\dfrac{1}{6}[2a_0 - 5a_2 - 18a_1]$

$$= -\frac{2}{\pi}\int\limits_1^\infty \frac{1+3z}{(z^2-1)^2}\,\mathrm{Im}F^1(z+i\varepsilon)\,dz + \frac{4}{\pi}\int\limits_1^\infty \frac{dz}{(z-1)(z+1)^2}. \qquad (4.17)$$
$$\cdot\,\mathrm{Im}\left[\tfrac{1}{3}F^0 - \tfrac{5}{6}F^2\right].$$

Olsson (1967) used a recast form of (4.16) to yield a sum rule for a_1:

$$\frac{1}{3v}\,\mathrm{Re}\,A^1(v) = \frac{2(1+v)}{3\pi}\,\mathrm{P}\int\limits_0^\infty \frac{dv'(1+2v')}{\sqrt{v'(v'+1)}}\,\frac{1}{v'+v+1}$$
$$\cdot\left[\frac{\mathrm{Im}F^1(v')}{\sqrt{v'(v'+1)}} - \frac{\mathrm{Im}F^1(v)}{\sqrt{v(v+1)}}\right] \qquad (4.18)$$
$$+ \frac{1}{6\pi}\,\mathrm{P}\int\limits_0^\infty dv'\,\frac{\mathrm{Im}\left[\tfrac{2}{3}F^0(v') + F^1(v') - \tfrac{1}{3}F^2(v')\right]}{(v'+1)(v'+v+1)(v'-v)}.$$

Before discussing the result of applying the forward relations, we shall
make a digression to discuss the application of dispersion relations for
the inverse amplitude.

Dispersion Relations for the Inverse Forward Amplitudes — The Goebel-Shaw Bounds

An alternative approach using forward amplitudes is provided by
dispersion relations for the inverse amplitudes. This will be illustrated
by a brief discussion of the important bounds on scattering lengths
derived *Goebel* and *Shaw* (1968). The technique is to treat the forward
amplitude for the elastic scattering of physical particles (Eq. (4.5)). This
guarantees positivity of ImF on both cuts. One then writes down a sum
rule which takes the form (in our notation)

$$-(a^\pm)^{-1} = \frac{1}{\pi}\int\limits_1^\infty dz\left[\frac{R^\pm(z)}{z-1} + \frac{R^\mp(z)}{z+1}\right],$$

where $R^I = \dfrac{q}{w}\,\dfrac{\sigma_{tot}^I}{\pi\,d\sigma_{el}^I/d\Omega} \equiv (\mathrm{Im}F^I)^{-1}$. The form written here is for
the case where $a^\pm$ are both negative. (Where this is not the case, the
bounds apply a fortiori.) The idea of the method is that any lower bound
on the positive definite quantities $R^\pm$ leads to a lower bound on the a's.

Introducing very mild assumptions on the form of the $\pi\pi$ forward amplitude, *Goebel* and *Shaw* achieve a number of bounds which may be summarized as:

$$a_0 \gtrsim -0.5 m_\pi^{-1}\,. \tag{4.19}$$

(The authors emphasize how the concept of range enters through the appearance of the ratio $\sigma_{\text{tot}}/\pi \, d\sigma_{\text{el}}/d\Omega$ which expresses the "peaking" of the angular distribution (multiplied by the inelasticity).)

Shaw (1968) has applied a related technique to achieve upper bounds on the $\pi\pi$ coupling parameter λ, for example, with rather mild assumptions

$$\lambda < 0.25 \tag{4.20}$$

and with more specific assumptions especially in the rho region

$$\lambda < \lambda_m \sim 0.08 \quad \text{to} \quad 0.15\,. \tag{4.21}$$

Application of Forward Dispersion Relations

We now resume the discussion of forward relations and select some examples of their application.

Antoniou and *Palev* (1968) used forward dispersion relations with a sub-threshold zero explicitly built in. Varying the position of the zero they found

$$0 < a_0 < 0.14\,,$$

$$-0.08 < a_2 < -0.03\,, \tag{4.22}$$

$$a_1 \approx 0.044\,.$$

A similar value for the a_1 scattering length had been found earlier by *Olsson* (1967) who made a systematic evaluation of his sum rule for a_1 (4.18). His result was

$$\alpha_1 = 0.04 \pm 0.005\,. \tag{4.23}$$

A similar value was also obtained by *Pišút, Lichard,* and *Bona* (1966) where the Wolf phase shifts (*Wolf,* 1964) were fed into the sum rule (4.16). Dominant contribution came from the ϱ-meson and the result was

$$a_1 = 0.037 \pm 0.004\,. \tag{4.24}$$

An interesting approach based on weight functions and analytic extrapolations was proposed by *Bowcock* and *John* (1969), who succeeded in performing an analytic extrapolation from the left to the right hand cut. Large errors — inevitably present in an honest extrapolation — prevented qualitative conclusions. The qualitative result is that a broad ε is required just by the analytic properties of $\pi\pi$ amplitudes.

2*

Castoldi (1969) used FDR for the $\pi^-\pi^0$ amplitude, demanding consistency of the output and input phase shift δ_0^2 in the low energy region. Feeding in information on the δ_1^1 and δ_0^2 in the ϱ-region, he obtained as a preferred solution δ_0^2 of a "turn down" shape with $a_2 = 0.26$ and a zero in δ_0^2 at 457 MeV. Inequalities due to *Martin* and *Wanders* were then applied to predict the form of δ_0^0 and the related scattering length, giving $a_0 = 0.8$. The origin of the "turn-down" form for δ_0^2 may be traced to the input cross section employed in the calculation. This is based on the Chew-Low determination of $\sigma_{\pi-\pi^0}$ by *Baton, Laurens,* and *Reignier* (1967) for which the lowest energy point does have a larger value than the immediately succeeding ones. If this is a genuine effect, then turnover follows from analyticity.

Fulco and *Wong* (1967) used forward DR with F^0 and F^1 subtracted at threshold and F^2 at $s = 2$ (this subtraction constant being related by $s \leftrightarrow u$ crossing symmetry to the value of F^0 and F^1 at $s = 2$). They obtained a set of solutions depending on the value of the input scattering length a_0. Large scattering lengths, a_0, (compared to the prediction of *Weinberg*) were preferred, due to the requirement that the Adler $\pi\pi$ sum rule (*Adler*, 1965) be saturated. (But see Footnote 12 of *Weinberg*, 1966.) A typical solution had $a_0 = 0.8$, $a_2 = 0.10$ with a turn-down δ_0^2, and subtantial δ_0^0. Zeroes developed for δ_0^0 below $s = 0$ and for δ_0^2 in the range $0 < s < 16$. The two last features were common to all solutions.

The inverse dispersion relations (*Gilbert*, 1957) for forward $\pi\pi$ amplitudes were used by *Pišút* (1968). Demanding consistency of δ_0^0 and δ_0^2 with the DR, he found two solutions for each phase

$$I = J = 0 \quad \begin{array}{l} \text{i) } a_0 = -0.5 \text{ turnover at 450 MeV.} \\ \text{ii) } a_0 = 0.013 \text{ always positive.} \end{array}$$

In the former case, δ_0^0 tends to large values consistent with a broad ε, in the latter δ_0^0 prefers to be small up to 600 MeV.

$$I = 2, J = 0 \quad \begin{array}{l} \text{i) } a_2 = 0.2 \text{ turn-down of } \delta_0^2 \text{ at 450 MeV.} \\ \text{ii) } a_2 = -0.015 \text{ always negative.} \end{array}$$

In the former case, δ_0^2 reaches about -10^0 to -20^0 in the ϱ-meson region, while in the latter it again prefers to be small (less than a few degrees).

Now we shall come in some detail to the recent calculation by *Morgan* and *Shaw* (1968 and 1970) (In the following, we shall refer to these authors as SM to avoid confusion with *Malamud* and *Schlein* whose famous $\pi\pi$ phase determinations (*Malamud* and *Schlein*, 1967 and 1969) are often designated on graphs by the authors' initial letters).

The object of SM's work was to relate information and hypothesis on the $\pi\pi$ system in the rho region to the properties at threshold. To this end, forward dispersion relations were used in the approximate truncated form

$$\operatorname{Re} T^{0,2}(z) = \frac{2}{\pi} \int\limits_{1}^{z_2} dz' \frac{z' \operatorname{Im} T^{0,2}(z')}{z'^2 - z^2} + e_0^{0,2} + e_2^{0,2} z^2 , \qquad (4.25)$$

$$\operatorname{Re} T^1(z) = \frac{2z}{\pi} \int\limits_{1}^{z_2} dz' \frac{\operatorname{Im} T^1(z')}{z'^2 - z^2} + e_1^1 z + e_3^1 z^3 . \qquad (4.26)$$

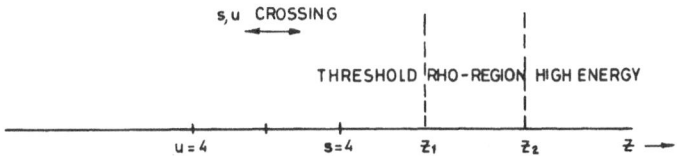

Fig. 9. Schematic for SM's Forward Dispersion Relation Method

The equations have explicit s, u crossing and the e's are subtraction constants which serve to parametrize the contribution to the real parts at low energies of the high energy absorptive parts $(z > z_2)$. The cut-off z_2 was taken to correspond to $M_{\pi\pi} = 1500$ MeV. The way the equations are used can be illustrated by reference to the picture (Fig. 9). One assumes an input for the phases in the rho-region. The six subtraction constants are then chosen so that this input is self-reproducing via the dispersion relations, also so that $A^{I_s=1}$ should vanish at threshold. The form of the phases at low energies is solved for iteratively using the dispersion relations to pass from $\operatorname{Im} T$ to $\operatorname{Re} T$ and unitarity to pass from $\operatorname{Re} T$ to $\operatorname{Im} T$. In this way, solutions are generated which have exact s, u crossing, approximate s, t crossing (from the threshold conditions), approximate unitarity and of specified form in the rho region.

Derivative relations i. e. dispersion relation for $\dfrac{d}{dt}[T^I]_{t=0, z \text{ fixed}}$ are also employed. These give information on the D waves which one needs in order to apply unitarity to the forward relations. They also confirm that F waves are negligible. One may sum up the procedure in a small table.

What Is Put In	What Comes Out
$m_\varrho, \Gamma_\varrho$	a_0, a_1, a_2
$\delta_0^2(m_\varrho^2)$ (value of $I = 2$. S wave at rho mass)	D waves
Form of δ_0^0 over rho-region i. e. $m_\varepsilon, \Gamma_\varepsilon$ } this is varied	Low energy form of phases
m_{f_0}, Γ_{f_0}	Subtraction constants.

Remark on the Subtraction Constants

In the SM method the subtraction constants are supposed to represent the effects of high energy contributions to the dispersion integrals. Therefore for each solution one should look at the resulting subtraction constants and see if they are reasonable. It turns out that the delicate one is e_2^0, the quadratic coefficient in the $I_t = 0$ amplitude. SM conclude that reasonable high energy effects would give $e_2^0 \lesssim 0.2$ in the units which they employ. They used this as a criterion for distinguishing among alternative solutions.

Philosophy of the Application and Remarks on the Data

SM applied their procedure to infer the low energy characteristics corresponding to alternative forms of the $\pi\pi$ phase shifts in the rho-region suggested by experiment. This topic has been discussed in Section 3. The important phase shifts are δ_1^1, dominated by the rho resonance, δ_0^2, the $I = 2\delta$ wave, always found experimentally to lie in the range $-20° \lesssim \delta_0^2(m_\varrho^2) \lesssim -10°$ and δ_0^0 with its up-down ambiguity at each energy across the rho-band. The functional form of δ_0^0 with energy is the outstanding problem. SM adopted a less pessimistic view as to the reliability of recent determinations of δ_0^0 than has sometimes been taken, (e. g. *Kane*, 1970). They generated solutions corresponding to alternative forms for δ_0^0 — Up-Down, Down-Up etc. and discussed them in relation to other data and theoretical criteria.

It should be emphasised that there is an unstated simplicity assumption, e. g. it was assumed that there exists at most one resonance in δ_0^0 below 850 MeV.

SM laid emphasis on the following pieces of data:

(i) Determination of δ_0^0 by extrapolation to the pion pole of the forward-backward asymmetry (*Scharenguivel et al.*, 1969a) (Fig. 7). SM discussed the influence of the values assumed for δ_0^2 and also compared the results with those of *Malamud* and *Schlein* (1969). They concluded that two distinct branches ("Up" and "Down") for δ_0^0 have been established above the rho mass but only a band of possibilities below. They therefore discussed solutions of the form not only "Up-Up", "Up-Down", "Down-Up", "Down-Down", but also "Between-Up", and "Between-Down",

(ii) Experimental Determination of a_0/a_2. Two recent determinations of this quantity were discussed in the previous section.

a) The relation between momentum transfer and mass for which the forward backward assymmetry in $\pi^+\pi^-$ production vanishes (*Gutay, Meiere* and *Scharenguivel*, 1969). Result $a_0/a_2 = -3.2 \pm 0.1$ from linear

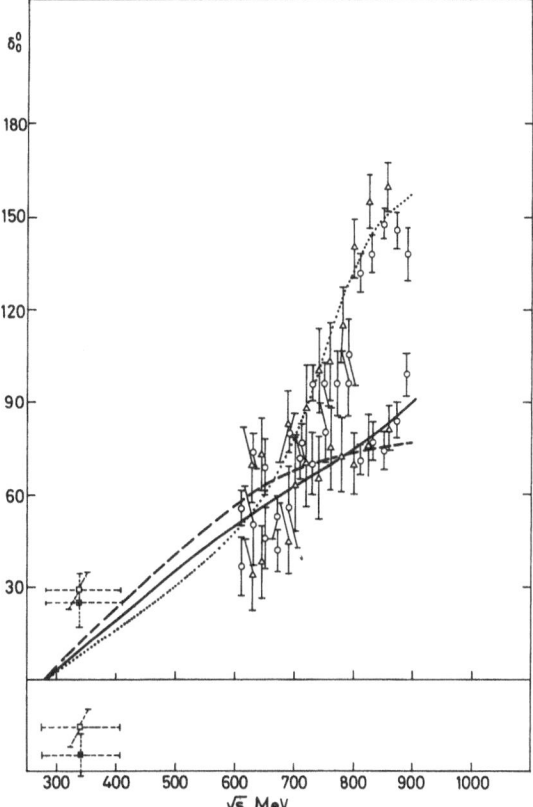

Fig. 10. Alternative Forms for δ_0^0 generated by SM compared with data (cf. legend to Fig. 7). Legend: ——— Between Down I (BDI), – – – Up-Down (UD), · · · · Between Up (BU)

formula fitted, but possible quadratic terms could lead to say $a_0/a_2 = -3.2 \pm 1.1$.

b) Discussion of charge branching ratios near threshold (*Cline et al.*, 1969a, 1969b). Result: $a_0/a_2 = -3.2 \pm 1.1$.

SM adopted the value

$$a_0/a_2 = -3.2 \pm 1.0. \tag{4.27}$$

(iii) Data on charge states $\pi^0\pi^0$ and $\pi^+\pi^+$, especially the charge ratios $R_1 = (\pi^0\pi^0)/(\pi^+\pi^+)$ and $R_2 = (\pi^0\pi^0)/(\pi^+\pi^-)$ emphasized by *Cline, Braun*, and *Scherer* (1969a) and (1969b), and the counter experiment of *Deinet et al.* (1969) on $\pi^-p \to \pi^0\pi^0 n$.

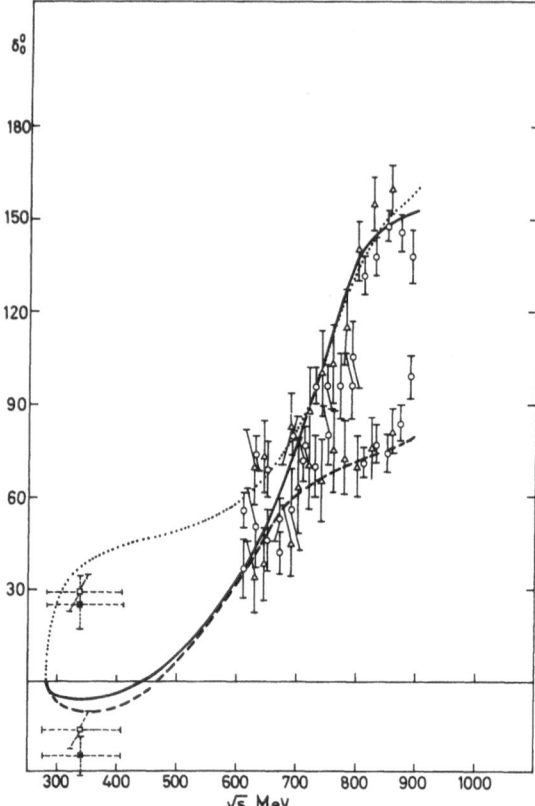

Fig. 11. Alternative Forms for δ_0^0 generated by SM compared with data (cf. legend to Fig. 7). Legend: ——— Down-Up (DU), - - - - Down-Down (DD), · · · · · · Up-Up (UU)

Results

We can now discuss the alternative forms for δ_0^0 generated by SM, and the associated low energy parameters*. The forms for δ_0^0 are shown in Fig. 10, 11, 12 and the associated parameters in Table 1. Keeping these forms in mind, one can now peruse the other available data with a view to resolving the ambiguity concerning δ_0^0.

a) The Threshold Region — The Universal a_0 v. a_2 Curve

The relation between the $I = 0$ and $2 S$ wave scattering lengths a_0 and a_2 for each of the forms treated by SM is shown in Fig. 13, together with the plots for roughly symmetrical epsilons with widths from 150

---

* When SM speak of alternative solutions, they are not alternative solutions in the sense of conventional phase shift analysis. The alternatives have been put in at the input stage.

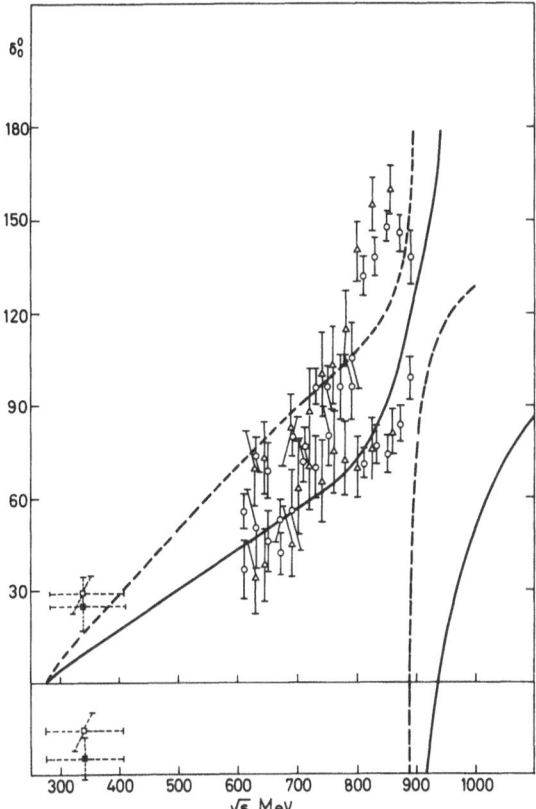

Fig. 12. Alternative Forms for δ_0^0 generated by SM compared with data (cf. legend to Fig. 7). Legend: ——— Between-Down II (BDII), – – – – "Lovelace" (LV) (This is not precisely Lovelace's solution, see *Morgan* and *Shaw*, 1970)

to 800 MeV and masses 600, 765 and 900 MeV. It is immediately seen that a_0 and a_2 are to good approximation related by a universal curve. It seems hard to pinpoint what exactly are the ingredients of this universal relation (but see the discussion by *Olsson*, 1970). It is clear that the position of the locus on the a_0, a_2 diagram would be altered by changing the assumed width of the rho. Taking the value $a_0/a_2 = -3.2 \pm 1.0$ referred to above SM deduced

$$\left.\begin{array}{rl} a_0 = & 0.16 \pm 0.04 \\ a_2 = & -0.05 \pm 0.01 \\ (L \equiv 2a_0 - 5a_2 = & 0.58 \pm 0.06) \end{array}\right\} \qquad (4.27)$$

Table 1. *Values of some important parameters corresponding to the various trial forms of SM discussed in the text (from Morgan and Shaw, 1970).*

Input forms	Threshold parameter					$\delta_0^0(M_K)$ Degrees	Subtraction Parameter e_2^0
	S-wave		P-wave	D-wave			
	a_0	a_2	a_1	c_0	c_2		
Down-Down (DD)	-0.31	-0.22	.032	.0016	.0001	5	0.61
Down-Up I (DUI)	-0.21	-0.17	.031	.0014	.0001	8	0.43
Down-Up II (DUII)	-0.27	-0.19	.030	.0013	$-$ .0001	2	0.47
Between-Up (BU)	0.12	-0.05	.033	.0014	.0001	29	0.21
Between-Down I (BDI)	0.16	-0.05	.035	.0015	.0002	34	0.25
Between-Down II (BDII)	0.16	-0.04	.034	.0015	.0002	30	0.22
Up-Down (UD)	0.24	-0.03	.037	.0017	.0004	40	0.27
Up-Up (UU)	1.49	0.19	.047	.0022	.0017	49	-0.53
'Lovelace' (LV)	0.29	-0.02	.037	.0018	.0006	49	0.23
Final solution	0.16 ± 0.04	-0.05 ± 0.01	0.035 ± 0.002	.0016 $\pm .0002$	0.0003 $\pm$ .0001	33 ± 5	

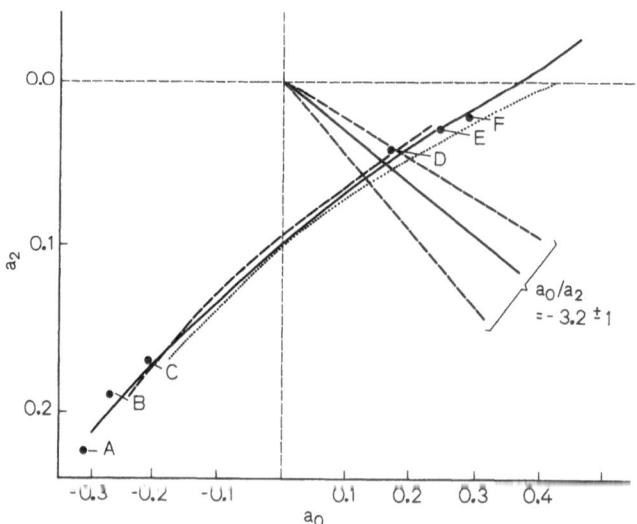

13. SM's "Universal Plot" of a_0 v. a_2. The Curves are for $M_\varepsilon = 900$ (dashed line), $M_\varepsilon = 765$ (full line) and $M_\varepsilon = 600$ (dotted line). The points corresponding to the alternative solutions shown in Fig. 10, 11 and 12 and Table 1 are also shown (A = Down-Down, B = Down Up II, C = Down Up I, D = Between Up, E = Up-Down, F = "Lovelace" (SM's Form, LV) Also shown is the allowed region for a_0/a_2 which is suggested by recent experiments (as discussed in the text)

in excellent agreement with the current algebra model of *Weinberg* (1966). The theoretical predictions for L are 0.54 or 0.69 according as the PCAC constant is extracted from the pion decay rate or the Goldberger Treiman relation. The P and D wave scattering lengths can similarly be extracted. Thus

$$\left.\begin{aligned}
a_1 &= 0.035 \pm 0.002\,, \\
c_0 &= 0.0016 \pm 0.0002\,, \\
c_2 &= 0.0003 \pm 0.0001\,.
\end{aligned}\right\} \qquad (4.28)$$

The quoted errors include the effects of modest changes in $\Gamma_\varrho (\pm 10\text{ MeV}$ about the assumed value of 120 MeV) and in $\delta_0^2 (m_\varrho^2)\,(\pm 5^0)$.

b) δ_0^0 in the Rho-Region

SM go on to discuss possible ways of discriminating among alternative solutions. The nine conditions of *Auberson, Piguet,* and *Wanders* (1968) embodying rigorous consequences of crossing, analyticity and positivity were found to be satisfied by all the alternative solutions, as might have been expected. It is possible that further constraints which have recently been published may prove more decisive. The subtraction constant argument does discriminate (see Table 1 and recall that e_2^0 is expected to be in the range $e_2^0 \lesssim 0.2$). Thus, the solutions "Between-Down", "Between-Up", and "Up-Down" are favoured on this argument, although the latter is ruled out if one takes $a_0/a_2 = -3.2 \pm 1.0$. Note, however that there has been a recent analysis tending to a more negative value of a_0/a_2 (*Scharenguivel et al.,* 1970a). In Fig. 8, predictions for the forward backward asymmetry in $\pi^+\pi^-$ elastic scattering corresponding to a selection of SM's theoretical solutions are compared with a reconstruction by one of us (D. M.) of the experimental values (from data of the Purdue Group, *Loos et al.,* 1970). The solutions with the largest a_0 fit this data best.

Thus far, the argument selects the "Between" Branch below the rho leaving open the behaviour at higher masses. A further narrowing down of possibilities is achieved if one employs the results from the other charge states $\pi^0\pi^0$ and $\pi^+\pi^+$.

Following *Cline et al.* (1969a, 1969b), SM compared the ratios $R_1 = (\pi^0\pi^0)/(\pi^+\pi^+)$ and $R_2 = (\pi^0\pi^0)/(\pi^+\pi^-)$ corresponding to their alternatives for δ_0^0. The comparison with the data considered by *Cline et al.* (1969a) is shown in Fig. 14. The result favours the Down Branch above the rho. The fit to R_1 is best if one assumes $\delta_0^2(m_\varrho^2) = -15^\circ$ rather than -10° or -20°.

Further information is provided by the spark chamber experiment of *Deinet et al.* (1969) on the process $\pi^- p \to \pi^0 \pi^0 n$. The statistics are good enough to render feasible a Chew-Low extrapolation to the total

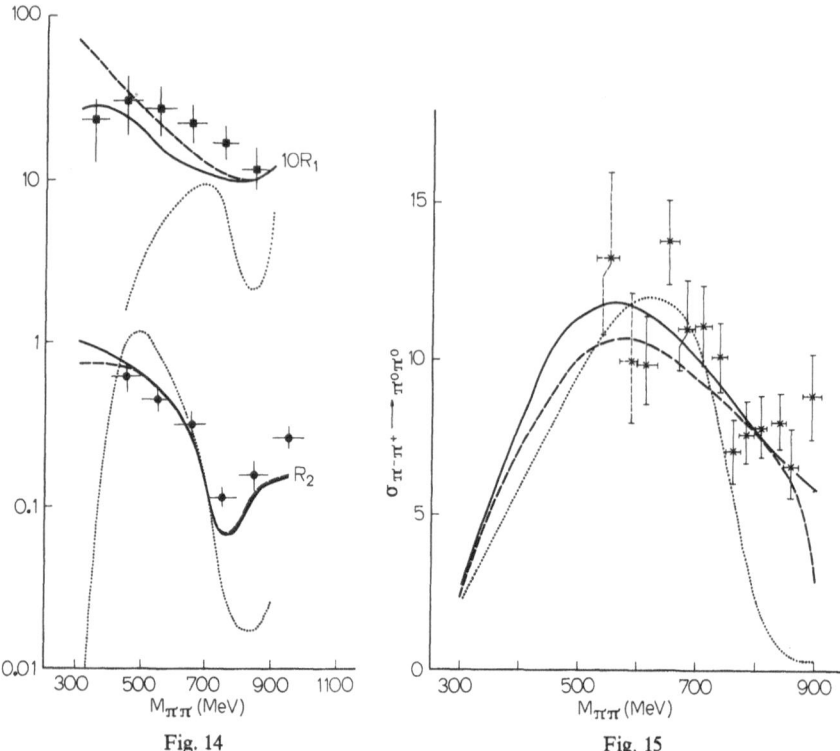

Fig. 14   Fig. 15

Fig. 14. Alternative Predictions for $R_1 = \sigma(\pi^0\pi^0)/\sigma(\pi^+\pi^+)$ and $R_2 = \sigma(\pi^0\pi^0)/\sigma(\pi^+\pi^-)$ with data taken from *Cline, Braun,* and *Scherer* (1969a, 1969b). The curves correspond to SM's Solutions BDI (full line), DUI (dotted line) and UD (dashed line)

Fig. 15. Comparison of SM's predictions for $\sigma(\pi^+\pi^- \to \pi^0\pi^0)$ with the results of *Deinet et al.* (1969) for the forms BDI (full line), BDII (dash-dot line) and BU (dashed line)

cross-section for $\pi^+\pi^- \to \pi^0\pi^0$. The results of this experiment are compared with the SM solutions BDI, BDII, and BU in Fig. 15. The Down branch is clearly favoured (as stressed by the authors using the Malamud and Schlein alternatives (1969)). (Note, however, that *Sondereger* and *Bonamy* (1969) claim to find something between the Up and Down Branches above the rho (1969).)

SM concluded that on the present evidence the Between-Down form of solution is the favoured one.

All the arguments for BDI (with no second resonance) apply equally well to the BDII form (with a zero and second resonance). (See *Bizarri et al.* (1969); also *Hyams* et (1968), and *Hoang* (1969).) Although the choice between these alternatives is of great interest, it appears to have little effect on the discussion below say 850 MeV.

§ 5. Pion-Nucleon Dispersion Relations

Crossing symmetry relates by analytic continuation amplitudes of the three processes

$$(s): \quad \pi N \rightarrow \pi' N', \tag{5.1}$$

$$(u): \quad \pi' N \rightarrow \pi N', \tag{5.2}$$

$$(t): \quad \pi \bar{\pi}' \rightarrow \bar{N} N', \tag{5.3}$$

where the Mandelstam variables s, u and t are used to denote the respective channels.

The unitarity condition applied to the t-channel above the $\pi\pi$ threshold implies

$$f_\pm^{JI}(t + i\varepsilon) = |f_\pm^{JI}| e^{i\delta_J^I(t)}, \quad 4\mu^2 < t < 4M^2. \tag{5.4}$$

Here δ_J^I is the pion-pion phase shift and the $f_\pm^{JI}$ denote the helicity amplitudes for the reaction (5.3). The symbol $\pm$ refers to the relative helicity states of $\bar{N}$ and N', I is the isospin and J the angular momentum (notation of *Frazer* and *Fulco*, 1960). Using the Ommes formula, (the dispersion relation for $\log f$), one can express $f_\pm^{JI}$ (and consequently the πN scattering amplitudes) in terms of δ_J^I in the t-channel for $4\mu^2 < t < 4M^2$. In this way, pion-pion phase shifts are related to the singularities of partial wave- or backward amplitudes of πN scattering. Thus study of the corresponding dispersion relations enables one to gain information on the pion-pion interaction.

The first method along these lines was developed by Hamilton and co-workers*.

The principle of the method is as follows. Consider a partial wave $f_{l\pm}^I(s)$ for πN scattering. The conventional normalisation is

$$f_{l\pm}^I(s) = q^{-2l-1} e^{i\delta_{l\pm}^I(s)} \sin \delta_{l\pm}^I(s) \tag{5.5}$$

where q is the c.m. momentum. The function $f_{l\pm}^I$ is analytic in the complex s-plane with cuts as shown in Fig. 16. It follows from the Cauchy Theorem that for s on the right hand cut

$$\mathrm{Re}\, f(s) - \frac{\mathrm{P}}{\pi} \int_{\mathrm{P}} \frac{\mathrm{Im}\, f(s' + i\varepsilon)}{s' - s} \, ds' - \frac{1}{\pi} \left[\int_{\mathrm{CR}} + \int_{\mathrm{S}} \right] \frac{\mathrm{Im}\, f(s' + i\varepsilon)}{s' - s} \, ds'$$

$$= \frac{1}{\pi} \int_{\mathrm{L}} \frac{\mathrm{Im}\, f(s' + i\varepsilon)}{s' - s} \, ds' + \frac{1}{\pi} \int_{\mathrm{C}} \frac{\Delta f(s')}{s' - s} \, ds' \equiv \Delta(s). \tag{5.6}$$

Here we have suppressed spin and isospin indices, denoted cuts as indicated in Fig. 16 and used the symbol $\Delta f(s)$ for the discontinuity

---

    * *Hamilton* and *Spearman* (1961); *Hamilton, Menotti, Spearman*, and *Woolcock* (1961); *Hamilton, Spearman*, and *Woolcock* (1962a); *Hamilton, Menotti, Oades*, and *Vick* (1962b); see also *Spearman* (1963).

of $f(s)$ along the circle C. A similar relation may be written for s on CR. In the parlance of *Hamilton et al.* the important quantity $\Delta(s)$ in Eq. (5.6) is called the discrepancy. All quantities on the left hand side of Eq. (5.6) can in principle be determined from experimental data, or more precisely from πN phase shifts on P and CR and from πN coupling constants (Born terms) on s. (One should of course bear in mind, that in regard to the estimating of errors, the taking of the principal value of an integral is rather delicate.) The short range contribution (arising from the cut

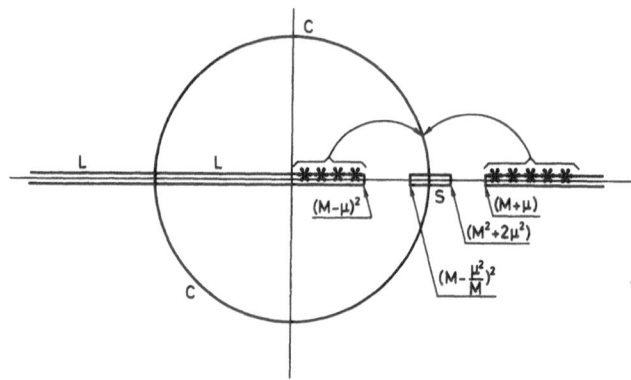

Fig. 16. Analytic properties of pion-nucleon partial wave amplitudes in the s-plane. The "flowers" indicate data coming from low energy πN scattering, the arrows show the extrapolation (a rather complicated $B \to B$ type) leading to information about $\pi\pi$ phase shifts

L) is much farther from the πN threshold than the front part of the circle. This leads one to expect that the integral over L can be lumped into a constant or a pole term for s near the πN threshold. In this way, one can isolate the contribution from the front part of the circle, which contains the $\pi\pi$ phase-shifts at low energies. To obtain $\Delta f(s)$, or what amounts to the same thing, to deduce $\pi\pi$ phase shifts, one has to solve the equation

$$\frac{1}{\pi} \int_{C'} \frac{\Delta f(s')}{s' - s} \, ds' = \Delta_{\pi\pi}(s), \tag{5.7}$$

where C' is the front part of the circle and $\Delta_{\pi\pi}$ is what remains of Δ after contributions from L and the non-frontal part of the circle are subtracted.

    Hamilton et al. (1962b) (HMOV) found their solutions in a natural way. They calculated the discrepancies $\Delta_s^{(+)}(s)$ and $\Delta_{p,j}^{(+)}(s)$ for S and P waves in πN scattering. The index $(^+)$ refers to the isospin combination $(\frac{1}{3})(F^{T=\frac{1}{2}} + 2F^{T=\frac{3}{2}})$ which crosses to $I=0$ in the t-channel. The other index $j = \frac{1}{2}$ or $\frac{3}{2}$ indicates the value of j for a given p-wave combination. HMOV investigated the following alternative parametrisations for δ_0^0.

a) A one pole form in the N/D method.

b) A linear form for N in a variable η, where $\eta = \eta(v)$ is a conformal mapping of the v-plane, cut along $-\infty < v < -1$ onto the unit circle*.

In both cases, the two free parameters were determined by comparison of the "experimental" values of the discrepancies with the calculated results. The solutions for δ_0^0 using parametrisation (a) have a large scattering length $a_0 \approx 1.3 - 2.0$ (in units m_π^{-1}) reach about $30°$ at $m_{\pi\pi} \approx 370 - 400$ MeV and then slowly decrease. Under the parametrisation (b) using N linear in the conformal variable $\eta(v)$, two solutions of a different shape were found, the former has a positive scattering length $a_0 \approx 0.6 \, m_\pi^{-1}$ and δ_0^0 rises to a plateau of about $30°$ in the region $m_{\pi\pi} \approx 470 - 700$ MeV.

The latter solution is of a "turn-over type". δ_0^0 starts with a negative scattering length $a_0 \approx -0.6$, reaches a minimum of about $-15°$, then starts to increase, passing through zero at 400 MeV, and attaining the value $60°$ at $m_{\pi\pi} \approx 700$ MeV.

The results of HMOV were sensitive to the P-wave πN scattering lengths. Some acceptable alternatives for these (from πN analysis) were incompatible with the $\pi\pi$ solution with negative a_0.

It is clear that a more refined parametrisation of the phase shifts δ_0^0 would allow more complicated behaviour.

In summary, the work by *Hamilton et al.* demonstrated clearly that the interaction in the state $I = J = 0$ is mainly attractive and that δ_0^0 is predominantly positive in the low energy region.

Even if the data were more precise it would be difficult to exclude "turnover" solutions in favour of δ_0^0's which are always positive.

The problem is basically that of an analytic extrapolation. In fact one extrapolates from a segment (along which the $\Delta_{\pi\pi}(s)$ is given) to the boundary. It has recently become clear that such an extrapolation is stable**only if one extrapolates not to the value of a function at a point on the boundary but instead to average values over an arc. To make a distinction between δ_0^0 always positive and δ_0^0 of the turn-over type, one should calculate the average of δ_0^0 (or of some quantity simply related thereto) over an arc, say, 280 to 400 MeV in $m_{\pi\pi}$ and carefully estimate the errors. This has not so far been done.

There are two other calculations along these lines. *Spearman* (1963) calculated the discrepancies in s-waves after having divided the partial wave amplitude by square-root factors obtaining thus a modification of Gilbert dispersion relations. This form enhances the contributions

---

* This type of mapping has been used by *Frazer* (1961) and *Ciulli* and *Fischer* (1961).

** *Bowcock* and *John* (1964), *Pišút*, *Presnajder* and *Fischer* (1969).

to the discontinuity from regions near to the πN threshold. The data was not sufficient to determine two parameters in δ_0^0 and the additional constraint of being in accord with the ABC anomaly led to a large scattering length $a_0 \sim 1.6$.

The problem was recently reconsidered by *Ellison* and *Humble* (1968). These authors found that a turn-over type of δ_0^0 is well accommodated into the discrepancy equation in the form of *Spearman* (1963) (i. e. for s-waves when one parametrises δ_0^0 in terms of the N/D method with two poles to represent the left hand cut.)

Another method, similar in spirit but different in technique, was proposed by *Atkinson* (1962), who considered dispersion relations for πN *backward scattering*. In this case, the s and t channels are related by crossing. The invariant amplitudes for πN scattering $A^\pm$, $B^\pm$ are then analytic functions in the complex v-plane (v is the square of the momentum in the s-channel and is related to the s-variable by the relation

$$s = [(v + M^2)^{\frac{1}{2}} + (v + \mu^2)^{\frac{1}{2}}]^2 ,$$

with two cuts

$$-\infty < v < -1 , \tag{5.8}$$

$$0 < v < \infty . \tag{5.9}$$

The former corresponds to the physical region of the t-channel, the latter to that of the s-channel. It is the amplitude $A^{(+)}(v, \theta_s = \pi)$ which is important for the determination of the δ_0^0 phase shift. For $v > 0$, A^+ is given by the usual partial wave expansion and, for $v < -1$ one can express the t-channel helicity amplitudes as

$$A^{(+)}(v) = \frac{4\pi}{\sqrt{6(v + M^2)}} f_+^{00}(-4v) + D\text{-waves} + \cdots . \tag{5.10}$$

The notation for helicity amplitudes is as before, and the argument $t = -4v$ gives the Mandelstam variable in the t-channel in terms of v.

The phase of $f_+^{00}(t)$ is related by Eq. (5.4) to δ_0^0. The problem to be solved is depicted in Fig. 17.

The function $A^+(v)$ is given on the (low energy part of) the R. H. cut. One has to determine the phase of A^+ on the low energy part of the L. H. cut. Atkinson proposed the following solution. On the basis of the Cauchy theorem he obtained the relation

$$\frac{A^+(v) - A^+(0)}{v} - \frac{1}{\pi} \int_0^\infty dv' \frac{\operatorname{Im} A^+(v' + i\varepsilon)}{v'(v' - v)}$$

$$= \frac{1}{\pi} \int_{-\infty}^{-1} dv' \frac{\operatorname{Im} A^+(v' + i\varepsilon)}{v'(v' - v)} \equiv E(v), \tag{5.11}$$

where the function $E(v)$, the complex discrepancy is by construction analytic in the complex v-plane with only the left hand cut. Taking v on the right cut, one can evaluate $E(v)$ for real v $(0 < v < \infty)$ from experimental data on πN scattering. The problem is then to extrapolate the function $E(v)$ from a segment interior to the analyticity region up to the boundary. To solve this problem Atkinson conformally mapped

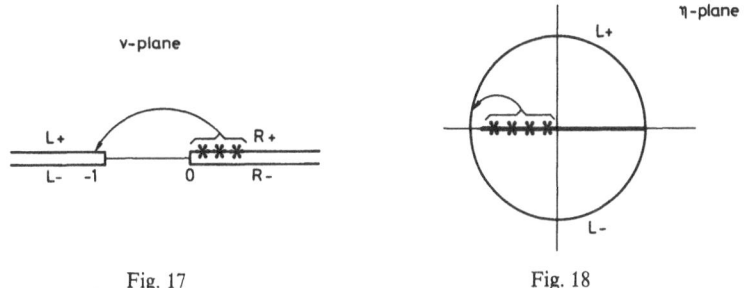

Fig. 17   Fig. 18

Fig. 17. Analytic properties of backward πN amplitudes. The "Flowers" indicate the data available and the arrow the extrapolation to be performed to obtain information about $\pi\pi$ phase shifts

Fig. 18. Conformal mapping of the situation in Fig. 17. This mapping was used by *Atkinson* (1962)

(see papers by *Ciulli* and *Fischer* and by *Fraser* quoted above) the v plane with l. h. cut onto the unit circle in the η-plane: see Fig. 18.

The function $E(\eta)$ is known from πN data on the segment R, it is developed into the Fourier series

$$E(\eta) = \sum_0 a_n \eta^n ,\qquad (5.12)$$

and the coefficients are determined from the fit on R. When the $\{a_n\}_0^N$ are known, it is easy to determine $E(v)$ on the unit circle in the η-plane, i. e. $E(v)$ on the l. h. cut in the v-plane. One has the relations $\operatorname{Im} E(v + i\varepsilon) = \operatorname{Im} \Lambda^+(v + i\varepsilon)$ on the l. h. cut and thence one can reconstruct the δ_0^0 phase shift. Using as input the phase shifts of *Woolcock* (1961) and *Hamilton* and *Woolcock* (1960, 1963), *Atkinson* obtained a form for δ_0^0 of the turn-over type. The phase starts off negative with scattering lengths $a_0 = -0.6$, reaches $-10°$ then turns over, passes through zero at about $m_{\pi\pi} = 470$ MeV and reaches $40° - 50°$ for 800 MeV $< m_{\pi\pi} < 1400$ MeV. As noted by *Atkinson*, the low energy behaviour of δ_0^0 is particularly sensitive to the $(+)$ combination of s-wave scattering lengths $(a_1 + 2a_3)$. The phase shifts of *Hamilton* and *Woolcock* (1960 and

1963) gave

$$a_1 + 2a_3 = 0.004 \pm 0.005 \ . \tag{5.13}$$

Even now there are large discrepancies in this particular s-wave combination, thus *Lovelace* (1969b, 1967) gives 0.056 and 0.069, *Zovko* (1969) obtained 0.045, while *Hamilton* (1966) found, in agreement with Eq. (5.13), (-0.002 ± 0.008). Apparently the low energy behaviour of δ_0^0 cannot at present be resolved in this way for want of precision in πN phase determinations stemming from shortcomings in the experiments.

Lovelace, Heinz, and *Donnachie* (1966), in a very influential paper, reapplied *Atkinson*'s method using the greatly extended πN phase information available to them (*Auvil, Lovelace, Donnachie,* and *Lea,* 1964; see also *Donnachie,* 1966). Instead of writing the backward dispersion relation for A^+ as *Atkinson* had done, *Lovelace, Heinz,* and *Donnachie* used the backward no flip amplitudes which are subject to smaller experimental errors. They obtained[*] a huge number of different solutions (depending on the parametrization used) which could be classified into two groups:

a) those with a positive scattering length $a_0 \approx 0.7$ with a broad resonance at 430 ± 70 MeV and width 400^{+400}_{-150} MeV.

b) turn-over type of δ_0^0 with a negative scattering length, passing through 0^0 at about 350 MeV, with a resonance at 680 ± 85 MeV of width 750 ± 50 MeV.

Their paper furnished in fact the first strong evidence for the existence of a broad ε-resonance. Shortly afterwards, phases δ_0^0 of this type were shown to be consistent with data on peripheral dipion production by *Malamud* and *Schlein* (1967).

The variety of the results of *Lovelace, Heinz,* and *Donnachie* illustrates the basic point common to all analytic extrapolations on to the boundary of the analyticity domain. Averages over arcs are "good", values at separate points are "bad".

The "proof" of the broad ε presented by *Lovelace et al.* is therefore most likely to survive any kind of a mathematical scrutiny, while any attempt to force conclusions about the low energy behaviour of δ_0^0 would be unrealistic.

In conclusion, one can make the following general statement about the results of study of δ_0^0 based on πN dispersion relations·

(i) the low energy $\pi\pi$ interaction in the $T = J = 0$ state is predominantly attractive.

(ii) the phase shift δ_0^0 exhibits a broad ε-resonance.

---

[*] The method used by *Lovelace et al.* differed in technical points from that of *Atkinson.* Discrepancies were estimated after having applied a conformal mapping techniques to both l. h and r. h. cuts separately. A detailed description of the method is given by *Donnachie* (1966).

(iii) the low energy behaviour of δ_0^0 remains uncertain, due partly to inaccuracies of the πN scattering lengths and basically due to the nature of the analytic extrapolation itself.

Note: A very recent and exhaustive paper on the determination of δ_0^0 from πN backward dispersion relations is that of *Nielsen, Petersen,* and *Pietarinen* (1970). As a result of this analysis, the authors prefer the "Down" branch for δ_0^0 above the rho.

§ 6. Information from $K e_4$ Decays and $K_L^0 \to 2\pi$ Decays

(i) *$K e_4$ Decays*
This reaction

$$K^+ \to \pi^+ \pi^- e^+ v$$

affords the opportunity to study a system with two pions as the only strongly interacting particles in the final state. Unfortunately the Branching Ratio $(3.3 \pm 0.3 \times 10^{-5})$ for this decay mode is very small and, to date, the experiment with the largest statistics (*Ely et al.,* 1969) has only 269 events in a heavy liquid bubble chamber*. The events are analysed under the following assumptions:

(i) Local current-current interaction of $V - A$ type.
(ii) Time-Reversal Invariance.
(iii) Only S and P wave for dipion system.
(iv) Only $I = 0$ and $I = 1$ for dipion system ($\Delta I = \frac{1}{2}$ rule).

The local current assumption means that the decay intensity can depend on five variables — dipion mass, S_π, dilepton mass S_e, and three angles θ_π, θ_e and ϕ (θ_π is the angle of the π^+ in the dipion CM with respect to the dipion line of flight, θ_e the analogous quantity for e^+ and ϕ the angle between the planes defined by $\pi^+ \pi^-$ and $e^+ v$). The distribution function $I(S_\pi, S_e, \theta_\pi, \theta_e, \phi)$ has an explicit structure in the variables θ_e and ϕ (from assumption (1)).

$$I = \sum_{i=1}^{q} I_i(s_\pi, s_e, \cos \theta_\pi) Z_i(\theta_e, \phi)$$

where the I are bilinear functions of the weak interaction form factors for the process — one vector form factor h and two axial form factors of f and g (There is in principle a third axial form factor r but this is un-important for $K e_4$ where $m_e \ll$ the energy release). The "f" coupling

---

* The following account is based on the paper of *Ely et al.,* 1969.

can involve either S or P wave dipions and g and h only P waves. Thus

$$f = f_S e^{i\delta_S} + f_P e^{i\delta_P} \cos\theta_\pi ,$$

$$g = g_P e^{i\delta_P} ,$$

$$h = h_P e^{i\delta_P} .$$

There thus exists the possibility of measuring S, P interference in the $\pi^+\pi^-$ final state. In principle, with large statistics (and no biases) a relatively model free analysis can be made (*Pais* and *Treiman*, 1968). With the existing statistics further assumptions have to be made. In the experiment of *Ely et al.*, fits are made with the quantities f_s, f_p, g_p and h_p assumed either to be constant (aside from necessary centrifugal factors), or allowing f_S to be "enhanced", with a Watson enhancement factor

$$f_S = f_S^0 \sin\delta_S \times [\text{kinematic factors}] . \tag{7.1}$$

The result is a measure of $\langle\delta_S - \delta_P\rangle$ or essentially $\langle\delta_0^0\rangle$ averaged over the dipion spectrum. Unfortunately, as pointed out by *Berends, Donnachie,* and *Oades* (1968) there is a two way ambiguity in the solution leading to values of $\langle\delta_0^0\rangle$ of opposite sign. This is resolvable in principle by observation of the linear coefficient in the $\cos\theta_e$ distribution but this turns out to be small:

The results are summarised below:

	A	B	
No form factor	-25 ± 8^0	$+25 \pm 9^0$	$M_{\pi\pi} \approx 330 \pm 50$ MeV .
Form factor	-16 ± 7^0	29 ± 6^0	

(Summary of Results on $\langle\delta_0^0\rangle$ from Analysis of Ke_4 Decays — *Ely et al.,* 1969.)

(ii) $K_L^0 \to 2\pi$ *Decays*

For a detailed review, the reader is referred to *Kabir* (1968, 1969). Here we give a skeleton of $K_L^0 \to 2\pi$ phenomenology as it bears on $\pi\pi$ phase shifts at the K mass.

The measureable effects are conventionally described (*Wu* and *Yang,* 1964), by the complex parameters η_{+-} and η_{00} (the ratios of the decay amplitude of the longlived to that of the short lived K^0 to the dipion charge state indicated). It may be that $\eta_{+-} = \eta_{00}$ (e. g. super weak theory), otherwise (by TCP) $\arg(\eta_{+-} - \eta_{00}) = \delta_0^2 - \delta_0^0 + \pi/2$.

The latest results due to *Chollet et al.* (1970) are displayed in Fig. 19, with the prediction from the strong interaction analysis reported in

§4 superimposed. It is clear that the difference $(\eta_{+-} - \eta_{00})$, if non-zero, is certainly small and therefore its direction is hard to determine. The two determinations of the phase are clearly compatible at the present level of precision.

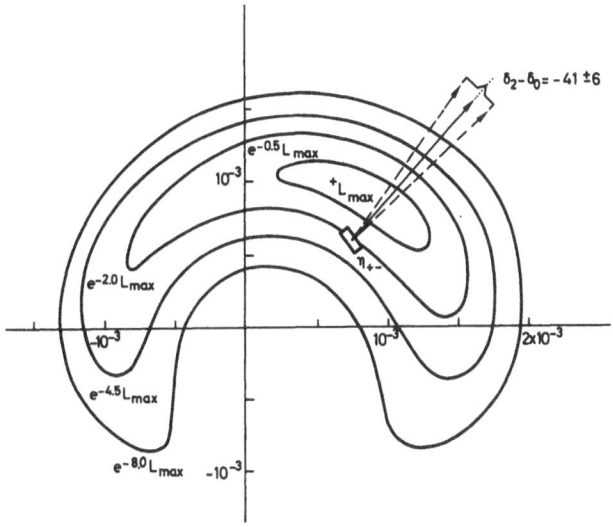

Fig. 19. $K_L^0 \to 2\pi$ Decay. Likelihood Contour for η_{00} from *Chollet et al.* (1970), with the received value for η_{+-} (*Kabir*, 1969) drawn on. Also shown is a representative *strong interaction prediction* for the phase arg $(\eta_{+-} - \eta_{00}) = \delta_0^2 - \delta_0^0 + \pi/2$ based on $\delta_0^2 - \delta_0^0 = -41 \pm 6^0$ (*Morgan* and *Shaw*, 1970)

7. Comments and Conclusions

Talks at a meeting on coupling constants and low energy parameters should end up with a table of scattering lengths and figures of low energy phase shifts. In the field of pion-pion scattering any clear-cut statement of this kind would be risky and most probably shortlived.

Still, the following general features of the low energy pion-pion scattering seem to be well established at present:

(i) the P-wave is dominated by the rho meson at about 765 MeV and of width 110–150 MeV. The phase shifts appears to behave in a "smooth" way below the resonance and can be described there by a Breit-Wigner formula with a reasonable energy dependent width (threshold behaviour and perhaps a barrier penetration factor being built in). The scattering length is, making a conservative estimate within the limit

$$0.03 m_\pi^{-3} < a_1 < 0.10 m_\pi^{-3}$$

with the lower values apparently preferred. There is no evidence for a zero in δ_1^1 below the ϱ-meson mass. Above the rho mass the behaviour of δ_1^1 is less clear. In particular, there is the question as to whether the phase goes on rising passing through π and on to a second resonance (CDD pole type solution), or whether the phase is just of an uncompli-cated Breit Wigner Form. The dipion mass distribution in $\pi N \to \pi \pi N$ can be fitted by either type of behaviour (*Pišút* and *Roos*, 1968). The Veneziano ansatz of *Lovelace* (1969) calls for the "CDD" type of behaviour. One awaits further data.

(ii) The $I = 2 S$-wave is indicated from decay angular distributions in $\pi N \to \pi \pi N$ to be somewhere in the range $-10°$ to $-20°$ at the rho-mass (see Section 3). Experimental evidence below this region is not conclusive, since the peripherality of dipion production in $\pi N \to \pi \pi N$ reaction is not pronounced below, say, 600 MeV. Any dramatic behaviour of δ_0^2 below 600 MeV, would, however, have been seen if it were present.

The most probable low-energy behaviour of δ_0^2 is described by a negative and small δ_0^2, corresponding to a small negative scattering length in the range -0.0 to -0.3. This type of phase shift is consistent with the Veneziano-Lovelace-Wagner model (*Lovelace*, 1969), with current algebra predictions (Section 2), and with the results of dispersion relations calculations (Section 4). In the latter connection we consider the most reliable result to be the "universal curve" (*Morgan* and *Shaw*, 1970, and *Olsson*, 1970) shown in Fig. 13. If one accepts the experimental results on the ratio a_0/a_2 of *Gutay*, *Meiere*, and *Scharenguivel* (1969) and *Cline*, *Braun*, and *Scherer* (1969a, 1969b) the "universal curve" leads to the value

$$a_2 = -0.05 \pm 0.01$$

which is quite close to the prediction of *Weinberg* (1966). With some reservation for unexpected suprises this value should be considered as a favoured one at the present level of information.

What the δ_0^2 does above the rho-region is an open question. Does it perhaps resonate in defiance of the quark model? The only thing which is clear is that it does not resonate very soon, above the rho mass.

(iii) The behaviour of the δ_0^0 phase shift has for years been a contro-versial subject. Over the past few years our understanding has been influenced above all by three advances:

a) the work by *Lovelace, Heinz*, and *Donnachie* (1966) on πN backward dispersion relations, predicting a broad sigma-resonance.

b) Progress in the analysis of peripheral $\pi N \to \pi \pi N$ data, notably by *Malamud* and *Schlein* (1967) who emphasized the ambiguities of δ_0^0 determinations and showed that the evidence admitted solutions with a broad resonanace above the rho mass.

c) *Weinberg*'s predictions bases on current algebra which gave S-wave scattering lengths much smaller than had been believed.

The allowed domain for theoretical speculation has been significantly de-limited by the Goebel-Shaw bound forbidding large negative scattering lengths.

The most probable value of a_0 is, in our opinion,

$$a_0 = 0.16 \pm 0.04$$

which is based directly on the experimental results on a_0/a_2 referred to above, and on the "universal curve" following from the dispersion relations.

Errors quoted in determinations of a_0, a_1, and a_2 should be taken with a grain of salt. In fact, no reliable errors on the determination of low energy phase shifts are available for two reasons:

(i) experimental data on $\pi\pi$ interactions contain apart from statistical also hidden systematic errors based on assumptions that the quantities under consideration behave in a "smooth" way.

(ii) Likewise all calculation based on dispersion relations include the same "smoothness" assumptions, mostly via the form of the parametrisation used.

The behaviour of δ_0^0 in the rho-meson region is dominated by a broad resonance, whose existence should be now considered, in our opinion as well established. The width of the ε is not known and this is reflected in the up-down ambiguities of δ_0^0 below and above the rho-region. Dispersion relation calculations (*Morgan* and *Shaw*, 1968) indicate that a narrow ε (of a width say less than 400) implies a negative a_0, so that to be consistent we should adhere to a very broad epsilon i. e. an "Up-Down" type of phase shift.

Such a phase-shift has been derived by *Lovelace* from a form of single channel unitarization of the Veneziano model and employed with success in phenomenological discussions of most of the available data by *Wagner* and *Roberts*. (When systems with three hadrons are treated, these authors like everybody else have to make additional assumptions; notwithstanding, the overall success is impressive (References from *Lovelace*, 1969).)

As we have seen such a phase shift is also preferred by the dispersion relation approach (§ 4), by πN backward scattering (*Nielsen, Lyng, Petersen,* and *Pietarinen,* 1970) and by *Chew-Low* extrapolation results of *Deinet et al.* (1969) on the reaction $\pi^- p \rightarrow \pi^0 \pi^0 n$.

The "Between-Down" form of δ_0^0 seems therefore to fit most of the present evidence, when this is taken at its face value. However, in view of the many assumptions overt and tacit made in all the analyses, it should be viewed as a tentative assignment – a sort of "best buy" – subject to confirmation or revision as further evidence becomes available.

How will our understanding be advanced? Above all, by the accumulation of data. High statistics data on $\pi^- p \to \pi^+ \pi^- n$ will enable the background to OPE to be properly explored and Chew-Low extrapolation to be performed with much greater confidence. In this connection, and in all treatment of data, the use of more refined extrapolation techniques with realistic and reliable error estimates is highly desirable. Further data on $\pi^- p \to \pi^0 \pi^0 n$ and on $\pi - p \to K \bar{K} n$ will extend our knowledge of charge exchange and inelasticity in $\pi\pi$ interactions. Exploration in greater precision of the region $M_{\pi\pi} > 900$ MeV is of the greatest interest. There is plenty of work for the future.

The views of one of the authors (D.M.) on the $\pi\pi$ problem were developed in collaboration with *Graham Shaw* to whom he is greatly indebted. The other of us (J.P.) is indebted to *Matts Roos* for valuable and helpful discussions and thanks Professor *J. Prentki* and *W. Thirring* for the hospitality extended to him at the CERN Theory Division.

We should like to thank Professors *Höhler* and *Pilkuhn* and all the Karlsruhe group for their hospitality at Ruhestein.

References

Adler, S. L.: Phys. Rev. **140**, B 736 (1965).
Antoniou, N. G., Palev, C.: Phys. Letters **26** B, 301 (1968).
Argonne: Proceedings of the Conference on $\pi\pi$ and πK Interactions, Argonne National Laboratory, 1969.
Atkinson, D.: Phys. Rev. **128**, 1908 (1962).
Auberson, G., Piguet, O., Wanders, G.: Phys. Letters **28** B, 41 (1968).
Augustin, J. E., et al.: Phys. Letters **28** B, 508 (1969a).
— *et al.:* Orsay Preprint (1969b).
Auslander, V. L. et al.: Novosibirsk Preprint (1969).
Auvil, P., Lovelace, C., Donnachie, A., Lea, A. T.: Phys. Letters **12**, 76 (1964).
Baton, J. P., Laurens, G., Reignier, J.: Nucl. Phys. B **3**, 349 (1967).
    (also **Phys. Letters 26** B, 471 (1968), Phys. Letters **25** B, 419 (1967).)
Berends, F. A., Donnachie, A., Oades, G. C.: Phys. Rev. **171**, 1457 (1968).
Bizzarri, R., et al.: Nucl. Phys. B **14**, 169 (1969).
Bowcock, J. E., John, G.: Nucl. Phys. B **11**, 659 (1969).
— *Cottingham, W. N., Williams, J. G.:* Nucl. Phys. B **3**, 95 (1967).
Castoldi, P.: Nucl. Phys. B 12, 567 (1969).
Chew, G. F., Mandelstam, S.: Phys. Rev. **119**, 467 (1960).
Cho, C. F., Sakurai, J. J.: Phys. Letters **30** B, 119 (1969).
Chollet, J., et al.: Phys. Letters **31** B, 658 (1970)
Ciulli, S., Fischer, J.: Nucl. Phys. **24**, 456 (1961).
Cline, D., Braun, K. J., Scherer, V. R.: Univ. Of Wisconsin, preprint (1969a).
— — — (1969b). *Argonne* (1969) p. 179.
Colton, E., Malamud, E., Schlein, P. E. (1969). Argonne (1969), p. 93.
Deinet, W. et al.: Phys. Letters **30** B, 359 (1969).

Donnachie, A.: Particle Interactions at High Energies (Scottish Universities Summer School, 1966), ed. *Preist, Vick.* London: Oliver, Boyd 1966.

Ellison, G. W., Humble, S.: Phys. Rev. **173**, 1563 (1968).

Ely, R. P.,et al.: Phys. Rev. **180**, 1319 (1969).

Frazer, W., Fulco, J. R.: Phys. Rev. **117**, 1603 (1960).

— Phys. Rev. **123**, 2180 (1961).

Froggatt, C. D., Morgan, D.: Phys. Rev. **187**, 2044 (1969).

— — RHEL preprint (in preparation) (1970).

Fulco, J. R., Wong, D. Y.: Phys. Rev. Letters **19**, 1399 (1967).

Gilbert, W.: Phys. Rev. **108**, 1078 (1957).

Goebel, C. J., Shaw, G.: Phys. Letters **27** B, 291 (1968).

Gounaris, G. J., Sakurai, J. J.: Phys. Rev. Letters **21**, 244 (1968).

Gutay, L. J., Meiere, F. T., Scharenguivel, J. H.: Phys. Rev. Letters **23**, 431 (1969a).

Gutay, L. J., et al.: Nucl. Phys. B **12**, 31 (1969b).

Haissinski, J.: (1969). *Argonne*, (1969), p. 373.

Hamilton, J., Woolcock, W. S.: Phys. Rev. **118**, 291 (1960).

— *Spearman, T. D.:* Ann. Phys. (N. Y.) **12**, 172 (1961).

— *Menotti, P., Spearman, T. D., Woolcock, W. S.:* Nuovo Cimento **20**, 519 (1961).

— *Spearman, T. D., Woolcock, W. S.:* Ann. Phys. (N.Y.) **17**, 1 (1962a).

— *Menotti, P., Oades, G. C., Vick, L. L. J.:* Phys. Rev. **128**, 1881(1962b).

— *Woolcock, W. S.:* Rev. Mod. Phys. **35**, 737 (1963).

— Phys. Letters **20**, 687 (1966).

Hoang, T. F., Nuovo Cimento **64** A, 55 (1969).

Hyams, B. D., et al.: Nucl. Phys. B **7**, 1 (1968).

Kabir, P. K.: The *C P* Puzzle. London-New York: Academic Press 1968.

— Springer Tracts Mod. Phys. **52**, 91 (1969).

Kane, G. L., Ross, M.: Phys. Rev. **177**, 2353 (1969).

— Proceedings of Conference on Experimental Meson Spectroscopy, Philadelphia, May, 1970: Columbia University Press 1970 (to be published).

Katz, W. M., et al.: Univ. of Rochester preprint. UR-875-282 (1969).

Loos, L. L., Fuchs, N. H., Gutay, L. J., Scharenguivel, J. H.: Purdue preprint 1970.

Lovelace, C., Heinz, R. M., Donnachie, A.: Phys. Letters **22**, 332 (1966).

— Proceedings of the Heidelberg Conference, 1967, Ed. *Filthuth.* Amsterdam: North Holland 1967.

— Phys. Letters **28** B, 264 (1968).

— Argonne, (1969) p. 562.

— Pion-Nucleon Scattering. Ed. *Shaw* and *Wong.* New York-London: J. Wiley 1969b.

Malamud, E., Schlein, P. E.: Phys. Rev. Letters **19**, 1056 (1967).

— Argonne (1969), p. 93.

Morgan, D., Shaw, G.: Nucl. Phys. B **10**, 1387 (1968).

— — Phys. Rev. D **2**, 520 (1970).

Nielsen, H., Lyng Petersen, J., Pietarinen, E.: Nordita Preprint (1970).

Olsson, M. G.: Phys. Rev. **162**, 1338 (1967).

— Univ. of Wisconsin preprint COO-270, January,(1970).

Osborn, H., Nuovo Cimento Letters **1**, 513 (1969).

Pais, A., Treiman, S. B.: Phys. Rev. **168**, 1858 (1968).

Particle Data Group: Rev. Mod. Phys. **42**, 1 (1970).

Pišút, J., Lichard, P., Bóna, P.: Nucl. Phys. **87**, 433 (1966).

— Nucl. Phys. B **8**, 159 (1968).

— *Roos, M.:* Nucl. Phys. B **6**, 325 (1968).

— *Presnajder, P., Fischer, J.:* Nucl. Phys. B **12**, 110 and 586 (1969).

Scharenguivel, J. H., et al.: Phys. Rev. **186**, 1387 (1969a).
— Argonne, (1969), p. 306.
— *et al.*: Purdue University preprint COO-1428-154 (1970a).
— — Phys. Rev. Letters **24**, 332 (1970b).
Shaw, G.: Phys. Letters **28** B, 44 (1968).
Sondereger, P., Bonamy, P.: Saclay Preprint (1969).
Spearman, T. D.: Phys. Rev. **129**, 1847 (1963).
Wagner, F.: Nuovo Cim. **64** A, 189 (1969).
Walker, W. D., et al.: Phys. Rev. Letters **18**, 603 (1967).
Weinberg, S.: Phys. Rev. Letters **17**, 616 (1966).
Williams, P. K.: Phys. Rev. D **1**, 1312 (1970).
Wolf, G.: Phys. Letters **19**, 329 (1964).
Woolcock, W. S.: Cambridge University Thesis (unpublished) (1961).
Wu, T. T., Yang, C. N.: Phys. Rev. Letters **13**, 380 (1964).
Zovko, I.: Nucl. Phys. B **11**, 231 (1969).

Prof. Dr. *D. Morgan*
Science Research Council
Rutherford High Energy Laboratory
Chilton, Didcot/England
and
Dr. *J. Pišút*
CERN, Genf
and Bratislava

Analytic Extrapolations and the Determination of Pion-Pion Phase Shifts*

JÁN PIŠÚT **

Contents

Abstract 43
1. Introduction 43
2. A Classification of Analytic Extrapolations 44
   2.1 $I \rightarrow I$ Extrapolation. 44
      2.1.1 The Chew-Low-Goebel Extrapolation 46
      2.1.2 The Estimate of Errors in $I \rightarrow I$ Extrapolation 50
   2.2 $I \rightarrow B$ Extrapolation 52
   2.3 $B \rightarrow I$ Extrapolation 53
   2.4 $B \rightarrow B$ Extrapolation 54
3. Ambiguities in the Determination of Phase Shifts from Final State Enhancement
   Factors 56
4. Comments 59
Acknowledgements 59
References 59

Abstract

Analytic extrapolations directly or implicitly involved in obtaining the information on the pion-pion scattering are briefly discussed.

A qualitative description is given of ambiguities and instabilities arising in some of the methods.

1. Introduction

Owing to the lack of direct experimental evidence, a reliable picture of pion-pion interaction can only emerge as a consistent result of various indirect methods. These methods are, in general, making substantial use of analyticity properties of amplitudes — either of those for pion-pion

---

* An extended Appendix to the companion review by *D. Morgan* and the author on the pion-pion interactions at low energies.

** On leave of absence from the Department of Theoretical Physics, Comenius University, Bratislava, Czechoslovakia.

scattering or of others in which the pion-pion interaction plays an important role. On a technical level this amounts to saying that information on pion-pion scattering has to be obtained by some kind of analytic extrapolation.

The instabilities inherent in some of the types of extrapolation are probably the reason why even phenomenological models of pion-pion scattering led to a broad spectrum of frequently contradictory results.

The purpose of the present talk is to give a brief review of various types of analytic extrapolations, and to indicate their connection with some of the methods used to obtain the information about $\pi\pi$ scattering. We shall concentrate mainly on the qualitative description of instabilities and ambiguities.

In the next section we shall give a classification of different types of analytic extrapolation and mention briefly some of the methods in which they are involved. In Section 3 we shall discuss, on the same qualitative level, the ambiguities in the loosely related problem of the determination of phase shifts from enhancement factors in final-state interactions.

2. A Classification of Analytic Extrapolations

The boundary of the analyticity region differs in many respects from the interior of the analyticity region. It is therefore natural to classify the analytic extrapolations according to the following scheme*.

2.1 $I \rightarrow I$ Extrapolation

In this case the extrapolation is performed from a segment inside the analyticity region to a point inside the analyticity region (in a shorthand notation I stands for Inside). The problem is displayed in a schematic way in Fig. 1 for both the typical case of a cut plane and for the unit circle. The mapping of a cut plane onto a finite region frequently simplifies the discussion and provides an insight into the problem**.

This type of extrapolation is stable in the following sense.

Let $f(z)$ denote the function to be extrapolated; let errors of the data be of the order ε; and let $|f(a)|$ be bounded by a constant M in the whole analyticity region. Then the error of extrapolation to the point a is given as

$$\Delta f(a) \sim \varepsilon^{\omega(a)} M^{1-\omega(a)},\tag{1}$$

* Unless specified otherwise, we have in mind the case of a simply connected region.
** The mapping due to *Cutkosky* and *Deo* [1] and *Ciulli* [2], optimal from the point of view of polynomial expansions, will be mentioned shortly.

where $\omega(z)$ is a so-called harmonic measure, i. e. a real function which satisfies the equation

$$\frac{\partial^2}{\partial x^2}\,\omega(z) + \frac{\partial^2}{\partial y^2}\,\omega(z) = 0$$

in the analyticity region (with an additional cut along the segment on which the data are given) and the boundary conditions

$$\omega(z) = 1 \quad \text{for } z \text{ in "experimental region"}$$
$$\omega(z) = 0 \quad \text{for } z \text{ on the boundary.}$$

(2)

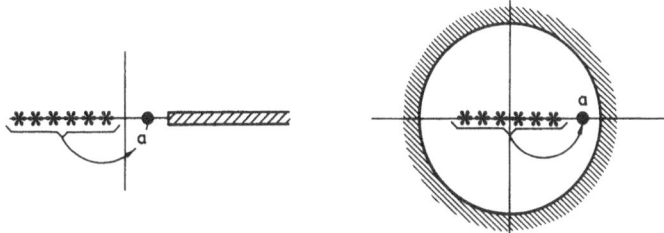

Fig. 1. Extrapolation from data (denoted by flowers) inside the analyticity region to a point a inside the analyticity region. The extrapolation is referred to as $I \rightarrow I$ type. The left half shows a typical case of a cut plane, the right half is the conformal mapping of the cut plane onto the unit circle

Instead of being more precise in statements contained in Eqs. (1) and (2) or reproducing the proofs*, we shall stress here the most important points and give a simple illustrative example.

Eq. (1) and (2) show that any estimate of the error requires either information or a hypothesis about the bound of the extrapolated function $(|f(z)| < M)$ in the whole analyticity region. To show this in a simple way, suppose that the analyticity region is the unit circle, the experimental region is, say, the segment $\langle -c, c \rangle, c < 1$ and one extrapolates to the point a, a real, $c < a < 1$.

Consider now the function

$$\chi(z) = \varepsilon' \exp\left\{\left[z - \frac{a+c}{2}\right]\frac{2}{a-c}\,\frac{K}{\varepsilon'}\right\},$$

(3)

---

* The Eqs. (1) and (2) are the Nevanlinna principle known from the theory of functions. Its first application to problems in particle physics is apparently due to *Bertero* and *Viano* [3]. The derivation of the Nevanlinna principle is given, for example in Refs. [2] and [3].

where K and ε' are positive and otherwise arbitrary. By construction, $\chi(z)$ is analytic in the unit circle,

$$|\chi(z))| < \varepsilon' \quad \text{for} \quad z \in (-c, c), \qquad (4)$$

$$\chi(a) = K.$$

Choosing $\varepsilon' \ll \varepsilon$, we can add $\chi(z)$ to any solution of the extrapolation problem without spoiling the fit [since errors on $(-c, c)$ are supposed to be of the order of ε]. Because of Eq. (4) we change at the same time the result of extrapolation by K (and so far K is arbitrary). Requiring $|f(z)|$ to be bounded by M in the whole unit circle, one restricts K, and as a consequence restricts also the error of extrapolation. The resulting error of course agrees with Eqs. (1) and (2). A somewhat more complicated example which just "saturates" the bounds given by the Nevanlinna principle is given by *Bertero* and *Viano* [3].

A glance at Eqs. (1) and (2) (or the preceding example) shows that the extrapolation to a point on the boundary is basically instable since the error of extrapolation becomes proportional to the bound M on $|f(z)|$ and not to the experimental arror ε. This is the reason why the extrapolation to the boundary is a more delicate problem which has to be treated separately.

The most important application of the $I \to I$ extrapolation to $\pi\pi$ physics is the *Chew-Low-Goebel* [4, 5] extrapolation, which we shall now discuss together with methods suitable for this type of extrapolation.

2.1.1 The Chew-Low-Goebel Extrapolation

The problem can roughly be described as follows: a function $f(t)$ is analytic in the complex t-plane with a cut $-\infty < t < -9$ (pion mass units, the cut is due to the 3π intermediate state in the $N\bar{N} \to 3\pi$ channel). At $t = -1$, $f(t)$ is simply related to the pion-pion cross-section or to one of the coefficients in the Legendre polynomial decomposition of the angular distribution for the dipion decay. The experimental data are available along the segment $0 < t < t_0$. The situation is displayed in Fig. 2a.

In applications of the CLG method performed so far, $f(t)$ has been constructed as a polynomial in t and the coefficients determined by the fit to the data in the experimental region*.

It is, however, well known that such a procedure is not the best thing to follow. About ten years ago *Frazer* [6] and *Ciulli* and *Fischer* [7] proposed to use conformal mapping techniques for phenomenological

---

* The error is then simply derived from statistical errors of these coefficients. We hope that the example given above indicates clearly that this estimate of error is not conservative enough.

extrapolation. *Frazer* dealt with the problem of determining the coupling constant of a particle which is exchanged in a given reaction by extrapolation in the variable $z = \cos\theta$, where θ is the scattering angle. He proposed to fit the data not by a polynomial in the original variable z but by a polynomial in $v = v(z)$, where $v(z)$ is a function which maps conformally the original analyticity region onto the unit circle (see Fig. 2b). He demonstrated by a numerical example that a fit by a polynomial of the fourth order in the original variable may be no better than a fit by a polynomial of the second order in a new variable v. The procedure is, of course, equivalent to performing first the conformal mapping $v = v(z)$ and then expanding the experimental data into a polynomial in the variable v.

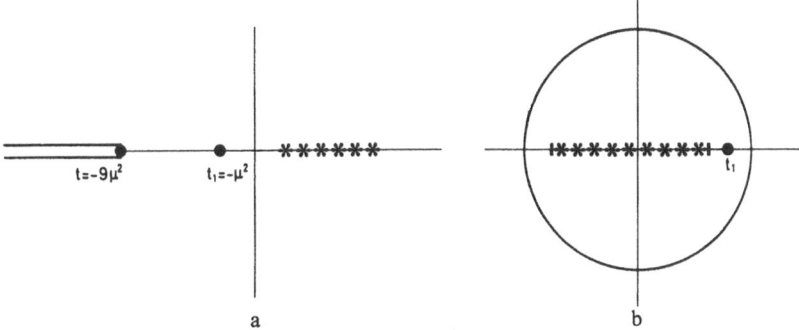

Fig. 2. a Analytic properties of functions extrapolated by the Chew-Low-Goebel Method. The region where data are available is denoted by flowers. b The conformal mapping of the cut t-plane of Fig. 2a onto the unit circle

Recently, *Cutkosky* and *Deo* [1] and *Ciulli* [2] improved the original proposal of *Frazer* [6] and of *Ciulli* and *Fischer* [7], and found the conformal mapping leading to the optimal asymptotic convergence of polynomial expansions. Even if the formal mathematical proofs assure that one has only the asymptotically optimal solutions, the method, as already exemplified by *Frazer*, may lead to marked improvements in fits by low-order polynomials. The use of these methods in future practical extrapolations is to be highly recommended.

In a rough but hopefully transparent way, the idea of the optimal mapping can be described as follows★: suppose that instead of knowing the data on a segment (as it is always in realistic cases) we know that a function $f(z)$ is analytic in a region shown in Fig. 3a and we have an exact information about the first N-derivatives of $f(z)$ in the origin.

★ For a full account of relevant mathematics, the reader is referred to original papers by *Cutkosky* and *Deo* [1] and *Ciulli* [2], and to the excellent book by *Walsh* [8].

Denoting

$$C_n = f^{(n)}(0)[n!]^{-1}$$

we can write approximate expression for $f(a)$ as

$$f(a) \approx f_N(a) = \sum_0^N C_n a^n . \qquad (5)$$

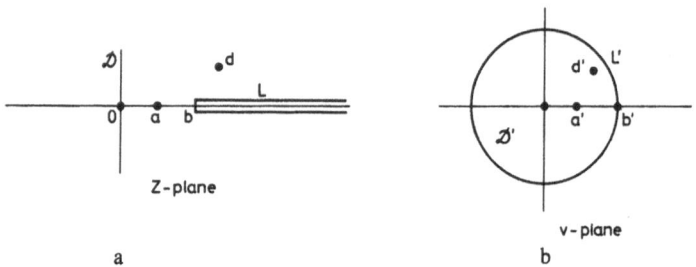

Z-plane

v-plane

a   b

Fig. 3a and b. A simple example illustrating the idea of the optimal mapping

If $f(z)$ is bounded by M in the analyticity region $\mathcal{D}$ and on the boundary L, we have from Cauchy theorem

$$C_m = \frac{1}{2\pi i} \int \frac{f(z')}{z'^{m+1}} \, dz' , \qquad (5')$$

and hence

$$|C_n| < M b^{-n} , \qquad (6)$$

where b is the distance from the origin to the nearest singularity. For the remainder of the approximation in Eq. (5) we can easily obtain the estimate

$$|f(a) - f_N(a)| \leq M \frac{b}{b-a} \left(\frac{a}{b} \right)^{n+1} \qquad (7)$$

Suppose now that a mapping $v = v(z)$ is performed [such that $v(0) = 0$]. In the new variable one can again construct the Taylor series and find the remainder similar to that given by Eq. (7). The new variable will be "asymptotically better" if the remainder becomes asymptotically smaller than in the original case. This is equivalent to the requirement that $|a'/b'|$ be smaller than a/b, where a' is the image of a and b' is the distance from the origin (in the v-plane) to the nearest singularity.

It is easy to believe* that the "asymptotically optimal" mapping for this case is the one shown in Fig. 3b, e. g. the mapping of the cut z-plane onto the unit circle. This mapping has two important properties:

---

* From the symmetry arguments.

i) The Taylor series converges in the new variable for any point inside the analyticity region (compare points d and d' in Figs. 3a and 3b).

ii) The upper bound of the remainder given in Eq. (7) is asymptotically the smallest in the v-plane.

The oversimplified point in the preceding example lies in the fact that the information (or the data) were assumed to be given in only one point. In a more realistic case when data are given on a segment, the optimal mapping was given by *Cutkosky* and *Deo* [1] and by *Ciulli* [2]. In *Ciulli*'s formulation one makes an additional cut along the "experimental region" (see Fig. 1) and maps the whole analyticity region (with two cuts) shown in Fig. 1 onto the annular region shown in Fig. 4a. If, as in most cases, the function is real in the data region in the original variable, it will be real on the unit circle in Fig. 4a. The appropriate expression for expanding the amplitude is then

$$w = \tfrac{1}{2}(v + v^{-1}) \tag{8}$$

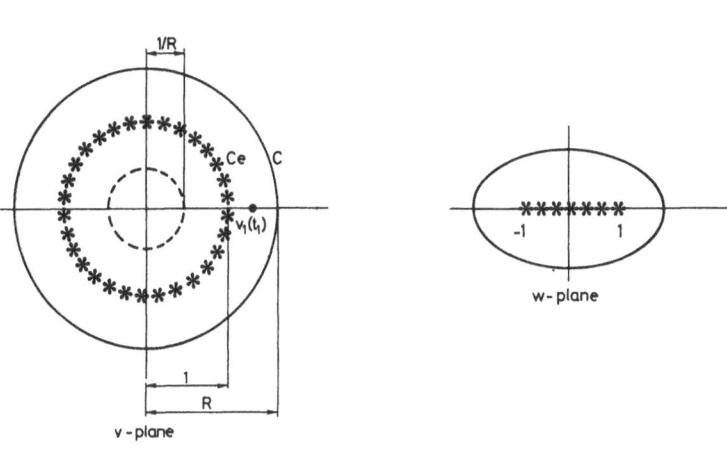

Fig. 4. a *Ciulli*'s form of the optimal mapping for cases similar to the Chew-Low-Goebel extrapolation. b *Cutkosky* and *Deo*'s form of the optimal mapping

which is by construction real for $|v| = 1$. Interpreting Eq. (8) as a conformal mapping, one obtains the form of the optimal mapping given by *Cutkosky* and *Deo* [2] (shown in Fig. 4b). *Cutkosky* and *Deo*'s and *Ciulli*'s mappings are completely equivalent, and both have the virtues (i) and (ii) mentioned above. For practical applications the *Cutkosky* and *Deo* mapping is perhaps more convenient since the amplitude is then decomposed (fitted) on the segment $-1 \leqq w \leqq 1$ by more familiar orthogonal polynomials.

As an alternative to polynomial expansions, a method based on using the weight functions was proposed in Ref. [9]. The idea goes as follows: denote as C_e the "experimental region" shown in Fig. 4a and as C the boundary of the analyticity region. Let $f(v)$ be the function which should be extrapolated from C_e to the point v_1. Let $g(v)$ be any function analytic for $1 \leq |v| \leq R$ and such that $g(v_1) = 1$.

On the basis of the Cauchy theorem, one has

$$f(v_1) = \frac{1}{2\pi i} \int_{C_e} \frac{f(v)\,g(v)}{v - v_1}\,dv + \frac{1}{2\pi i} \int_{C} \frac{f(v)\,g(v)}{v - v_1}\,dv. \qquad (9)$$

Choosing now $g(v)$ so that it is small on C [to suppress our ignorance of the function $f(v)$ on the boundary] we can obtain the estimate of $f(v_1)$. A simple example of a weight function $g(v)$ with this property is

$$g(v) = (v_1/v)^n \qquad (9')$$

with n being a positive integer. The optimal weight function of this kind is given in Refs. [10, 24]

2.1.2 The Estimate of Errors in $I \to I$ Extrapolation

The most serious pitfall of analytic extrapolations lies in the estimate of errors. Let us start again with a simple example. Suppose that $f(z)$ is analytic in the unit circle and that the derivatives at the origin are known with experimental errors up to the order N, so that one can write

$$C_n \equiv f^{(n)}(0)\,[n!]^{-1} = C_n' + \varepsilon_n, \qquad n = 0,1,...,N.$$

Here C_n are the true values, C_n' the experimental data, and ε_n the errors. For a value of $f(z)$ in a point a one then obtains

$$f(a) - f_N'(a) = \sum_0^N \varepsilon_n a^n + \sum_{N+1}^\infty C_n a^n, \qquad (10)$$

where

$$f_N'(a) = \sum_0^N C_n' a^n, \qquad (10')$$

is the estimate of $f(a)$ obtained from the "data", e. g. from the information about C_n' for $n = 0,1,...N$. The first term on the right-hand side of Eq. (10) can be treated as a statistical error, while the second term is a typical extrapolation error. As shown in Eqs. (5), (6), and (7) one can estimate the extrapolation error if $f(z)$ is known to be bounded on the boundary by a constant M. If such information is not available one has to guess the behaviour of the coefficients C_n for $n > N$ from the behaviour of C_n with $n \leq N$.

A proposal by *Cutkosky* and *Deo* [1] consists essentially in estimating the constant M in Eq. (6) from the coefficients C_n with $n \leq N$ and using this M also for $n > N$. Even if this is a guess it is apparently the best that can be done when *a priori* information about $f(z)$ on the boundary is not available.

The situation gets a bit worse in a realistic situation when the data are given on a segment and not idealized as a set of derivatives in a point. The discussion of the estimates of errors is, however, similar to the simplified example just described, and an interested reader is referred to *Bertero* and *Viano* [3], for example, or to the last section of Ref. [9] for details★.

An important question connected with estimating the error of extrapolation can be formulated as follows:

When are the data good enough for performing
the Chew-Low-Goebel extrapolation?★★

In performing a CLG extrapolation one actually separates the contribution of the one-pion exchange from that of other mechanisms. It is therefore advisable to choose only reactions which are known (from the study of t-dependence of cross-sections and from the behaviour of density matrix elements) to be dominated by OPE★★★. A quantitative insight into the question whether the data are able to reveal the presence of the pion pole and give some information about its residue can be obtained along the following lines.

Coming back to a simple example. Suppose that $f(z)$ is analytic in the unit circle; we know the coefficients C_n in its Taylor expansion in the origin and we suspect that the function has a pole in a point b, $b < 1$. According to Eq. (6) it holds that

$$|C_n| < M \tag{11a}$$

if there is no singularity inside the unit circle, and

$$|C_n| < M b^{-n} \tag{11b}$$

if there is a singularity in a point b. If the information about C_n is good enough to favour at a pragmatic level the alternative (11b), it implies

---

★ The troubles with estimating errors of analytic extrapolations come, in general, from difficulties connected with merging statistical concepts with analyticity. A pioneering step in this direction was recently taken by *Cutkosky* [11].

★★ The author is indebted to *F. James* for valuable discussions concerning this point.

★★★ The related questions are discussed in detail in excellent reviews by *G. Kane* [12].

that the data "feel" the presence of the pole and the extrapolation makes sense. In a more realistic situation when the data are given along a segment, the argument is unchanged; one only needs to study the dependence on n of coefficients in polynomial representation of the data, instead of C_n's described here.

2.2 $I \rightarrow B$ Extrapolation

The extrapolation from a segment inside the analyticity region to the boundary is shown in a schematic way in Fig. 5. The nature of the problem is illustrated by the following example. Suppose that $f(z)$ is analytic in the unit circle, data are given on the segment $(-\frac{1}{2}, \frac{1}{2})$, and we wish to extrapolate to the point $z = 1$. Let errors of the data be of the order ε.

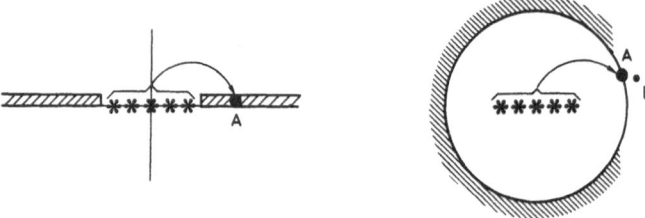

Fig. 5. Extrapolation from data inside the analyticity region to the boundary. The case is referred to as $I \rightarrow B$ extrapolation

Adding a function

$$\chi(z) = K \frac{\varepsilon'}{1 + \varepsilon' - z} \tag{12}$$

to any solution of the extrapolation problem does not spoil the fit to the data on the segment $(-\frac{1}{2}, \frac{1}{2})$, provided that $\varepsilon' < \varepsilon$, but changes the result of extrapolation to the point $z = 1$ by an arbitrary* number K.

The derivatives of $\chi(z)$ for $z = 1$ are given by the relation

$$\chi^{(n)}(1) = \frac{K}{(\varepsilon')^n}. \tag{13}$$

This relation illustrates one possible way of avoiding the instability of the extrapolation to the boundary. If the *derivatives* of the extrapolated function are required to exist on the boundary up to a given order n_0 and

---

* If $f(z)$ is known to be bounded by M for $|z| \leqq 1$, $|K|$ should naturally be less than M.

be bounded by a constant, this in turn restricts the value of K in Eq. (13) and assures stability. A method for extrapolation to the boundary along these lines was discussed by *Ciulli* [2]. In cases when bounds on derivatives are not available, one has to use another approach* which consists in extrapolating, *not* to values of a function at points on the boundary, but to *average* values over arcs on the boundary. The reason why the extrapolation to average values over arcs is stable is intuitively clear from the example of $\chi(z)$ as given by Eq. (12). Even if $\chi(1) = K$, any average value of $\chi(z)$ along an arc longer than ε' would be of the order of $(\varepsilon' K)$.

The results of Refs. [13], [14] and [10] show that the stability of extrapolation increases with the increase of the length of the arc over which the average value is calculated.

Instances where $I \rightarrow B$ extrapolation is employed, in the context of $\pi\pi$ scattering are

i) models for $\pi\pi$ scattering above threshold incorporating the constraints of current algebra (hard-pion theory):

ii) models for $\pi\pi$ scattering based on *Martin* inequalities and other rigorous constraints valid below the threshold.

2.3 $B \rightarrow I$ Extrapolation

This is shown in Fig. 6. The methods applicable to this situation are described in Refs. [15–18]. The error of the extrapolation is formally given by Eq. (1) where the harmonic measure is this time given by the conditions

$$\omega(z) = 1 \quad \text{on the "experimental" part of the boundary,}$$
$$\omega(z) = 0 \quad \text{on the "unknown" part of the boundary.}$$

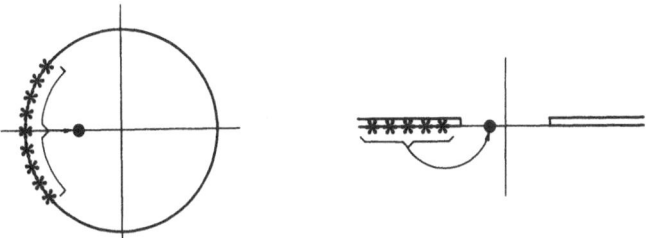

Fig. 6. Analytic extrapolation from a part of the boundary to a point inside the analyticity region (referred to as $B \rightarrow I$ extrapolation)

---

* This is actually the approach used by *Bowcock et al* [13] and in a different formulation in Refs. [14, 10].

The extrapolation becomes perfectly stable (in the sense that the error of extrapolation is proportional to the error of the data) when the "experimental" region is identical to the whole boundary. This is the case of various sum rules expressing coupling constants (or values of amplitudes at a particular point) as integrals of scattering amplitudes over the whole physical region.

The $B \rightarrow I$ problem is just the inverse of $I \rightarrow B$ extrapolation. Thus $B \rightarrow I$ extrapolation can, for example, be used for testing consistency of $\pi\pi$ phase shifts with *Martin* inequalities or with current algebra predictions.

2.4 $B \rightarrow B$ Extrapolation (see Fig. 7)

Mathematically the problem of the stability is similar to that of the $I \rightarrow B$ case. The solution for the case where restrictions on derivatives are imposed on the boundary is most probably feasible, but to the best of our knowledge has not, so far, been discussed in the literature. The optimal solution for extrapolation to average values along arcs on the boundary is given by *Prešnajder et al.* [10].

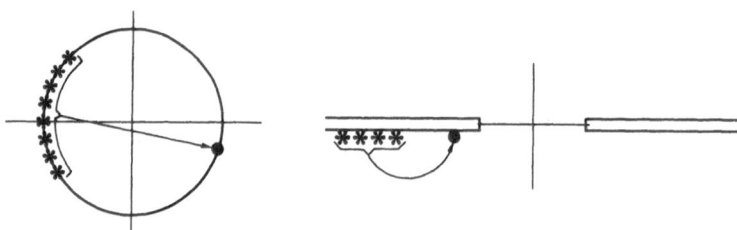

Fig. 7. Extrapolation from a part of the boundary to the rest of the boundary ($B \rightarrow B$ extrapolation)

In pratice, one uses suitable weight functions [14–19], and one applies the Cauchy theorem to the product of the weight function and the function to be extrapolated. The results [10] can be summarized as follows. The region shown in Fig. 7 is mapped onto the one shown in Fig. 8. Then if the error in the experimental region is of the order ε, the error of the mean value over the segment $\langle -1,1 \rangle$ in Fig. 8, obtained with the optimal weight function, is proportional to $\varepsilon^{(\alpha/\pi)}$. Note that for $\alpha = \pi$ the situation (as expected) is perfectly stable.

Actual extrapolations to the boundary (both cases $I \rightarrow B$ and $B \rightarrow B$) have, so far, been performed almost exclusively in the following way: the amplitudes were first written in forms containing a few free parameters which were determined later on from fits in the "experimental region". Due to the instabilities inherent in the problem, results are expected to be very sensitive to the form of the parametrization chosen.

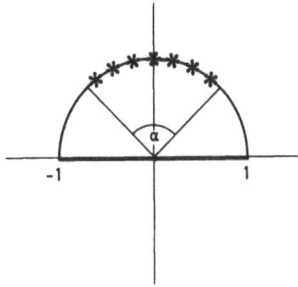

Fig. 8. The conformal mapping used in extrapolation to average values along arcs on the boundary in the $B \rightarrow B$ case

An honest solution to these complications was given by *Lovelace*, *Heinz*, and *Donnachie* [20] in their determination of δ_0^0 from πN backward dispersion relations*. Fig. 9 shows a limited sample of the set of phase shifts obtained by them in different parametrizations. It is clear that average values of corresponding amplitudes over the range from threshold to about 1 GeV are all more or less the same, while averages taken over smaller intervals, say from threshold up to 350 MeV, are still dependent on the parametrization chosen. As an alternative to this type of solution, one could calculate directly the average values of the $I = J = 0$ $\pi\pi \rightarrow N\bar{N}$ partial wave amplitude.

The problems where one is faced with a $B \rightarrow B$ type of extrapolation are numerous. For instance:

a) the determination of the left-hand cut singularities from the right-hand cut ones (*Bowcock* and *John* [21]);

b) predictions of $\pi\pi$ phase shifts from πN backward scattering;

c) predictions of $\pi\pi$ phase shifts from πN partial wave dispersion relations;

d) the determination of $\pi\pi$ low-energy behaviour from forward dispersion relations. In this case the extrapolation ($B \rightarrow B$ type) is not the full story, since at low energies s-waves dominate and the requirement of the unitarity will probably reduce the ambiguities to something like CDD-pole ambiguities. At present there is, however, no mathematical theory for this problem.**

---

* For other references pertaining to this approach, see the companion paper by *D. Morgan* and the author.

** The author is indebted to *D. Atkinson* for agreeable discussions about these questions.

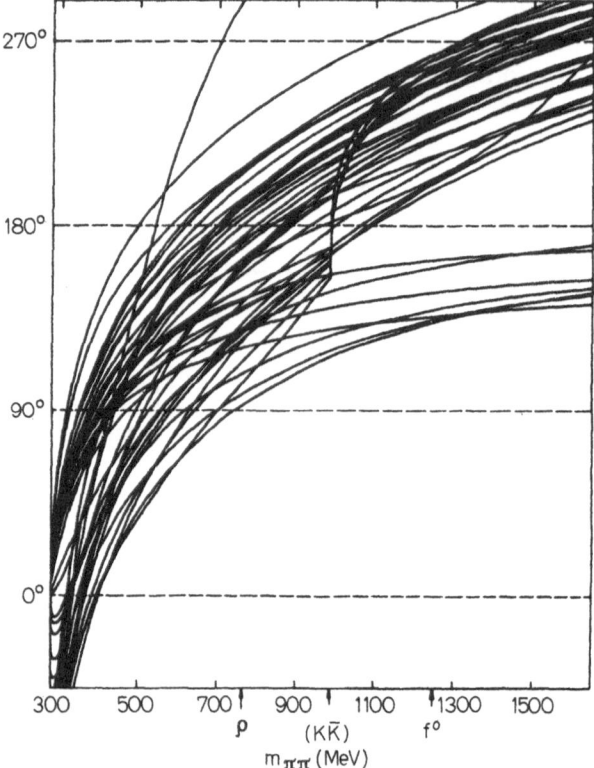

Fig. 9. $\pi\pi\ \delta_0^0$ phase shift obtained by *Lovelace, Heinz* and *Donnachie* from πN backward dispersion relations

In all these applications quite contradictory results can be found in the literature. The major explanation lies probably in the fact that the extrapolation to the boundary is, in a local sense, basically instable, and as a consequence the results of calculations are strongly dependent on the parametrizations used.

3. Ambiguities in the Determination of Phase Shifts from Final State Enhancement Factors*

Suppose, for simplicity, that there are only two strongly interacting particles in the final state. A common procedure is to make an ansatz for the phase shift δ (the two particles are supposed to be in a state

---

    * Various enhancement factors and their relations are discussed in, for example, *Jackson* and *Kane* [22], and *Jacob, Mahoux*, and *Omnès* [23].

with given angular momentum) containing a few free parameters, then to construct an enhancement factor

$$E(s) = \exp\{R(s) + i\delta(s)\} \tag{14}$$

with

$$R(s) = \frac{s - s_0}{\pi} \, \mathrm{P} \int\limits_{s_0}^{\infty} \frac{\delta(s')}{(s' - s_0)(s' - s)} \, ds' \tag{15}$$

and insert it into the amplitude*. If there is no other final-state interaction, the invariant mass distribution of the two strongly interacting particles is given as

$$\frac{d\sigma}{dm} \sim |E(s)|^2 = \exp\{2R(s)\} . \tag{16}$$

Using Eq. (16) one fits the data in a finite region, say $s_0 < s < s_1$, determines the free parameters, and obtains the phase shift $\delta(s)$. It is natural to ask wether the procedure leads to an unambiguous result, or whether the result may be biased by the parametrization chosen.

An obvious ambiguity is exhibited by the transformation

$$\delta(s) \to \delta'(s) = \delta(s) + C\sqrt{s - s_0} \tag{17}$$

since the principal value integral with the square root term explicitly gives zero. The example may be considered as unrealistic since $\delta'(s) \to \infty$ for $s \to \infty$. It is, however, easy to construct a general class of functions $\delta(s)$ which will lead to the same $R(s) = |E(s)|^2$ for $s_0 < s < s_1$. Let us first perform the conformal mapping $z = z(s)$ shown in Fig. 10 and take any function of the form

$$\Delta(s) = \mathrm{Re}\left\{ \sum_0^N a_{2n} z^{2n} \right\}. \tag{18}$$

The corresponding part of the enhancement factor is then given as**

$$R(s) = -\mathrm{Im}\left\{ \sum_0^N a_{2n} z^{2n} \right\}. \tag{19}$$

Taking the coefficients a_{2n} as real, it obviously follows that

$$\Delta(s) = \sum_0^N a_{2n} [z(s)]^{2n} \quad \text{for } s_0 < s < s_1 \tag{20}$$

---

* The expression in Eq. (14) corresponds to $D^{-1}(s)$ in the N/D method. Other enhancement factors contain also effects of the left-hand cut or other, slowly varying functions of s.

** Note that $\Delta(s)$ and $R(s)$ are the imaginary and the real parts of an analytic function for $s_0 < s < s_1$.

and

$$R(s) = 0 \qquad\qquad \text{for } s_0 < s < s_1. \qquad (21)$$

The addition of $\Delta(s)$ to the phase shift $\delta(s)$ does not change $R(s) \equiv |E(s)|^2$ on the interval $s_0 < s < s_1$, while it can change the phase shift in a substantial way★. Note further that $\Delta(s)$ is finite by construction for $s \to \infty$, although it shows oscillations, which can become physically unacceptable for large N. In a fitting procedure the phase shift $\delta(s)$ may not approach the true phase shift but one of the class

$$[\delta_{\text{true}}(s) + \Delta(s) + c\sqrt{s - s_0}],$$

depending on the way in which $\delta(s)$ was parametrized.

From this point of view the results on $\pi\pi$ phase shifts coming from the analysis of $\bar{n}p \to 3\pi$ are preferable to those from $\pi N \to \pi\pi N$ below the threshold for $\Delta(1238)$ production, since in the former case the phase shifts δ_l^I are responsible for the interference of various amplitudes, and the preceding argument does not apply. On the other hand, a three-body final state offers better possibilities of neglecting important three-body diagrams.

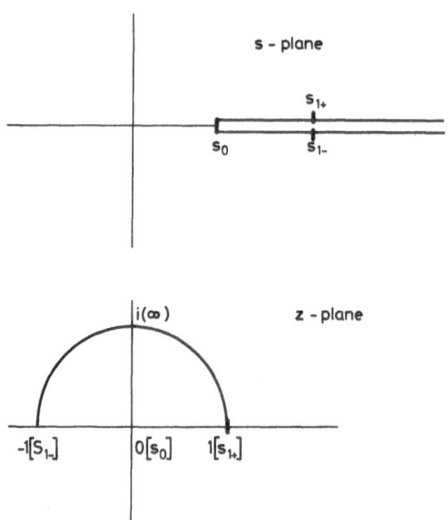

Fig. 10. Ambiguities in determining phase shifts from enhancement factors. The upper part shows the original s-plane, the lower part its mapping onto the semicircle in the z-plane. The enhancement factor is supposed to be known experimentally for $s_0 < s < s_{1+}$

★ The whole ambiguity is due to the trivial fact that an analytic function is not given in an unambiguous way by values of its real part along a *part* of the boundary.

4. Comments

The complications that arise when obtaining reliable information on low-energy parameters by explicit or tacit analytic extrapolations are not limited to the pion-pion case. They are, however, most pronounced here since a clear-cut resolution of ambiguities by experimental data is difficult and perhaps impossible owing to the lack of direct evidence. A critical attitude towards some of the results obtained is probably necessary if a consistent picture is finally to emerge. In the present brief review of pitfalls of extrapolations, we have restricted ourselves only to the description of a few basic points; for details about the extrapolation methods the reader should consult the references quoted above, and for illustrative examples from the field of pion-pion interactions the companion lecture by *D. Morgan* and the author on pion-pion scattering at low energies.

Acknowledgements

The author is indebted to Drs. *Peter Prešnajder* (Bratislava), *A. Nogová* (Bratislava), *J. Fischer* (Prague), *Matts Roos* (CERN), *David Morgan* (Rutherford Lab.), *Fred James* (CERN), *F. Mueller* (CERN), and *D. Drijard* (CERN) for valuable discussions about analytic extrapolations and their applications.

References

1. *Cutkosky, R. E., Deo, B. B.:* Phys. Rev. **174**, 1859 (1968); — Phys. Rev. Letters **22**, 1272 (1968).
2. *Ciulli, S.:* Nuovo Cimento **61** A, 787 (1969); **62** A, 301 (1969).
3. *Bertero, M., Viano, G. A.:* Nuovo Cimento **38**, 1915 (1965).
4. *Chew, G. F., Low, F. E.:* Phys. Rev. **113**, 1640 (1959).
5. *Goebel, C.:* Phys. Rev. Letters **1**, 337 (1958).
6. *Frazer, W. R.:* Phys. Rev. **123**, 2180 (1961).
7. *Ciulli, S., Fischer, J.:* Nucl. Phys. **24**, 465 (1961).
8. *Walsh, J. L.:* Interpolation and Approximation by Rational Functions in the Complex Domain, Amer. Math. Soc. Coll. Publ. Vol. XX, published by the Amer. Math. Soc., Providence, Rhode Island, 1956.
9. *Pišút, J., Prešnajder, P.:* Nucl. Phys. B **12**, 110 (1969).
10. *Prešnajder, P., Pišút, J.:* CERN preprint TH. 1158 (1970).
11. *Cutkosky, R. E.:* Ann. Phys. (N. Y.) **54**, 350 (1969).
12. *Kane, G. L.:* Proc. Argonne Conf. on $\pi\pi$ and πK Scattering, 1969. Also, talk given at the Philadelphia Conference on $\pi\pi$ and πK Scattering, 1970 (Michigan preprint).
13. *Bowcock, J. E., Cottingham, W. N., Williams, J. G.:* Nucl. Phys. B3, 95 (1967).
14. *Pišút, Prešnajder, P., Fischer, J.:* Nucl. Phys. B **12**, 586 (1969).
15. *Carleman, T.:* Sur les fonctions quasianalytiques, Paris 1926.
16. *Lavrentiev, M. M.:* On Some Incorrect Problems of Mathematical Physics (Russian), published by Acad. of Sci., Novosibirsk 1962 (see also Springer Tracts, Vol. 11).

17. *Prešnajder, P., Pišut, J.*: Nucl. Phys. B **14**, 489 (1969).
18. *Ciulli, S., Fischer, J.*: NORDITA, preprint 1970.
19. *Fischer, J., Pišút, J., Prešnajder, P., Šebesta, J.*: Czech. J. Phys. **19**, 1486 (1969).
20. *Lovelace, C., Heinz, R. M., Donnachie, A.*: Phys. Letters **22**, 332 (1966).
21. *Bowcock, J. E., John, G.*: Nucl. Phys. B **11**, 659 (1969).
22. *Jackson, J. D., Kane, G. L.*: Nuovo Cimento **23**, 444 (1962).
23. *Jacob, M., Mahoux, G., Omnès, R.*: Nuovo Cimento **23**, 838 (1962).
24. *Nenciu, G.*: Lett. Nuovo Cim. **4**, 96 (1970).

Dr. *Ján Pišút*
CERN (European Organization for Nuclear Research)
CH-1211 Genf 23

Coulomb Corrections in the Analysis of πN Experimental Scattering Data

G. C. Oades

Contents

1. Introduction 61
2. General Form of the Scattering Amplitude in the Simultaneous Presence of the
      Nuclear and Electromagnetic Interactions 62
3. Review of Various Treatments of the πN Coulomb Corrections 65
4. Influence of Coulomb Corrections on Low Energy Data 70
References 71

1. Introduction

Our knowledge of the πN interaction comes mainly from the study of the scattering processes

$$\pi^+ p \to \pi^+ p \,,$$

$$\pi^- p \to \pi^- p \,,$$

$$\pi^- p \to \pi^0 n \,.$$

In all three cases the initial states involve charged particles as do the final states of the first two processes and it is clear that a complete analysis of such experiments must take into account both the electromagnetic and the strong interactions of the particles involved. It is the purpose of this review to describe those electromagnetic effects which can most easily be included when analysing the basic scattering data. These are the effects which, in a potential model, would occur when the Coulomb potential is included in addition to the nuclear potential. It is also desirable to include the $\pi^\pm - \pi^0$ and $p - n$ mass differences and the radiative capture process $\pi^- p \to \gamma n$ when analysing this basic scattering data although here the situation is less well defined. This does not, of course, exhaust the electromagnetic effects and other types of corrections will be discussed by *Steiner* (1970).

The problem of Coulomb corrections is not confined to πN scattering and the contents of this review can easily be extended to other processes. For a discussion of the corrections usually applied in NN scattering see *Breit* and *Haracz* (1967) and in $\bar{K}N$ scattering see *Dalitz* and *Tuan* (1960).

2. General Form of the Scattering Amplitude in the Simultaneous Presence of the Nuclear and Electromagnetic Interactions

In the absence of electromagnetic effects and making the usual assumption that the πN strong interaction is of short range, the asymptotic form of the outgoing wave function for a given angular momentum state can be written for a $\pi^+ p$ or $\pi^- p$ incoming state as

$$\pi^+ p \text{ incoming state} : \left\{ \sin(\varrho - l\pi/2) + \frac{1}{2i}(S_{N,l\pm}^{++} - 1)\,e^{i(\varrho - l\pi/2)} \right\} |\pi^+ p\rangle ,$$

$$\pi^- p \text{ incoming state} : \left\{ \sin(\varrho - l\pi/2) + \frac{1}{2i}(S_{N,l\pm}^{--} - 1)e^{i(\varrho - l\pi/2).} \right\} |\pi^- p\rangle \quad (1)$$

$$+ \left\{ \frac{1}{2i} S_{N,l\pm}^{0-}\, e^{i(\varrho - l\pi/2)} \right\} |\pi^0 n\rangle ,$$

where, neglecting mass differences, $\varrho = qr$, q being the c. m. momentum and r the relative radial coordinate (units $\hbar = c = 1$). Assuming charge independence the purely nuclear S-matrix elements have the usual form

$$S_{N,l\pm}^{++} = e^{2i\delta_{l\pm}^{3/2}} ,$$

$$S_{N,l\pm}^{--} = \frac{1}{3} e^{2i\delta_{l\pm}^{3/2}} + \frac{2}{3} e^{2i\delta_{l\pm}^{1/2}} , \quad (2)$$

$$S_{N,l\pm}^{0-} = \frac{\sqrt{2}}{3}(e^{2i\delta_{l\pm}^{3/2}} - e^{2i\delta_{l\pm}^{1/2}}) ,$$

where $\delta_{l\pm}^{3/2}$ and $\delta_{l\pm}^{1/2}$ are the isotopic spin 3/2 and 1/2 (in general complex) scattering phases for the angular momentum states $l\pm$.

When the long range Coulomb force is present and mass differences are included the asymptotic forms (1) become modified to (neglecting the γn channel)

$\pi^+ p$ incoming state $: \left\{ \sin(\varrho - l\pi/2 + \sigma_l - \eta \ln 2\varrho) \right.$

$$+ \frac{1}{2i} (S^{++}_{\text{TOT},l\pm} - 1) \, e^{i(\varrho - l\pi/2 + \sigma_l - \eta \ln 2\varrho)} \bigg\} | \pi^+ p \rangle ,$$

(3)

$\pi^- p$ incoming state $: \left\{ \sin(\varrho - l\pi/2 - \sigma_l + \eta \ln 2\varrho) \right.$

$$+ \frac{1}{2i} (S^{--}_{\text{TOT},l\pm} - 1) \, e^{i(\varrho - l\pi/2 - \sigma_l + \eta \ln 2\varrho)} | \pi^- p \rangle$$

$$+ \left\{ \frac{1}{2i} S^{0-}_{\text{TOT},l\pm} e^{i(\varrho_0 - l\pi/2)} \right\} | \pi^0 n \rangle ,$$

where now $\varrho = qr$, q being the $\pi^\pm p$ c. m. momentum and $\varrho_0 = q_0 r$, q_0 being the $\pi^0 n$ c. m. momentum. The Coulomb parameter η is defined by

$$\eta = \alpha m/q \quad (\alpha = 137.036^{-1}),$$

m being the reduced mass of the $\pi^\pm p$ system, and the Coulomb phases are given by

$$\sigma_l = \arg \Gamma(l + 1 + i\eta).$$

The connection between the asymptotic forms (3) and the scattering amplitude proceeds in the usual way (see Chapter 3 of *Mott* and *Massey* (1965) for the treatment of the spinless case and *Oades* and *Rasche* (1970) for a summary of the extension to spin 1/2 − spin 0 scattering). The result is that the scattering amplitudes have the form

$$F = f + i\boldsymbol{\sigma} \cdot \hat{\boldsymbol{n}} g$$

where

$$\hat{\boldsymbol{n}} = \hat{\boldsymbol{q}} \times \hat{\boldsymbol{q}}'$$

$\hat{\boldsymbol{q}}$ and $\hat{\boldsymbol{q}}'$ being unit vectors along the incoming and outgoing pion directions.

For $\pi^\pm p \to \pi^\pm p$ the non-spin-flip amplitude has the form

$$f^{\pi^\pm p \to \pi^\pm p} = f^{\pi^\pm p \to \pi^\pm p}_{\text{Coulomb}} + \sum_l \{ (l+1) f^{\pm\pm}_{l+} + l f^{\pm\pm}_{l-} \} \, P_l(\cos\theta) \tag{4}$$

and the spin-flip amplitude has the form

$$g^{\pi^\pm p \to \pi^\pm p} = g^{\pi^\pm p \to \pi^\pm p}_{\text{Coulomb}} + \sum_l \{ f^{\pm\pm}_{l+} - f^{\pm\pm}_{l-} \} \, P^1_l(\cos\theta) \tag{5}$$

where θ is the c. m. scattering angle and where

$$f_{l\pm}^{\pm\pm} = e^{\pm 2i\sigma_l}\left(\frac{S_{\text{TOT},l\pm}^{\pm\pm}-1}{2iq}\right).$$ (6)

Also $f_{\text{Coulomb}}^{\pi^\pm p\to\pi^\pm p}$ and $g_{\text{Coulomb}}^{\pi^\pm p\to\pi^\pm p}$ are the electromagnetic scattering amplitudes which would be present in the absence of strong interactions. Non-relativistically these have the form

$$f_{\text{Coulomb}}^{\pi^\pm p\to\pi^\pm p} = \frac{\mp\eta}{2q\sin^2\theta/2}\exp(\mp i\eta\ln\sin^2\theta/2\pm 2i\sigma_0),$$ (7)

$$g_{\text{Coulomb}}^{\pi^\pm p\to\pi^\pm p} = 0.$$

(It should be noted that most authors multiply the total scattering amplitude by the unobservable phase factor $e^{\mp 2i\sigma_0}$ thus removing it from the Coulomb amplitude). Several authors have treated the problem of including magnetic moment and relativistic effects in the electro-magnetic scattering amplitudes at which stage a contribution to $g_{\text{Coulomb}}^{\pi^\pm p\to\pi^\pm p}$ is obtained. A convenient review is given by *Roper* (1965) although it should be noted that some authors prefer to express these magnetic moment and relativistic effects as modifications to the Coulomb phases e. g. *Vik* and *Rugge* (1963).

For the charge exchange process the corresponding forms for the non-spin-flip and spin-flip amplitudes are

$$f^{\pi^- p\to\pi^0 n} = \sum_l \{(l+1)f_{l+}^{0-} + lf_{l-}^{0-}\}\, P_l(\cos\theta),$$ (8)

$$g^{\pi^- p\to\pi^0 n} = \sum_l \{f_{l+}^{0-} - f_{l-}^{0-}\}\, P_l^1(\cos\theta),$$ (9)

where

$$f_{l\pm}^{0-} = e^{-i\sigma_l}(S_{\text{TOT},l\pm}^{0-}/2iq).$$ (10)

These formulae are independent of any particular model, corre-sponding only to the extension of the usual partial wave decomposition to include long range Coulomb effects. The obvious disadvantage is that the three processes,

$$\pi^+ p\to\pi^+ p,$$
$$\pi^- p\to\pi^- p,$$
$$\pi^- p\to\pi^0 n,$$

are now described for each angular momentum state in terms of three, in general complex, S-matrix elements rather than in terms of two, again in general complex, charge independent phases. In principal it would be desirable to analyse the experimental data for each process in terms of the $S_{\mathrm{TOT},l\pm}$ and quote values and errors for these quantities, thus clearly distinguishing between the experimental uncertainties in determining the $S_{\mathrm{TOT},l\pm}$ and the theoretical uncertainties in obtaining the $S_{N,l\pm}$ from these values. At present the increased number of parameters renders this impossible and it is already necessary at the experimental analysis stage to relate the $S_{\mathrm{TOT},l\pm}$ to the $S_{N,l\pm}$ so that the various processes can be simultaneously analysed in terms of the charge independent phases $\delta_{l\pm}^{3/2}$ and $\delta_{l\pm}^{1/2}$. It is here that model dependent treatments become necessary and in the next section work on this problem is reviewed.

3. Review of Various Treatments of the πN Coulomb Corrections

In all phase shift analyses, with the exception of some early work below 100 MeV, the data is analysed in the approximation of replacing $S_{\mathrm{TOT},l\pm}$ by the charge independent values $S_{N,l\pm}$ (e. g. see *Donnachie*, 1967). While this may be a good approximation when dealing with data having errors of 5% or more, it can be dangerous when analysing data of higher accuracy. As the accuracy increases it is no longer possible to simultaneously fit elastic $\pi^+ p$, elastic $\pi^- p$ and $\pi^- p$ charge exchange data in this charge independent approximation (*Bugg*, 1970). This does not necessarily mean that charge independence has broken down since the usual definitions of charge independence include a phrase of the type "in the absence of electromagnetic effects". It means that the approximation of replacing $S_{\mathrm{TOT},l\pm}$ by $S_{N,l\pm}$ is no longer good enough and Coulomb corrections must be included.

It is hoped that, in the low energy region, these Coulomb corrections are dominated by those effects which can be calculated in a potential model by including the Coulomb potential in addition to the charge independent nuclear potentials. It is also hoped that the importance of these effects decreases with energy so that their inclusion at higher energies is not necessary. These assumptions are open to both experimental and theoretical investigation and further work on this problem is very desirable. From now on only these potential model Coulomb corrections will be discussed in detail although, as mentioned in the introduction, *Steiner* (1970) will discuss other possible effects.

The problem of Coulomb corrections in πN scattering was first discussed by *Van Hove* (1952). He assumed that at each energy the πN radial wave functions obeyed the Schrödinger equation with separate charge independent nuclear potentials for each angular momentum state. Since the treatment is at fixed energy it is not necessary to assume energy independent nuclear potentials and so the model is rather general. *Van Hove* went on to make the more restrictive assumption that the Coulomb potential vanished within the nuclear interaction region and he then obtained expressions for the s and p wave S-matrix elements, $S_{\text{TOT}, 0+}$ and $S_{\text{TOT}, 1\pm}$, for $\pi^{\pm} p \to \pi^{\pm} p$ and for $\pi^- p \to \pi^0 n$. He also indicated how first order relativistic effects could be included by suitably modifying the Coulomb parameter but he did not include mass differences in his treatment.

The corrections obtained by *Van Hove* are now called the outer Coulomb corrections and one serious problem is that they depend strongly on the value chosen for the nuclear interaction radius. It was later shown by *Hamilton* and *Woolcock* (1960) that if further inner Coulomb corrections were calculated by including the Coulomb potential within the nuclear interaction region then this dependence on the nuclear radius was greatly reduced. It should be noted that it is at this stage that charge structure effects also enter into the problem, the Coulomb potential departing from the point charge potential at small distances due to the finite spatial charge distributions.

In their work *Hamilton* and *Woolcock* estimated values for the inner Coulomb corrections to the s-wave scattering lengths. They used a real nuclear potential for $\pi^+ p \to \pi^+ p$ and a complex nuclear potential for the $\pi^- p$ processes, the real part of this potential determining $\pi^- p \to \pi^- p$ and the imaginary part $\pi^- p \to \pi^0 n$. *Schnitzer* (1963) later extended this treatment to obtain the inner Coulomb corrections to the scattering amplitudes at finite energies. In this work he did not, however, take into account the coupling between the $\pi^- p$ and $\pi^0 n$ channels due to the Coulomb potential and so his results are only applicable to $\pi^+ p \to \pi^+ p$. *Auvil* (1968) obtained similar results to those of *Schnitzer* but again only for the single channel $\pi^+ p \to \pi^+ p$ problem.

Recently *Sauter* (1969) has reconsidered the problem of Coulomb corrections to $\pi^+ p \to \pi^+ p$ from a different standpoint. He uses the techniques of *Dashen* and *Frautschi* (1964) rather than the non-relativistic Schrödinger equation and he compares the s-wave inner and outer Coulomb corrections obtained by the two methods, finding reasonable agreement in the non-relativistic region. This work is very interesting since it represents the first attempt to develop a method of treatment which can be extended into the relativistic region.

As pointed out above, the methods of *Schnitzer* or *Auvil* are only applicable to the single channel $\pi^+ p \to \pi^+ p$ problem. To remedy this defect *Oades* and *Rasche* (1970) have made a similar treatment of the coupled channel $\pi^- p \to \pi^- p$ and $\pi^- p \to \pi^0 n$ processes. At present this work corresponds to the extension *Van Hove*'s treatment of these processes to include inner Coulomb corrections. As in the other cases this ignores mass differences and also the effects of the radiative capture process, $\pi^- p \to \gamma n$, but these effects will be included later.

To summarise the conclusions of all these potential model treatments of low energy Coulomb corrections, in the case of $\pi^+ p \to \pi^+ p$ the effect of these corrections is to modify the charge independent phases $\delta_{l\pm}^{3/2}$ to new values $\tau_{l\pm}^+$ so that

$$S_{\mathrm{TOT},l\pm}^{++} = e^{2i\tau_{l\pm}^+} \tag{11}$$

where

$$\underset{\eta \to 0}{\mathrm{Lt}} \, \tau_{l\pm}^+ = \delta_{l\pm}^{3/2} . \tag{12}$$

In the case of $\pi^- p \to \pi^- p$ and $\pi^- p \to \pi^0 n$ the situation is more complicated since the Coulomb potential acts only on the $\pi^- p$ state and so couples the $T = 3/2$ and $T = 1/2$ states. Thus there is a mixing in isospin space and the S-matrix elements are expressed in terms of two phases and a mixing parameter in the form

$$
\begin{aligned}
S_{\mathrm{TOT},l\pm}^{--} &= \frac{(\sqrt{\tfrac{1}{3}} + \sqrt{\tfrac{2}{3}} \tan \omega_{l\pm})^2 \, e^{2i\tau_{l\pm}^{3/2}} + (\sqrt{\tfrac{2}{3}} - \sqrt{\tfrac{1}{3}} \tan \omega_{l\pm})^2 \, e^{2i\tau_{l\pm}^{1/2}}}{1 + \tan^2 \omega_{l\pm}}, \\
S_{\mathrm{TOT},l\pm}^{0-} &= \frac{(\sqrt{\tfrac{1}{3}} + \sqrt{\tfrac{2}{3}} \tan \omega_{l\pm})(\sqrt{\tfrac{2}{3}} - \sqrt{\tfrac{1}{3}} \tan \omega_{l\pm})(e^{2i\tau_{l\pm}^{3/2}} - e^{2i\tau_{l\pm}^{1/2}})}{1 + \tan^2 \omega_{l\pm}}.
\end{aligned}
\tag{13}
$$

Comparing these expressions with the charge independent values (2) gives the limiting behaviour

$$
\begin{aligned}
\underset{\eta \to 0}{\mathrm{Lt}} \, \tau_{l\pm}^{3/2} &= \delta_{l\pm}^{3/2} , \\
\underset{\eta \to 0}{\mathrm{Lt}} \, \tau_{l\pm}^{1/2} &= \delta_{l\pm}^{1/2} , \\
\underset{\eta \to 0}{\mathrm{Lt}} \, \tan \omega_{l\pm} &= 0 .
\end{aligned}
\tag{14}
$$

The derivation of expressions for the Coulomb corrections is discussed in detail by *Oades* and *Rasche* (1970) and will not be repeated here, only the formulae necessary for numerical calculation being summarised. In order to calculate these Coulomb corrections it is not necessary to directly specify the nuclear potential but it is sufficient to have a parametric form for the purely nuclear wave functions within the nuclear interaction region. These wave functions are normalised such that

$$R_{N,l\pm}^T = \cos \delta_{l\pm}^T \, \varrho j_l(\varrho) - \sin \delta_{l\pm}^T \, \varrho n_l(\varrho) , \qquad \varrho \geq \varrho_N \tag{15}$$

where j_l and n_l are the usual regular and irregular spherical Bessel functions. Also $\varrho_N = q r_N$ where r_N is the nuclear radius. The parameters of the wave functions can be determined from the phases $\delta_{l\pm}^T$ by imposing continuity at the nuclear boundary ϱ_N.

If it is assumed that charge structure effects are of finite range then the Coulomb potential is equal to the point charge potential outside some charge radius r_c i. e.

$$\tilde{V}_c^{\pi^\pm p} = \pm \frac{2\eta}{\varrho}, \qquad \varrho \ge \varrho_c, \tag{16}$$

where $\varrho_c = \varrho r_c$. Finally defining ϱ_0 by

$$\varrho_0 = \max(\varrho_N, \varrho_c)$$

the inner Coulomb corrections correspond to effects over the range $0 \le \varrho \le \varrho_0$ and the outer Coulomb corrections correspond to effects for $\varrho > \varrho_0$.

For $\pi^+ p \to \pi^+ p$ the inner Coulomb corrections modify the phase $\delta_{l\pm}^{3/2}$ to a new value $\tilde{\tau}_{l\pm}^+$ given by

$$\sin(\delta_{l\pm}^{3/2} - \tilde{\tau}_{l\pm}^+) = \int_0^{\varrho_0} \tilde{V}_c^{\pi^+ p}(\varrho) \, (R_{N,l\pm}^{3/2}(\varrho))^2 \, d\varrho + O(\eta^2). \tag{17}$$

The outer Coulomb corrections further modify the phase to a final value $\tau_{l\pm}^+$ given in terms of $\tilde{\tau}_{l\pm}^+$ by

$$\frac{\varrho_0 j_l(\varrho_0) - \tan \tilde{\tau}_{l\pm}^+ (\varrho_0 n_l(\varrho_0))}{(\varrho_0 j_l(\varrho_0))' - \tan \tilde{\tau}_{l\pm}^+ (\varrho_0 n_l(\varrho_0))'} = \frac{F_l(\eta, \varrho_0) + \tan \tau_{l\pm}^+ G_l(\eta, \varrho_0)}{F_l'(\eta, \varrho_0) + \tan \tau_{l\pm}^+ G_l'(\eta, \varrho_0)}, \tag{18}$$

where F_l and G_l are the usual regular and irregular radial Coulomb wave functions, j_l and n_l the regular and irregular spherical Bessel functions and the prime denotes differentiation with respect to ϱ. Substituting $\tau_{l\pm}^+$ into (11) then gives $S_{\text{TOT},l\pm}^{++}$.

For $\pi^- p \to \pi^- p$ and $\pi^- p \to \pi^0 n$ the inner and outer Coulomb corrections are not easily separated. Defining

$$\chi_{3\alpha} = \frac{1}{3} \int_0^{\varrho_0} \tilde{V}_c^{\pi^- p}(\varrho) \, (R_{N,l\pm}^{3/2}(\varrho))^2 \, d\varrho,$$

$$\chi_{1\alpha} = -\chi_{3\beta} = \frac{\sqrt{2}}{3} \int_0^{\varrho_0} \tilde{V}_c^{\pi^- p}(\varrho) \, R_{N,l\pm}^{3/2}(\varrho) \, R_{N,l\pm}^{1/2}(\varrho) \, d\varrho, \tag{19}$$

$$\chi_{1\beta} = -\frac{2}{3} \int_0^{\varrho_0} \tilde{V}_c^{\pi^- p}(\varrho) \, (R_{N,l\pm}^{1/2}(\varrho))^2 \, d\varrho,$$

wave functions $R_{\text{IN},l\pm}^{(-)\alpha}$ and $R_{\text{IN},l\pm}^{(-)\beta}$ are constructed according to the formulae

$$R_{\text{IN},l\pm}^{(-)\alpha} = \sqrt{\tfrac{1}{3}}\, R_{N,l\pm}^{3/2} + \frac{\sqrt{\tfrac{1}{3}}\chi_{3\alpha} R_{N,l\pm}^{1/2} - \sqrt{\tfrac{2}{3}}\chi_{1\alpha} R_{N,l\pm}^{3/2}}{\sin(\delta_{l\pm}^{3/2} - \delta_{l\pm}^{1/2})},$$

$$R_{\text{IN},l\pm}^{(-)\beta} = -\sqrt{\tfrac{2}{3}}\, R_{N,l\pm}^{1/2} + \frac{\sqrt{\tfrac{1}{3}}\chi_{3\beta} R_{N,l\pm}^{1/2} - \sqrt{\tfrac{2}{3}}\chi_{1\beta} R_{N,l\pm}^{3/2}}{\sin(\delta_{l\pm}^{3/2} - \delta_{l\pm}^{1/2})}. \tag{20}$$

In terms of the wave function $R_{\text{IN},l\pm}^{(-)\alpha}$ a phase $\tau_{l\pm}^{\alpha}$ and a constant $X_{l\pm}^{\alpha}$ are defined by

$$\tan\tau_{l\pm}^{\alpha} = \frac{F_l(-\eta,\varrho_0)\,R_{\text{IN},l\pm}^{(-)\alpha\prime}(\varrho_0) - F_l'(-\eta,\varrho_0)\,R_{\text{IN},l\pm}^{(-)\alpha}(\varrho_0)}{G_l'(-\eta,\varrho_0)\,R_{\text{IN},l\pm}^{(-)\alpha}(\varrho_0) - G_l(-\eta,\varrho_0)\,R_{\text{IN},l\pm}^{(-)\alpha\prime}(\varrho_0)},$$

$$X_{l\pm}^{\alpha} = \frac{R_{\text{IN},l\pm}^{(-)\alpha}(\varrho_0)}{\cos\tau_{l\pm}^{\alpha}\, F_l(-\eta,\varrho_0) + \sin\tau_{l\pm}^{\alpha}\, G_l(-\eta,\varrho_0)}. \tag{21}$$

Similar expressions define a phase $\tau_{l\pm}^{\beta}$ and a constant $X_{l\pm}^{\beta}$ in terms of the wave function $R_{\text{IN},l\pm}^{(-)\beta}$.

Writing

$$\tau_{l\pm}^{\alpha} = \delta_{l\pm}^{3/2} + \Delta_{l\pm}^{3/2}, \qquad \tau_{l\pm}^{\beta} = \delta_{l\pm}^{1/2} + \Delta_{l\pm}^{1/2},$$

$$X_{l\pm}^{\alpha} = \sqrt{\tfrac{1}{3}}(1+\varepsilon_{l\pm}^{3/2}), \qquad X_{l\pm}^{\beta} = -\sqrt{\tfrac{2}{3}}(1+\varepsilon_{l\pm}^{1/2}), \tag{22}$$

the total Coulomb corrections are given to first order in η by

$$\tau_{l\pm}^{3/2} = \delta_{l\pm}^{3/2} + \frac{\Delta_{l\pm}^{3/2} - 2\chi_{3\alpha}}{3} + O(\eta^2),$$

$$\tau_{l\pm}^{1/2} = \delta_{l\pm}^{1/2} + \frac{2\Delta_{l\pm}^{1/2} + \chi_{1\beta}}{3} + O(\eta^2), \tag{23}$$

$$\tan\omega_{l\pm} = \sqrt{2}\left\{ \frac{2\Delta_{l\pm}^{3/2} - \Delta_{l\pm}^{1/2}}{3\tan(\delta_{l\pm}^{3/2} - \delta_{l\pm}^{1/2})} + \frac{2\varepsilon_{l\pm}^{3/2} + \varepsilon_{l\pm}^{1/2}}{3} \right\} + O(\eta^2).$$

Substituting these values into (13) then gives $S_{\text{TOT},l\pm}^{--}$ and $S_{\text{TOT},l\pm}^{0-}$.

In these formulae for the inner and outer Coulomb corrections first order relativistic effects can be included by use of *Van Hove's* suggestion of modifying the Coulomb parameter to a new value η' given by

$$\eta' = \eta(1 + q^2/m^2)^{1/2}.$$

Finally in this section it should be pointed out that the approach described here is only suitable when the experimental data is analysed via partial wave decompositions. A different problem is encountered when one wishes to obtain information on the purely nuclear forward scattering amplitude by making measurements in the Coulomb interference region. This is particularly clearly discussed by *Rix* and *Thaler* (1966) who use a potential model and the J. W. K. B. approximation to conclude that neglect of inner and outer Coulomb corrections leads to an error of the order of 0.001% above 1 GeV.

4. Influence of Coulomb Corrections on Low Energy Data

Up to the present time the only application of these low energy πN Coulomb corrections has been in the extraction of s-wave scattering lengths from very low energy data. In their paper *Hamilton* and *Woolcock* (1960) estimate that the measured $\pi^+ p$ scattering length should be correct since the inner and outer Coulomb corrections happen to cancel but that the effect of the inner Coulomb corrections changes $a_1 - a_3$ by 0.02 (from 0.265 to 0.245). The $\pi^+ p$ data was reexamined by *Rasche* (1969) who concluded that the measured $\pi^+ p$ scattering length should be corrected by $+0.003$. It should perhaps be pointed out that the treatment of such low energy data requires care. In the case of $\pi^+ p \rightarrow \pi^+ p$ the measured s-wave phase τ_{0+}^+ does not obey the usual scattering length formula but rather the modified form (see *Mott* and *Massey*, 1965)

$$C^2 q \cot \tau_{0+}^+ + 2q\eta h(\eta) = \frac{1}{a_+} + \frac{1}{2} r_+ q^2 + \dots . \qquad (24)$$

where

$$C^2 = \frac{2\pi\eta}{e^{2\pi\eta} - 1}$$

and

$$h(\eta) = \mathrm{Re}\, \frac{\Gamma'(-i\eta)}{\Gamma(-i\eta)} - \ln(\eta).$$

Thus one procedure is to determine a_+ by fitting the data with (24) and then make the inner and outer Coulomb corrections to obtain a_3. An easier alternative is to assume some parametric form for the purely nuclear phase $\delta_{0+}^{3/2}$, from this calculate the $\pi^+ p$ phase τ_{0+}^+ and then adjust the nuclear parameters until τ_{0+}^+ fits the measured data.

At this point mass difference effects and the radiative capture process should again be mentioned since, as pointed out before, the formulae given above neglect these effects. The effect at the $\pi^- p$ threshold of the open $\pi^0 n$ and γn channels is to contribute an imaginary part to the scattering amplitude. *Woolcock* (1961) estimates that, in the case of $\pi^- p \rightarrow \pi^- p$, mass differences are responsible for a threshold imaginary part which is about 3% of the real part while the γn channel produces a threshold imaginary part of about 2% of the real part. In the case of $\pi^- p \to \pi^0 n$, apart from the usual phase space factor in the differential cross section formula, mass difference effects are negligible but the γn channel is responsible for a threshold imaginary part of about 1% of the real part.

Finally a warning is necessary concerning the low energy $\pi - N$ phases resulting from recent large scale phase shift analyses. As stated before these analyses approximate $S_{TOT,l\pm}$ by the charge independent values $S_{N,l\pm}$ and so the estimation of Coulomb corrections to their results is difficult. As an example of the effects that could occur, suppose that the $T = 3/2$ phases were determined entirely by $\pi^+ p$ data. In that case the values would really correspond to the $\pi^+ p$ phases, $\tau_{l\pm}^+$, and amplitudes reconstructed from them would still contain Coulomb effects. If the forward scattering amplitude is reconstructed from these phase it will differ from the purely nuclear forward scattering amplitude which would be obtained by use of the phases $\delta_{l\pm}^{3/2}$. Typical changes obtained using the CERN "1967 Theory" $T = 3/2$ s and p waves (Donnachie, 1968) are shown in the table.

Table. *Difference between the forward scattering amplitude calculated from $\delta_{l\pm}^{3/2}$ and from $\tau_{l\pm}^+$, expressed as a percentage of the latter value*

T_π	31 MeV	98 MeV	194 MeV
$\Delta \operatorname{Re} f(\theta = 0)$	+7.4%	+3.0%	+10.3%
$\Delta \operatorname{Im} f(\theta = 0)$	+8.7%	+6.0%	+ 0.1%

It should be noted that these estimates are probably too large since the $\pi^- p \to \pi^- p$ and $\pi^- p \to \pi^0 n$ data also play a role in determining the $T = 3/2$ phases. In the best case the weights of the various pieces of data and the signs of the corrections could have the effect of cancelling these deviations. This cannot, however, be checked and a reanalysis of the low energy data with the inclusion of Coulomb corrections is desirable. Until that time the use of these low energy phases requires care especially as regards conclusions about deviations from charge independence.

References

Auvil, P. R.: Phys. Rev. **168**, 1568 (1968).
Breit, G., Haracz, R. D.: High Energy Physics, Vol. 1 (E. H. S. Burhop, Ed.). New York: Academic Press, 1967.

Bugg, D. V.: Contribution to the Ruhestein Meeting, May 1970.

Dalitz, R. H., Tuan, S.: Ann. Phys. (N. Y.) **10**, 307 (1960).

Dashen, R. F., Frautschi, S. C.: Phys. Rev. **135**, B 1190 (1964).

Donnachie, A.: Particle Interactions at High Energies, (*T. W. Preist* and *L. L. J. Vick*, Ed.) Edinburgh: Oliver and Boyd 1967.

— *Kirsopp, R. G., Lovelace, C.:* Phys. Letters **26** B, 161 (1968).

Hamilton, J., Woolcock, W. S.: Phys. Rev. **118**, 291 (1960).

Mott, N. F., Massey, H. S. W.: The Theory of Atomic Collisions Oxford: University Press 1965.

Oades, G. C., Rasche, G.: Zürich preprint (1970).

Rasche, G.: Nuovo Cimento **62** A, 229 (1969).

Rix, J., Thaler, R. M.: Phys. Rev. **152**, 1357 (1966).

Roper, L. D., Wright, R. M., Feld, B. T.: Phys. Rev. **138**, B 190 (1965).

Sauter, E.: Nuovo Cimento **61** A, 515 (1969).

Schnitzer, H. J.: Nuovo Cimento **28**, 752 (1963).

Steiner, F.: Contribution to the Ruhestein Meeting, May 1970.

Van Hove, L.: Phys. Rev. **88**, 1358 (1952).

Vik, O. T., Rugge, H. R.: Phys. Rev. **129**, 2311 (1963).

Woolcock, W. S.: Cambridge University Ph. D. Thesis (1961).

Prof. Dr. *G. C. Oades*
Science Research Council
Rutherford High Energy Laboratory
Chilton, Didcot/England

Kaon-Nucleon Interactions below 1 GeV/c

B. R. Martin

Abstract. A review of kaon-nucleon interactions below 1 GeV/c is given, with emphasis on direct analyses of KN scattering data, and the phenomenological use of dispersion relations.

Contents

1. Introduction 74
2. Dispersion Relations 75
   2.1 Kinematics and Notations 75
   2.2 Fixed-t Dispersion Relations 77
       2.2.1 Forward Dispersion Relations 79
             a) Sign Subtraction 80
             b) Energy Subtractions 81
             c) Other Subtractions 82
       2.2.2 Superconvergence Relations and Finite Energy Sum Rules 83
             a) Finite Contour Relations 84
   2.3 Partial-Wave Dispersion Relations 85
3. Kaon Nucleon Scattering below 1 GeV/c 87
   3.1 $S = +1$, $T = 1$ Interaction 87
       3.1.1 Phase-Shift Analysis 88
             a) Analysis of *Goldhaber et al.* 90
             b) Analysis of *Lea et al.* 91
             e) Analysis of *Cutkosky* and *Deo* 93
             d) Other Analyses 94
   3.2 $S = +1$, $T = 0$ Interaction 96
       3.2.1 Phase-Shift-Analysis 97
   3.3 $S = -1$ Interaction 99
       3.3.1 Analyses below ~ 500 MeV/c 100
             a) Scattering Length Models 100
             b) Zero-Range K-Matrix Models 103
             c) Effective-Range K-matrix Models 105
             d) Other Analyses 107
       3.3.2 Analyses above ~ 500 MeV/c 109
4. Low-Energy Parameters 110
   4.1 $S = -1$ Scattering Lengths 110
   4.2 $S = +1$ Scattering Lengths 112
       4.2.1 $T = 0$ Scattering 112
       4.2.2 $T = 1$ Scattering 113
             c) s-Wave 113
             b) p-Waves 117

4.3 Kaon-Nucleon Coupling Constants 121
    4.3.1 Forward Dispersion Relations 121
        a) Sign-subtracted Sum Rule. 121
        b) Energy-subtracted Relations. 123
        c) Other Subtractions. 125
        d) Finite Contour Relations 126
    4.3.2 Consistency of Models for $\bar{K}N$ Scattering 127
    4.3.3 Other Determinations 130
        a) Work of *Cutkosky* and *Deo* 130
        b) Phenomenological Bounds 131
        c) Partial-Wave Dispersion Relations 133
4.4 Forward Scattering Amplitude 135
5. Summary and Outlook 136
References 138

1. Introduction

Our current understanding of the nature of the low-energy pion-nucleon interaction has been achieved largely by a combination of direct analyses of pion-nucleon scattering data, combined with the phenomenological use of dispersion relations. Two examples will suffice to show how fruitful this approach has been. Firstly, the careful evaluation of forward dispersion relations has enabled the value of the πNN coupling constant to be calculated with an error of only about 5 % (see e.g. *Moorhouse*, 1969), which is an impressive accuracy for a strong interaction quantity. Secondly, the use of information from partial-wave and fixed-s dispersion relations has helped in finding acceptable solutions in recent phase-shift analyses (*Lovelace*, 1969).

With the success of this approach for πN scattering it is not surprising that similar techniques should be tried in kaon-nucleon physics, although the analogous analyses have not been so fruitful, partly due to a lack of reliable data, but also due to intrinsic complications not possessed by the simpler πN interaction. These two difficulties are not, of course, independent. For example, one of the major differences between πN and KN dispersion relations is the existence in the latter of a large unphysical region due to the exothermic production of pion-hyperon states. Because this region is not directly accessible to experiment its evaluation requires the use of a model, the parameters of which are determined from analyses of data in the low-energy physical region. However, this region is not well-known experimentally, and hence the parameters of the models are not well determined. In particular this leads to uncertainties in quantities evaluated from forward dispersion relations.

This article is an attempt to review the existing state of our knowledge about the kaon-nucleon interaction in the low-energy region below

~ 1 GeV/c, concentrating on the approaches mentioned above, i.e. phenomenological analyses supplemented by the use of dispersion relations. Some of the topics discussed therefore overlap the work of two recent review articles (*Bransden*, 1969; *Queen et al.*, 1969 a). Section 2 gives a brief discussion of dispersion relations for the kaon-nucleon system with particular reference to forward dispersion relations. Phenomenological analyses of $K N$ scattering data below 1 GeV/c, are reviewed in Section 3, and in Section 4 a discussion is given of the evaluation of the low-energy parameters of the kaon-nucleon interaction. Finally, Section 5 is devoted to a summary of the existing situation and suggestions for further experiments and analyses.

2. Dispersion Relations

Since dispersion relations, and particularly forward dispersion relations, have played an important role in determining the low-energy parameters we will review briefly in this section the necessary formalism, while deferring until Section 4 discussion of practical calculations.

2.1. Kinematics and Notations

The kinematics of spin 0-spin $\frac{1}{2}$ scattering are standard (*Chew et al.*, 1957) and we will give here just those formulas that we will need in later sections.

The three channels that we shall consider are

$$K + N \to K + N, \quad (s)$$

$$\bar{K} + N \to \bar{K} + N, \quad (u)$$

$$\bar{K} + K \to \bar{N} + N, \quad (t)$$

with $p_i(q_i)$ the initial four-momentum of the nucleon (kaon), and $p_f(q_f)$ the corresponding final-state four-momentum. These three processes are described by scattering amplitudes which are functions of the usual invariants

$$s = (q_i + p_i)^2 = (q_f + p_f)^2,$$

$$t = (q_i - q_f)^2 = (p_i - p_f)^2,$$

$$u = (q_i - p_f)^2 = (q_f - p_i)^2,$$

where momentum conservation implies

$$s + t + u = 2(M^2 + m^2), \tag{2.1.1}$$

and M and m are the masses of the nucleon and kaon, respectively.

The S-matrix for these three processes may be written

$$S_{fi} = \delta_{fi} + i(2\pi)^4 M \delta^{(4)}(p_i + q_i - p_f - q_f)\,\tau_{fi}, \qquad (2.1.2)$$

where

$$\tau_{fi} = \bar{u}(p_f)\, T_{fi} u(p_i)\,.$$

Spin is treated by writing the T-matrix element T_{fi} in terms of two scalar invariant amplitudes A and B, which are assumed to satisfy the Mandelstam representation. Thus

$$T_{fi} = A + \tfrac{1}{2}\gamma(q_i + q_f)\,B\,.$$

In the centre-of-mass of the s-channel the differential cross-section may be written

$$\left(\frac{d\sigma}{d\Omega}\right)_{fi} = \sum \left| \langle f | f_1 + \frac{(\boldsymbol{\sigma}\cdot\mathbf{q}_i)(\boldsymbol{\sigma}\cdot\mathbf{q}_f)}{q_i q_f} f_2 | i \rangle \right|^2, \qquad (2.1.3)$$

where the matrix is taken between two-component spinors, and the expression is summed over final spin states, and averaged over initial spin states. The amplitudes f_1 and f_2 are related to the invariant amplitudes by

$$\frac{A(s,t)}{4\pi} = \left(\frac{W+M}{E+M}\right) f_1(s,t) - \left(\frac{W-M}{E-M}\right) f_2(s,t), \qquad (2.1.4)$$

$$\frac{B(s,t)}{4\pi} = \left(\frac{1}{E+M}\right) f_1(s,t) + \left(\frac{1}{E-M}\right) f_2(s,t), \qquad (2.1.5)$$

where E is the total centre-of-mass energy of the nucleon, and $W = \sqrt{s}$. They may be decomposed into partial-wave amplitudes by the expansions

$$f_1(s,t) = \sum_{l=0}^{\infty} f_{l+}(s)\, P'_{l+1}(x) - \sum_{l=2}^{\infty} f_{l-}(s)\, P'_{l-1}(x), \qquad (2.1.6)$$

$$f_2(s,t) = \sum_{l=1}^{\infty} [f_{l-}(s) - f_{l+}(s)]\, P'_l(x), \qquad (2.1.7)$$

where $x \equiv \cos\theta$ is the cosine of the s-channel c.m. scattering angle, related to t by

$$t = -2q^2(1 - \cos\theta), \qquad (2.1.8)$$

and

$$f_{l\pm}(s) = \frac{\exp[2i\delta_{l\pm}(s)] - 1}{2iq(s)}, \qquad (2.1.9)$$

is the partial-wave amplitude for scattering in a state of total angular momentum $J = l \pm 1/2$ with phase shift $\delta_{l\pm}(s)$.

Finally, to specify charge, the amplitudes A and B are decomposed into amplitudes $A^\pm$ and $B^\pm$ by

$$A(s, t) = A^+(s, t) + A^-(s, t)(\tau_N \cdot \tau_K),$$
$$B(s, t) = B^+(s, t) + B^-(s, t)(\tau_N \cdot \tau_K),$$

$$(2.1.10)$$

where τ_N and τ_K are the isospin operators for the nucleon and kaon, respectively. In the s-channel, the amplitudes $A^\pm$ are related to amplitudes for a definite isospin state $T = 0, 1$ by

$$A_s^0(s, t) = A^+(s, t) - 3 A^-(s, t),$$
$$A_s^1(s, t) = A^+(s, t) + A^-(s, t),$$

$$(2.1.11)$$

and similarly for $B_s^T(s, t)$. The analogous relations in the t-channel are

$$A_t^0(s, t) = 2 A^+(s, t),$$
$$A_t^1(s, t) = 2 A^-(s, t),$$

$$(2.1.12)$$

and similarly for $B_t^T(s, t)$. Crossing symmetry is expressed most easily in terms of the $(\pm)$ amplitudes. Thus, if $s \leftrightarrow u$ at fixed t,

$$A^+(s, t, u) = A^+(u, t, s),$$
$$A^-(s, t, u) = - A^-(u, t, s),$$

$$(2.1.13)$$

and

$$B^+(s, t, u) = - B^+(u, t, s),$$
$$B^-(s, t, u) = B^-(u, t, s).$$

$$(2.1.14)$$

2.2 Fixed-t Dispersion Relations

When writing dispersion relations at fixed momentum transfer it is usual to introduce the variable

$$\nu = \frac{P \cdot Q}{M},$$

where

$$P_\mu = \tfrac{1}{2}(p_i + p_f)_\mu; \quad Q_\mu = \tfrac{1}{2}(q_i + q_f)_\mu.$$

Then

$$\nu = \omega + t/4 M,$$

$$(2.2.1)$$

where ω is the total laboratory energy of the incoming kaon. The invariants s and u are related to ω by

$$s = M^2 + m^2 + 2Mv - t/2\,,$$
$$u = M^2 + m^2 - 2Mv - t/2\,,$$

and so under $s \leftrightarrow u$ crossing, $v \leftrightarrow -v$. It is also convenient to define the linear combinations

$$A_\pm(v, t) = A^+(v, t) \pm A^-(v, t)\,,$$
$$B_\pm(v, t) = B^+(v, t) \pm B^-(v, t)\,,$$
(2.2.2)

which correspond to amplitudes for $K \pm N$ scattering. In terms of these amplitudes crossing now becomes.

$$A_+(v, t) = A_-(-v, t)\,,$$
$$B_+(v, t) = -B_-(-v, t)\,.$$
(2.2.3)

Dispersion relations for fixed negative t may now be written down for the invariant amplitudes $A_\pm(v, t)$ and $B_\pm(v, t)$. Ignoring the problem of subtractions they are, for $\omega \geqq m$,

$$\operatorname{Re} A_\pm(v, t) = \frac{1}{\pi} \int_{v_0}^{\infty} dv' \, \frac{\operatorname{Im} A_+(v', t)}{v' \mp v} + \frac{1}{\pi} \int_{v_0}^{\infty} dv' \, \frac{\operatorname{Im} A_-(v', t)}{v' \pm v}$$
$$+ \frac{1}{\pi} \int_{\bar{v}}^{v_0} dv' \, \frac{\operatorname{Im} A_-(v', t)}{v' \pm v} - \frac{1}{2M} \sum_Y \frac{g_Y^2(M_Y - M)}{\omega_Y \pm v + t/4M}\,,$$
(2.2.4)

and

$$\pm \operatorname{Re} B_\pm(v, t) = \frac{1}{\pi} \int_{v_0}^{\infty} dv' \, \frac{\operatorname{Im} B_+(v', t)}{v' \mp v} - \frac{1}{\pi} \int_{v_0}^{\infty} dv' \, \frac{\operatorname{Im} B_-(v', t)}{v' \pm v}$$
$$- \frac{1}{\pi} \int_{\bar{v}}^{v_0} dv' \, \frac{\operatorname{Im} B_-(v', t)}{v' \pm v} - \frac{1}{2M} \sum_Y \frac{g_Y^2}{\omega_Y \pm v + t/4M}\,,$$
(2.2.5)

where

$$v_0 = m + t/4M\,,$$
$$\bar{v} = \bar{\omega} + t/4M\,,$$
$$\bar{\omega} = \omega_{\pi Y}\,,$$
$$\omega_i = (M_i^2 - M^2 - m^2)/(2M)\,,$$
(2.2.6)

and g_Y^2 is the rationalized pseudoscalar Watson-Lepore coupling constant for $KYN(Y = \Lambda, \Sigma)$. One of the first two integrals in Eqs. (2.2.4) and (2.2.5) is to be interpreted in a principal-valued sense depending on whether the $K^+ N$ or $K^- N$ amplitude is evaluated.

The question of possible subtractions in these relations has been considered in detail by *Hamilton* and *Woolcock* for the analogous πN cases (*Hamilton* and *Woolcock*, 1963). In practice we shall need only the dispersion relation for $B_+(\nu, 0)$ which will converge in its unsubtracted form if *Pomeranchuk's* arguments hold (*Pomeranchuk*, 1956).

2.2.1 Forward Dispersion Relations

Setting $t = 0$ in Eqs. (2.2.4) and (2.2.5), and using (2.2.1) and (2.2.6) we have the following forward dispersion relations

$$
\begin{aligned}
\operatorname{Re} A_\pm(\omega, 0) = {} & \frac{1}{\pi} \int_m^\infty d\omega' \frac{\operatorname{Im} A_+(\omega', 0)}{\omega' \mp \omega} + \frac{1}{\pi} \int_m^\infty d\omega' \frac{\operatorname{Im} A_-(\omega', 0)}{\omega' \pm \omega} \\
& + \frac{1}{\pi} \int_{\bar\omega}^m d\omega' \frac{\operatorname{Im} A_-(\omega', 0)}{\omega' \pm \omega} - \frac{1}{2M} \sum_Y \frac{g_Y^2(M_Y - M)}{\omega_Y \pm \omega},
\end{aligned}
\tag{2.2.7}
$$

and

$$
\begin{aligned}
\pm \operatorname{Re} B_\pm(\omega, 0) = {} & \frac{1}{\pi} \int_m^\infty d\omega' \frac{\operatorname{Im} B_+(\omega', 0)}{\omega' \mp \omega} - \frac{1}{\pi} \int_m^\infty d\omega' \frac{\operatorname{Im} B_-(\omega', 0)}{\omega' \pm \omega} \\
& - \frac{1}{\pi} \int_{\bar\omega}^m d\omega' \frac{\operatorname{Im} B_-(\omega', 0)}{\omega' \pm \omega} - \frac{1}{2M} \sum_Y \frac{g_Y^2}{\omega_Y \pm \omega}.
\end{aligned}
\tag{2.2.8}
$$

The forward scattering amplitude in the laboratory system is given by

$$
f(\omega) = \frac{1}{4\pi} [A(\omega, 0) + \omega B(\omega, 0)],
\tag{2.2.9}
$$

and is related to the total cross-section $\sigma(\omega)$ by

$$
\operatorname{Im} f(\omega) = \frac{k}{4\pi} \sigma(\omega),
\tag{2.2.10}
$$

where k is the laboratory momentum of the incoming kaon. The forward amplitude in the c.m. system is

$$
f_c(\omega) = \frac{M}{4\pi W} [A(\omega, 0) + \omega B(\omega, 0)],
\tag{2.2.11}
$$

and since

$$
\frac{k}{q} = \frac{W}{M},
$$

we have

$$
\frac{f(\omega)}{k} = \frac{f_c(\omega)}{q},
\tag{2.2.12}
$$

and, by the optical theorem,

$$\operatorname{Im} f_c(\omega) = \frac{q}{4\pi}\,\sigma(\omega). \tag{2.2.13}$$

Dispersion relations for the forward amplitude may be constructed from Eqs. (2.2.7), (2.2.8), and (2.2.9). Using the optical theorem, the basic unsubtracted relations are

$$\operatorname{Re} f_\pm(\omega) = \frac{P}{4\pi^2} \int_m^\infty d\omega' k' \left[\frac{\sigma_+(\omega')}{\omega' \mp \omega} + \frac{\sigma_-(\omega')}{\omega' \pm \omega} \right]$$
$$+ \frac{1}{\pi} \int_0^m d\omega' \frac{\operatorname{Im} f_-(\omega')}{\omega' \pm \omega} + \sum_Y \frac{R_Y}{\omega_Y \pm \omega}, \tag{2.2.14}$$

where

$$R_Y = \left(\frac{g_Y^2}{4\pi} \right) \left[\frac{(M_Y - M)^2 - m^2}{4 M^2} \right]. \tag{2.2.15}$$

Eq. (2.2.14) are not expected to converge as they stand, even if the Pomeranchuk theorem holds, and so for practical calculations either substractions have to be made, or the finite contour relations of Section 2.2.2 below are used.

These subtractions can be made in a number of ways, and below we will mention some of the dispersion relations which have been used in practical analyses, although the details of the actual calculations will be deferred until Section 4.

a) Sign Subtraction

The simplest way to improve the convergence of (2.2.14) is to write a dispersion relation for the quantity

$$F^-(\omega) \equiv \tfrac{1}{2}[f_-(\omega) - f_+(\omega)].$$

This gives

$$\operatorname{Re} F^-(\omega) = \omega \sum_Y \frac{R_Y}{\omega_Y^2 - \omega^2} + \frac{\omega}{\pi} \int_0^m d\omega' \frac{\operatorname{Im} f^-(\omega')}{\omega'^2 - \omega^2}$$
$$+ \frac{\omega}{4\pi^2} P \int_m^\infty d\omega' \frac{[\sigma_-(\omega') - \sigma_+(\omega')]\,k'}{\omega'^2 - \omega^2}. \tag{2.2.16}$$

These integrals now converge provided the Pomeranchuk Theorem holds, and it is perhaps worthwhile pointing out that there exists no direct experimental evidence against this theorem. Eq. (2.2.16), evaluated at threshold $\omega = m$, has been extensively studied as a sum rule for the KYN coupling constants.

b) Energy Subtractions

If we wish to study the behaviour of $f_+(\omega)$, or $f_-(\omega)$, as a function of ω separately then we must make conventional energy subtractions in Eq. (2.2.14). One subtraction made at $\omega = \omega_0$ gives

$$\mathrm{Re}\, f_\pm(\omega) - \mathrm{Re}\, f_\pm(\omega_0) =$$

$$\pm \frac{(\omega - \omega_0)}{4\pi^2} \, \mathrm{P} \int\limits_m^\infty d\omega' k' \left[\frac{\sigma_+(\omega')}{(\omega' \mp \omega)(\omega' \mp \omega_0)} - \frac{\sigma_-(\omega')}{(\omega' \pm \omega)(\omega' \pm \omega_0)} \right]$$

$$\mp \frac{(\omega - \omega_0)}{\pi} \int\limits_{\bar{\omega}}^m d\omega' \, \frac{\mathrm{Im}\, f_-(\omega')}{(\omega' \pm \omega)(\omega' \pm \omega_0)} \mp (\omega - \omega_0) \sum_Y \frac{R_Y}{(\omega_Y \pm \omega)(\omega_Y \pm \omega_0)} .$$

$$(2.2.17)$$

For $K^+ N$ scattering the subtraction has the effect of suppressing the integral due to the unphysical region (and also the $K^- N$ physical integral) at the expense of enhancing the $K^+ N$ physical region. A convenient value of the subtraction energy is $\omega_0 = m$, where $\mathrm{Re}\, f_\pm(\omega_0)$ is related to the real parts of the s-wave $K^\pm N$ scattering lengths $a_s^\pm$ by

$$\mathrm{Re}\, f_\pm(m) = \left(\frac{M + m}{M} \right) a_s^\pm . \qquad (2.2.18)$$

If we set $\omega_0 = 0$ in Eq. (2.2.17) we have

$$\mathrm{Re}\, f_\pm(\omega) = K \pm \frac{\omega}{4\pi^2} \, \mathrm{P} \int\limits_m^\infty d\omega' \, \frac{k'}{\omega'} \left[\frac{\sigma_+(\omega')}{\omega' \mp \omega} - \frac{\sigma_-(\omega')}{\omega' \pm \omega} \right]$$

$$(2.2.19)$$

$$\mp \frac{\omega}{\pi} \int\limits_{\bar{\omega}}^m d\omega' \, \frac{\mathrm{Im}\, f_-(\omega')}{\omega'(\omega' \pm \omega)} + \sum_Y \frac{R_Y}{\omega_Y \pm \omega} ,$$

where

$$K = \mathrm{Re}\, f_\pm(0) - \sum_Y (R_Y/\omega_Y) . \qquad (2.2.20)$$

This subtraction produces convergent integrals without enhancing any particular region, at the expense of introducing the constant K, which is not directly measurable and must be treated as a free parameter in fits to $\mathrm{Re}\, f_\pm(\omega)$.

If necessary, further combinations of energy and sign subtractions can be performed on the basic relation (2.2.14). For example, if we subtract once at the K^+N threshold $\omega = m$, and a second time at the crossed K^-N threshold $\omega = -m$, we have

$$\operatorname{Re} f_\pm(\omega) - \frac{1}{2} \operatorname{Re} f_-(m)\left(1 \mp \frac{\omega}{m}\right) - \frac{1}{2} \operatorname{Re} f_+(m)\left(1 \pm \frac{\omega}{m}\right)$$

$$= \frac{k^2}{4\pi^2} \operatorname{P} \int\limits_m^\infty d\omega' \frac{1}{k'}\left[\frac{\sigma_-(\omega')}{\omega' \pm \omega} + \frac{\sigma_+(\omega')}{\omega' \mp \omega}\right] \tag{2.2.21}$$

$$+ \frac{k^2}{\pi} \int\limits_{\bar\omega}^m d\omega' \frac{\operatorname{Im} f_-(\omega')}{k'^2(\omega' \pm \omega)} + \sum_Y \frac{k^2 R_Y}{(\omega_Y \pm \omega)(\omega_Y^2 - m^2)},$$

which has the effect of strongly suppressing the high-energy integrals, which now converge even if asymptotically $\sigma_\pm(\omega') \to c_\pm$ with $c_+ \neq c_-$.

c) Other Subtractions

The energy-subtracted and sign-subtracted dispersion relations above only involve the real part of the forward amplitude at isolated values of ω. An alternative method of subtraction which involves a knowledge of both the real and imaginary parts of the amplitude in a continuous range of energies is obtained by considering the amplitude

$$F(\omega; \beta) = \frac{f_-(\omega)}{(\omega - m)^{1-\beta}(\omega - \omega_0)^\beta} \equiv \frac{f_-(\omega)}{\gamma(\omega)}, \tag{2.2.22}$$

where β is a parameter in the range $0 < \beta < 1$, and ω_0 is chosen so that $\bar\omega < \omega_0 < m$. The denominator is defined to be positive for $\omega > m$, and negative for $\omega < \omega_0$, with a cut joining the two branch points. An unsubtracted dispersion relation for $F(\omega; \beta)$ evaluated at the crossed channel threshold $\omega = -m$ is

$$\frac{\operatorname{Re} f_+(m)}{|\gamma(-m)|} = \sum_Y \frac{R_Y}{|\gamma(\omega_Y)|(\omega_Y + m)} + \frac{1}{\pi} \int\limits_{\bar\omega}^{\omega_0} d\omega' \frac{\operatorname{Im} f_-(\omega')}{|\gamma(\omega')|(\omega' + m)}$$

$$+ \frac{1}{\pi} \int\limits_{\omega_0}^m d\omega' \frac{[\cos\pi\beta \cdot \operatorname{Im} f_-(\omega') + \sin\pi\beta \cdot \operatorname{Re} f_-(\omega')]}{(m - \omega')^{1-\beta}(\omega' - \omega_0)^\beta(\omega' + m)} \tag{2.2.23}$$

$$- \frac{1}{4\pi^2} \operatorname{P} \int\limits_m^\infty d\omega'\left[\frac{k'\sigma_+(\omega')}{\gamma(\omega')(\omega' + m)} - \frac{k'\sigma_-(\omega')}{|\gamma(-\omega')|(\omega' - m)}\right].$$

This relation is an example of a so-called "broad-area" subtraction, so named by Adler who first used the technique for πN scattering (*Adler*, 1965).

By varying β and ω_0 one can sample different parts of the unphysical region with different weights. As $\beta \to 1$, Eq. (2.2.23) reduces to the conventional relation, Eq. (2.2.17), evaluated at $\omega = -m$, and as $\beta \to 0$ it reduces to the same relation with the subtraction at $\omega_0 = m$.

2.2.2 Superconvergence Relations and Finite Energy Sum Rules

It was emphasized by *de Alfaro et al.* (1966) that if an analytic function $f(z)$ satisfied a dispersion relation

$$f(z) = \frac{1}{\pi} \int_{-\infty}^{\infty} dz' \frac{\operatorname{Im} f(z')}{z' - z}, \qquad (2.2.24)$$

and is bounded such that

$$|f(z)| < z^{\alpha}, \qquad \alpha < -1 \qquad (2.2.25)$$

as $z \to \infty$, then it must also satisfy the so-called superconvergence relation

$$\int_{-\infty}^{\infty} dz \operatorname{Im} f(z) = 0, \qquad (2.2.26)$$

which may be derived from (2.2.24) by multiplying by z and taking the limit $z \to \infty$. For amplitudes that are odd under crossing $(z \leftrightarrow -z)$ Eq. (2.2.26) take the form

$$\int_{0}^{z} dz \operatorname{Im} f(z) = 0, \qquad (2.2.27)$$

but for amplitudes which are even under crossing the superconvergence relation is trivially satisfied. However, one can then construct superconvergence relations for $z^n f(z)$, where n is a positive integer, provided that condition (2.2.25) still holds.

If the asymptotic behaviour (2.2.25) does not hold, but the high-energy behaviour is known, then one can always subtract off that part of the amplitude violating the bound and write a superconvergence relation for the remainder. This procedure leads to the so-called finite-energy sum rules (FESRs) (*Dolen et al.*, 1968) which relate integrals of the full amplitude over a finite low-energy region to a known form which will vary depending on the assumed high-energy behaviour. Thus, for example, if we consider an amplitude $f(\nu, t)$ at fixed t, and assume that for small negative t and large $\nu > \nu_c$ the asymptotic behaviour is given by the Regge-pole model, i.e.

$$f_{\text{Regge}}(\nu, t) = \frac{\beta_i(t) \, \nu^{\alpha_i(t)}}{\sin \pi \alpha_i(t)} \left[\pm 1 - e^{-i \pi \alpha_i(t)} \right], \qquad (2.2.28)$$

where $\alpha_i(t)$ is the Regge trajectory of the ith pole, $\beta_i(t)$ its residue, and ± 1 denotes the signature, then FESRs take the form

$$\frac{1}{v_c^{n+1}} \int_0^{v_c} dv \, \mathrm{Im}\, f(v, t)\, v^n = \sum_i \frac{\beta_i(t)\, v_c^{\alpha_i(t)}}{\alpha_i(t) + n + 1} . \tag{2.2.29}$$

As a particular example of a FESR for KN scattering we can consider the work of *Restignoli et al.* (1966) who represent the forward $K^\pm N$ amplitude by a sum of Regge poles of the form

$$f_\pm(\omega) = \sum_l \gamma_l[i - \cot(\pi \alpha_l/2)]\, (k/k_0)^{\alpha_l}$$

$$\pm \sum_j \gamma_j[i + \tan(\pi \alpha_j/2)]\, (k/k_0)^{\alpha_j}, \tag{2.2.30}$$

where $l = P, P'$ and A_2 poles, and $j = \varrho$ and ω; k_0 is an arbitrary scale factor. If this form is used for the amplitude in the region $\omega > \omega_c$ then one of the relations that can be derived is the following

$$\mathrm{Re}\, f_+(m) = \frac{1}{\pi} \int_{\omega_{\pi A}}^{\omega_c} d\omega' \left[\frac{\mathrm{Im}\, f_+(\omega')}{\omega' - m} + \frac{\mathrm{Im}\, f_-(\omega')}{\omega' + m} \right]$$

$$+ \sum_Y \frac{R_Y}{\omega_Y + m} + H(\omega_c), \tag{2.2.31}$$

where

$$H(\omega_c) = \frac{-2}{\pi} \left[\sum_l \frac{\gamma_l}{\alpha_l} \left(\frac{k_c}{k_0} \right)^{\alpha_l} + \frac{m}{k_0} \sum_j \frac{\gamma_j}{\alpha_j - 1} \left(\frac{k_c}{k_0} \right)^{\alpha_j} \right], \tag{2.2.32}$$

which is a family of sum rules as a function of the parameter ω_c.

a) Finite Contour Relations

Sum rules of the finite energy type such as (2.2.31) can be derived as a Cauchy integral around some suitable contour in the complex ω-plane. For example, if we consider the function $f_+(\omega')/(\omega' - \omega)$ integrated around the contour ι shown in Fig. (2.1) then we have, by Cauchy's theorem

$$\mathrm{Re}\, f_+(\omega) = \sum_Y \frac{R_Y}{\omega_Y + \omega} + \mathrm{Re}\, \frac{1}{2\pi i} \int_{\mathrm{Circle}} d\omega' \frac{f_+(\omega')}{\omega' - \omega}$$

$$+ \frac{P}{\pi} \int_{\omega_{\pi A}}^{\omega_c} d\omega' \left[\frac{\mathrm{Im}\, f_+(\omega')}{\omega' - \omega} + \frac{\mathrm{Im}\, f_-(\omega')}{\omega' + \omega} \right]. \tag{2.2.33}$$

If $f_\pm(\omega)$ converged fast enough for large ω then we could let $\omega_c \to \infty$ and arrive at the standard form of the unsubtracted dispersion relation, Eq. (2.2.14). However, we do not have to take this limit provided we have an analytic form for $f_+(\omega)$ that can be continued into the complex ω-plane. Such a form is the Regge expansion (2.2.30), and using it in (2.2.33) for $\omega = m$ leads directly to Eq. (2.2.31). Eq. (2.2.33) is an example of a so-called Finite Contour Dispersion Relation (*Barger* and *Phillips*, 1970).

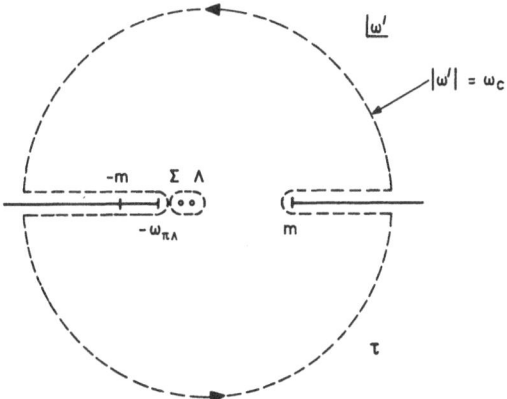

Fig. 2.1. The contour τ used to derive the finite contour relation, Eq. (2.2.33)

2.3 Partial-Wave Dispersion Relations

Consider firstly KN scattering. The singularities of the partial wave amplitudes $f_{l\pm}(s)$ of Eq. (2.1.9), as functions of s, were first given by *MacDowell* (1959), and are shown in Fig. (2.2). Exchanging a particle of mass $m_t = \sqrt{t}$ in the t-channel produces four branch points in the

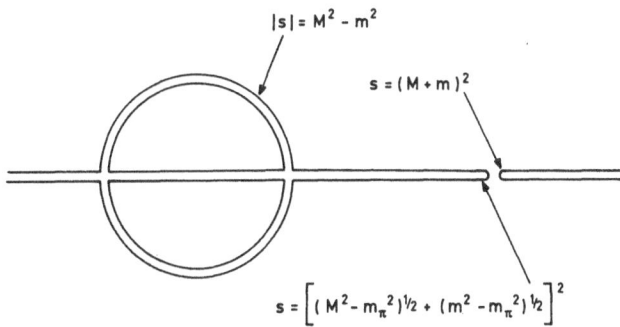

Fig. 2.2. Singularities of the KN partial-wave amplitudes $f_{l\pm}(s)$ as a function of s

s-plane, 0, $-\infty$ and s_+ where

$$s_\pm = [(M^2 - t/4)^{1/2} \pm (m^2 - t/4)^{1/2}]^2 \,.$$

For some values of t, $s_\pm$ are complex and obey

$$s_+ = s_-^* \,.$$

For the exchange of either a ϱ or ω meson in the t-channel $4m_\pi^2 < t < 4m^2$, and so $s_\pm$ are both real. For ϕ meson exchange, however, $s_\pm$ are complex, and hence this process contributes only to the circle and real-axis cuts to the left of the circle. Exchanging a particle of mass $m_u = \sqrt{u}$ in the u-channel also produces four branch points, 0, $-\infty$, s_1 and $s_2 > s_1$ where

$$s_2 = 2(M^2 + m^2) - u \,,$$
$$s_1 = (M^2 - m^2)^2/u \,.$$

For $m_u^2 < 2(M^2 + m^2)$ the cut along the real axis nearest the physical region may be taken between s_1 and s_2. The nearby left-hand singularities due to some exchange processes are shown in Fig. (2.3).

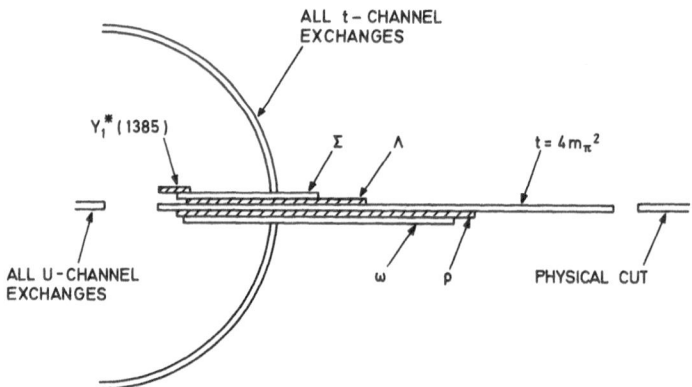

Fig. 2.3. Nearby singularities of the KN partial-wave amplitudes $f_{l\pm}(s)$ as a function of s. The arrows indicate which processes contribute to each singularity

Given the analytic structure described above, partial-wave dispersion relations for $f_{l\pm}(s)$ may be constructed.

The left-hand singularites of $\bar{K}N$ scattering amplitudes due to exchanges in the t-channel will be the same as those above. However, the u-channel exchanges have no hyperon terms. This simplification is obtained at the expense of greater complications in the s-channel which has simple poles at the masses of the Λ and Σ, and branch points at

$s = (M_\Lambda + m_\pi)^2, (M_\Sigma + m_\pi)^2$ and $(M + m)^2$. Partial wave dispersion relations for the $\bar{K}N$ system must therefore be considered in a multichannel framework.

3. Kaon Nucleon Scattering below 1 GeV/c

3.1 $S = +1$ $T = 1$ Interaction

The $S = +1$, $T = 1$ KN interaction is directly accessible in the form of $K^+ p$ scattering, and it is not surprising, therefore, that a considerable amount of work has been done on this system, both by way of experiments and phenomenological analyses. We will firstly describe briefly the main features of the data below ~ 1 GeV/c, and then in Section 3.1.1. discuss the results obtained by phase-shift analyses.

If we denote by $g(s, \vartheta)$ the spin-flip amplitude, and $f(s, \vartheta)$ the non-spin-flip amplitude for $K^+ p$ scattering, then these are given by

$$f(s, \vartheta) = \sum_{l=0}^{\infty} [(l+1) f_{l+}(s) + l f_{l-}(s)] P_l(\cos \vartheta), \qquad (3.1.1)$$

and

$$g(s, \vartheta) = i \sum_{l=1}^{\infty} [f_{l+}(s) - f_{l-}(s)] P_l^1(\cos \vartheta), \qquad (3.1.2)$$

where $P_l(\cos \vartheta)$ and $P_l^1(\cos \vartheta)$ are the ordinary and associated Legendre polynomials respectively.

The elastic differential cross-section for an unpolarized proton target is given by

$$\left(\frac{d\sigma}{d\Omega} \right) = |f(s, \vartheta)|^2 + |g(s, \vartheta)|^2. \qquad (3.1.3)$$

There are at present measurements of this quantity at 20 different momenta below 1 GeV/c; at 140, 175, 205, 235, 265, 355, 520 and 642 MeV/c. (*Goldhaber et al.*, 1962); 522 MeV/c. (*Kycia et al.*, 1960); 735 MeV/c. (*Barrelet*, 1968); 778 MeV/c. (*Focardi et at.*, 1967); 810 MeV/c. (*Stubbs et al.*, 1961) 864 and 969 MeV/c. (*Bland et al.*, 1969); 865 and 971 MeV/c. (*Andersson et al.*, 1969 a); 900 and 970 MeV/c. (*Giucomelli et al.*, 1970); 910 MeV/c. (*Hirsch* and *Gidal*, 1964); and 970 MeV/c. (*Cook et al.*, 1963). In the region below ~ 650 MeV/c, the angular distributions are found to be consistent with isotropy (after allowing for the Coulomb interaction), but at somewhat higher momenta ~ 800 MeV/c a marked departure from isotropy is observed, although significant forward peaking only starts to occur at the highest momenta.

The recoil proton polarization from an unpolarized target is given by

$$P(s, \vartheta) = \frac{2 \operatorname{Re}[f^*(s, \vartheta) \, g(s, \vartheta)]}{(d\sigma/d\Omega)} \, \hat{n} \, , \qquad (3.1.4)$$

where

$$\hat{n} = \frac{(\boldsymbol{q} \wedge \boldsymbol{q}')}{|\boldsymbol{q} \wedge \boldsymbol{q}'|} \, ,$$

and $q(q')$ is the centre-of-mass momentum of the initial (final) kaon. Until recently, measurements of P existed at only two momenta, 778 MeV/c (*Femino et al.*, 1967), and 910 MeV/c (*Hirsch* and *Gidal*, 1964), and these consisted of only three or four points at each momentum, all with very large errors. However, there are now measurements at two other momenta, 865 and 971 MeV/c (*Andersson et al.*, 1969 a), which cover the full angular range and with considerably better statistics. The polarization is positive across the whole angular range and is rather large.

The total cross-section is given by

$$\sigma(s) = \frac{4\pi}{q} \sum_{l=0}^{\infty} [(l+1) \operatorname{Im} f_{l+}(s) + l \operatorname{Im} f_{l-}(s)] \, . \qquad (3.1.5)$$

There are no published direct measurements of the total nuclear cross-section below ~ 600 MeV/c, but above this momentum several experiments exist, (*Bugg et al.*, 1968; *Bland et al.*, 1969; *Barrelet*, 1968; *Cook et al.*, 1963; *Burrowes et al.*, 1959; *Cool et al.*, 1966). An interesting feature of the total cross-section is a dip at ~ 700 MeV/c which comes from the data of *Bugg et al.* (*Bugg et al.*, 1968). The total inelastic cross-section

$$\sigma_{in}(s) = \sigma(s) - 4\pi \sum_{l=0}^{\infty} [(l+1) |f_{l+}(s)|^2 + l |f_{l-}(s)|^2] \, , \qquad (3.1.6)$$

has been measured only at a few momenta, (*Barrelet*, 1968; *Focardi et al.*, 1967; *Stubbs et al.*, 1961; *Bland et al.*, 1969; *Giacomelli et al.*, 1970). The first inelastic threshold is for single pion production at 525 MeV/c but σ_{in} is very small until about 800 MeV/c. where it has risen to ~ 1 mb. At 870 MeV/c. the production of $N^*(1236)$ can occur, and beyond this momentum the inelastic cross-section rises rapidly. The detailed behaviour of the inelastic cross-section in this region has been studied by *Bland* (1968) who has shown that most of the inelastic cross-section is indeed due to $N^*(1236)$ production initiated from p-waves in the K^+p channel.

3.1.1. Phase-Shift Analysis

The objective of phase-shift analysis is to obtain a form which represents all the data, and which can then be interpreted to give information about the scattering amplitudes themselves. This can clearly be done in a variety of ways but different methods fall into two basic classes, (a) multi-energy (or energy-dependent) and (b) single-energy (or energy-independent) analyses, and the most usual (but not the only) way of doing a phase-shift analysis of either type is to work with the partial-wave amplitudes.

In the energy-independent approach one analyzes all data at a common energy allowing, in principle, one complex number for each partial wave present. If one had a complete set of all physically observable data at a given energy, and if they had zero errors, then the analysis would yield a model-free set of partial-wave amplitudes at that energy. In practice, of course, there are disadvantages.

If one does not have a complete set of data, but lacks some measurement, then it is well-known that there exist transformations by which one can generate another exact solution from a given one. For example, without measurements of the polarization P this transformation gives rise to the Minami ambiguity, where a second solution can be produced from a given one by interchanging all partial waves of the same J but opposite parity. Other ambiguities also exist (see e.g. *Roper* and *Bailey*, 1967). Furthermore, all data have errors, and so at a given energy several solutions may appear which, from a purely statistical viewpoint, equally well fit the data.

In practice then one usually has a situation where at a given energy there are several solutions that fit the data, and the problem arises of how to find a continuous solution from amongst them. This is where the model-dependence enters. One takes all possible combinations of phase-shift-solutions at all energies and tests whether they are consistent with some continuity requirements. In practice, continuity models may be theoretically based, such as requiring that the amplitudes satisfy partial-wave dispersion relations, or they can be completely arbitrary, such as the so-called "shortest-path technique", which is the requirement that the distance along the arc of the amplitude when plotted in the complex plane is a minimum when summed over all partial waves. Both of these methods have been extensively used in πN work (see e.g. *Lovelace*, 1969), but little used in K^+p phase-shift analyses.

The second class of techniques is the energy-dependent analysis. Here one attempts to parameterize the scattering amplitudes for all values of energy analysed simultaneously. This immediately removes the problem of continuity, and also means that *all* data in the range can be used, and not just complete, or almost complete, sets. Furthermore, in practice, the number of ambiguous and multiple solutions is greatly reduced. However, the price one has to pay for these advantages is possibly a large one, that of potential bias, which occurs because, in practice, only a limited number of parameters can be determined from the data, and so one is forced to restrict the parametrization to rather simple forms.

Historically, K^+p analyses have followed the path of πN work. Thus, in the early days when there was little data and they were scattered throughout the momentum region, one was forced to use energy-dependent parameterizations. Then, as the data improved, energy-independent analyses became not only worthwhile but of practical necessity because of computer limitations. We are at the stage where nearly all current analyses above ~ 850 MeV/c. use energy-independent techniques, but below this momentum, where the data are still rather few, multi-energy analyses incorporating threshold properties are still profitable.

a) Analysis of *Goldhaber et al.*

Contrary to the outline above, the first important attempt to deduce low-energy phase-shifts from K^+p data used an energy-independent analysis.

Measurements of the elastic differential cross-section were made by *Goldhaber et al.* (1962) in 1962 at eight momenta from 140–642 MeV/c., and established that the angular distributions were consistent with isotropy, once due allowance was made for the Coulomb interaction. The simplest interpretations of K^+p scattering at these momenta are thus that it is (a) purely s-wave, (b) purely $p\frac{1}{2}$, or (c) an isotropic mixture of both $p\frac{1}{2}$ and $p\frac{3}{2}$ waves. By performing energy-independent phase-shift analyses at the eight available momenta, *Goldhaber et al.* deduced phase-shifts corresponding to the above three hypotheses. The p-wave solutions were rejected on the ground that they did not exhibit the characteristic q^3 threshold behaviour expected at these very low momenta. Finally, the sign of the pure s-wave solution was deduced to be negative by observing the effect of interference with the Coulomb interaction in the near forward direction for momenta below 350 MeV/c. The phase-shifts of this s-wave solution are shown in Table 3.1.

In subsequent phase-shift analyses at higher momenta (*Lea et al.*, 1966; *Lea et al.*, 1968; *Martin*, 1968a; *Andersson et al.*, 1969a; *Asbury et al.*, 1969; *Bland et al.*, 1969; *Giacomelli et al.*, 1970; *Kato et al.*, 1970); where frequently several possible solutions are found, it is common practice to reject those solutions which do not continue smoothly to the low-energy s-wave solution. Furthermore, these phase-shifts, and corresponding s-wave threshold parameters (see Section 4.3), have been widely used in theoretical work, and, in particular, the sign of the forward real part in the low-energy region is taken to be negative. Since so much other work rests on this analysis it is worthwhile considering the situation a little further.

Table 3.1. *The s-wave K^+p phase-shift solution of Goldhaber et al. (1962)*

k_L(MeV/c)	δ_s^1 (degrees)
140	$- 7.2 \pm 0.8$
175	-10.4 ± 0.9
205	-11.7 ± 0.9
235	-13.2 ± 0.9
265	-14.0 ± 1.1
355	-20.0 ± 1.1
520	-29.4 ± 1.7
642	-36.2 ± 1.4

b) Analysis of *Lea et al.*

The first attempt to give a systematic interpretation of all K^+p elastic scattering data was made by *Lea et al.* (1966). As a preliminary part of that analysis Lea performed a detailed study of the region below ~ 650 MeV/c using elastic s and p-waves parameterized according to the effective range form

$$q^{2l+1}\cot\delta_{l\pm}(s) = \frac{1}{a_{l\pm}} + b_{l\pm}q^2. \qquad (3.1.7)$$

The aim of the analysis was not to find a definitive set of phase shifts but merely to find feasible solutions for later work. Nevertheless, the results are interesting.

Fitting all the data below ~ 650 MeV/c, *Lea* found $\chi^2/N(N=$ number of degrees of freedom$)\sim 1.2$ for the dominant negative s-wave solution, ~ 3.0 for the dominant positive s-wave solution, ~ 1.2 for the dominant negative $p\frac{1}{2}$ solution, and ~ 3.4 for the dominant positive $p\frac{1}{2}$ solution. The difference in these fits is due almost entirely to the data below 355 MeV/c, whereas above this momentum there is little to choose between them. For the p-wave dominant solutions it is difficult to fit the data while maintaining the threshold behaviour of Eq. (3.1.7), while for the positive s-wave solution the increase in χ^2 is due to the fact that destructive interference with the Coulomb interaction produces a dip in the differential cross-section close to the forward direction, which is not exhibited by the data.

The possibility of an alternative solution at low momenta has been revived recently by *Carreras et al.* (1970a), who have performed phase-shift analyses of K^+p data constraining the partial-waves to lie close to values obtained in a previous Regge-pole analysis (*Carreras et al.*, 1970 b). The solution they obtain for $k \lesssim 350$ MeV/c has a positive s-wave showing the characteristic threshold behaviour; a negative $p\frac{1}{2}$ phase-shift of approximately the same magnitude (which cannot therefore have the correct threshold behaviour); and a small $p\frac{3}{2}$ positive phase shift. A typical fit to the data at 205 MeV/c is shown in Fig. (3.1). Although at all momenta below 350 MeV/c, where the dip is most prominent,

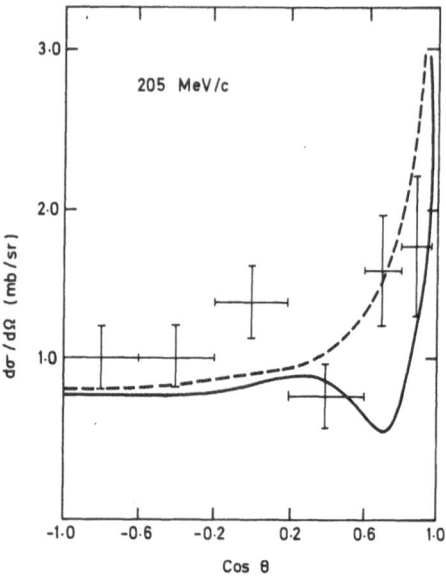

Fig. 3.1 Typical fit to the K^+p differential cross-section at 205 MeV/c from the solution of *Carreras et al.* (1970b). The broken line shows the fit *Lea et al.* (1968)

the cross-sections are systematically larger in the near forward region, *Carreras et al.* argue that since the discrepancy between the fit and the data is only at the average level of about two standard deviations this should not be taken as strong evidence against the solution. In Section 4.3 we will consider the possibility of rejecting such solutions on theoretical grounds, but throughout the rest of this section we will *assume* that the dominant negative s-wave solution is correct at low momenta.

The published analysis of *Lea at al.* (1968) was based on the following parameterizations

$$\operatorname{Re}\delta_{l\pm}(s) = q^{2l+1} \sum_{n=0}^{N} C_n^{l\pm} q^{2n},$$

$$\eta_{l\pm}(s) \equiv \exp[-2 \operatorname{Im}\delta_{l\pm}(s)] = [1 + Y_{l\pm}(s)]^{-1}, \qquad (3.1.8)$$

with

$$Y_{l\pm}(s) = \vartheta(s - s_T^{l\pm}) \left(\frac{s - s_T^{l\pm}}{s}\right)^{3/2} \sum_{m=0}^{M} D_m^{l\pm} \left(\frac{s - s_T^{l\pm}}{s}\right)^m ,$$

where the parameters are $C_n^{l\pm}$, $D_m^{l\pm}$ and $s_T^{l\pm}$. Typically, $N = 2$ and $M = 1$ were used, and s_T was constrained to lie above the threshold for single pion production. The threshold behaviour of η cannot in practice be determined from the data, but experience showed that the form used was satisfactory.

An analysis was made of all the data below 1.45 GeV/c using s, p and d-waves. The data were supplemented by values of the forward real part, calculated from Eq. (2.2.17) subtracted at threshold, but assigned large errors to prevent them biassing the solution. Because of the lack of polarization measurements at the time, four groups of solutions were found. Subsequently, a measurement of $P(\vartheta)$ at 1.22 GeV/c. (*Andersson et al.*, 1969 b) selected one of these solution (Group IV). This solution is characterized by a negative $p\frac{1}{2}$ amplitude of high elasticity and a positive $p\frac{3}{2}$ amplitude which rapidly becomes very inelastic above ~ 900 MeV/c. Also $|\delta_{p1}^1| > |\delta_{p3}^1|$, and so we will call this the Fermi set of phases.

The rapid onset of inelasticity in the $p\frac{3}{2}$ amplitude is consistent with the production of $N^*(1236)$, but since the initial $p\frac{1}{2}$ wave can also feed this channel it is somewhat surprising that the $K^+ p\, p\frac{1}{2}$ amplitude should remain rather elastic. Nevertheless this feature has also been found by other groups using energy independent techniques (*Asbury et al.*, 1969; *Andersson et al.*, 1969 a).

c) Analysis of *Cutkosky* and *Deo*

Cutkosky and *Deo* (1970) have recently made an analysis of data below 1.45 GeV/c. using a parametric form for the scattering amplitude which is based on the analytic properties of $A(s, t)$ and $B(s, t)$ as a function of $x = \cos\vartheta$. These authors have shown (*Cutkosky* and *Deo*, 1968a; see also *Ciulli*, 1969) that the most rapidly convergent polynomial expansion of the amplitudes is obtained by mapping the cut $\cos\vartheta$ plane onto the interior of a unifocal ellipse in which the physical region $-1 \leq x \leq 1$ is mapped on the line $-1 \leq z(x) \leq 1$, and the branch cuts are mapped onto the ellipse. The details of the mapping have been discussed by *Cutkosky* and *Deo*. The pole terms in A and B are introduced explicitly, as functions of the kaon-nucleon coupling constants, and the remainder is then expanded as a polynomial in z. This method has the advantage of including the correlations among the partial waves which are required by momentum-transfer analyticity properties, but the expansions are not guaranteed to satisfy unitarity term-by-term, although in practice this was found to give no trouble.

Above 650 MeV/c energy-independent analyses were made, but below this momentum the s and p-waves were parameterized by the expansion

$$q^{2l+1}\cot\delta_{l\pm}(s) = \frac{1}{a_{l\pm}} + \frac{r_{l\pm}}{2}p^2, \qquad (3.1.9)$$

where

$$p^2 = m^2\left[\cosh^{-1}\left(1 + \frac{q^2}{m^2}\right)^{1/2}\right]^2,$$

which converges in a greater region than the conventional effective-range expansion Eq. (3.1.7).

Assuming the negative s-wave solution at low-momenta, *Cutkosky* and *Deo* found two classes of solutions, corresponding to the Fermi-Yang ambiguity. Each of these solutions is a one-parameter family as a function of $G^2 \equiv g_A^2/4\pi + g_s^2/4\pi$, with χ^2 varying rather little with G^2. At 778 and 910 MeV/c no Fermi-Yang ambiguity is found and $\delta_{p1}^1 < 0$ is strongly suggested, at least up to 910 MeV/c. This clearly selects the Fermi set of phases, in agreement with the work of *Lea et al.* The s and p-wave phase shifts of this set are shown in Table 3.2 for $G^2 = 0.7$ and 14.

d) Other Analyses

Other analyses cited in references above have concentrated mainly on the region $k \gtrsim 850$ MeV/c where new data have become available, and are thus of little interest for our purpose, except insofar as they tend to confirm the results of lower-energy analyses in the cross-over region. Recently, however, as part of a new analysis of all K^+p data below 2.2 GeV/c, (*Lea et al.*, 1970a), the region below 800 MeV/c was reanalyzed using an energy-dependent parameterization based on the use of the inverse amplitude. The parameterizations are

$$q^{2l}\text{Re}[f_{l\pm}^{-1}(s)] = \frac{1}{a_{l\pm}} + \sum_{n=1}^{N} a_n^{l\pm}q^{2n},$$

and

$$q^{2l}\text{Im}[f_{l\pm}^{-1}(s)] = -q^{2l+1}\left[1 + \left(\sum_{m=1}^{M} b_m^{l\pm}q^m\right)^2\right],$$

which have the advantage of leading to rational functions for the real and imaginary parts of $f_{l\pm}(s)$. Only one solution giving an acceptable fit to the data was found, and this also has a negative $s\frac{1}{2}$ phase shift, a negative $p\frac{1}{2}$, and a positive $p\frac{3}{2}$, with $|\delta_{p\frac{1}{2}}^1| > |\delta_{p\frac{3}{2}}^1|$.

Table 3.2. The s and p-wave K^+p phase shifts from the analysis of Cutkosky and Deo (1970) showing the variation with G^2, the effective KN coupling constant

		Kaon laboratory momentum (MeV/c)										
		140	175	205	235	265	355	520	642	778	860	910	960
$s^{\frac{1}{2}}$	$G^2 = 0$	−7.8	−9.8	−11.5	−13.2	−15.0	−20.1	−29.0	−34.8	−42.9	−48.3	−60.7	−46.9
	$G^2 = 7$	−7.7	−9.6	−11.4	−13.1	−14.8	−19.9	−28.7	−34.5	−42.3	−49.0	−41.0	−49.2
	$G^2 = 14$	−7.8	−9.8	−11.4	−13.1	−14.8	−19.6	−27.6	−32.7	−39.6	−41.3	−31.8	−43.7
$p^{\frac{1}{2}}$	$G^2 = 0$	−0.05	−0.10	−0.15	−0.23	−0.32	−0.73	−2.0	−3.4	0.38	−2.7	−8.0	−8.1
	$G^2 = 7$	−0.10	−0.19	−0.30	−0.44	−0.62	−1.4	−3.9	−6.6	−3.0	−6.3	−14.2	−9.4
	$G^2 = 14$	−0.15	−0.30	−0.47	−0.69	−0.98	−2.2	−6.1	−10.3	−10.3	−11.7	−24.4	−19.5
$p^{\frac{3}{2}}$	$G^2 = 0$	0.04	0.07	0.11	0.17	0.24	0.53	1.5	2.5	0.66	5.5	2.9	10.1
	$G^2 = 7$	0.06	0.12	0.19	0.28	0.40	0.90	2.5	4.2	2.3	7.3	10.5	8.3
	$G^2 = 14$	0.09	0.18	0.29	0.43	0.60	1.4	3.8	6.4	6.6	11.5	15.6	10.9

3.2 $S=1$, $T=0$ Interaction

Until K^0 beams become more widely available the only practical way of gaining information about the $T=0$ KN interaction is to study K^+n scattering, where the neutron is bound in a deuteron. At low energies the three possible scattering processes for a K^+ and a deuteron are: elastic scattering

$$K^+ d \rightarrow K^+ d, \qquad (3.2.1)$$

and deuteron breakup, with or without charge-exchange,

$$K^+ d \rightarrow K^0 p p, \qquad (3.2.2)$$

$$K^+ d \rightarrow K^+ n p. \qquad (3.2.3)$$

The $K^+ d$ total cross-section is unknown below 590 MeV/c, but above this momentum it has been measured at 15 momenta below 1 GeV/c, (*Slater et al.*, 1961; *Bugg et al.*, 1968; *Cook et al.*, 1961; *Hirata et al.*, 1968; *Cool et al.*, 1966). Measurements of the differential cross-section for the deuteron breakup reaction with charge exchange have been made at 330, 530, 642, and 812 MeV/c (*Slater et al.*, 1961), 377 MeV/c (*Stenger et al.*, 1964), and 600 MeV/c (*Ray et al.*, 1969). The differential cross-section for the breakup reaction without charge-exchange has been measured at only two momenta, 377 and 530 MeV/c (*Stenger et al.*, 1964). The only polarization measurement is of the non-spectator proton in reaction (3.2.2), which has been made at 600 MeV/c (*Ray et al.*, 1969). Elastic $K^+ d$ scattering has never been measured directly at any momentum.

It is clear from the above that data on $K^+ d$ reactions are very sparse indeed, and of rather low quality. Moreover the extraction of $T=0$ KN information is further complicated by theoretical problems associated with the deuteron.

The scattering amplitude for three-body final states occurring in the deuteron breakup reactions can be expressed exactly in terms of off-energy-shell scattering amplitudes for each of the two-body sub-systems (*Faddeev*, 1961a; 1961b; 1963), and thus, in principle, information about the $T=0$ KN interaction may be obtained. However, in practice, approximations have to be made. We will describe briefly the nature of the approximations, without details. A more detailed discussion may be found in the article by *Bransden* (1969).

If the velocity of the incoming kaon is increased there will come a point at which the $K^+ d$ collision time is short compared with the characteristic time of nucleon motion within the deuteron. When this happens the binding forces between the nucleons can be neglected, and the interaction can be treated as the collision of a kaon with a free nucleon, but with a momentum distribution due to its Fermi motion. This is the impulse approximation (*Chew*, 1950; *Chew* and *Goldberger*, 1952; *Chew* and *Wick*, 1952), and has been widely used for $k \gtrsim 100$ MeV/c. However, certain corrections are, in principle, necessary to this simple model.

Firstly, there is the correction due to possible multiple scattering within the deuteron, which could be important below ~ 300 MeV/c, and many methods have been proposed to take account of these effects, all of which require two-body matrix elements off the energy shell. A very simple example of such models is that of *Chand* and *Dalitz* (1962) which neglects nucleon recoil but includes multiple scattering effects to all orders. Such a model could be suitable for very low momenta $\lesssim 100$ MeV/c. Another correction, which will occur irrespective of the momentum of the kaon, is the screening effect of the target nucleon by the spectator (*Glauber*, 1955), which makes measured cross-sections generally lower than free-nucleon cross-sections. At present there is no universally agreed way of making these various corrections in order to deduce $T = 0$ KN amplitudes.

3.2.1 Phase-Shift Analysis

The first attempt to obtain $T = 0$ KN phase-shifts from $K^+ d$ data was that of *Slater et al.* (1961) who analyzed their data on the charge-exchange breakup cross-section at 330, 530, 642, and 812 MeV/c in terms of $T = 0$ and $T = 1$ KN phase shifts plus Coulomb corrections. The simple impulse approximation was used and no corrections were made for pp final state interactions, or multiple scattering effects. Assuming the s-wave $T = 1$ phase shift solution of *Goldhaber et al.* (1962), as shown in Table 3.1, the data could only be fitted by using at least s and p-waves in the $T = 0$ state, with some slight evidence for a d-wave interaction. Later analyses of the same data *included* the effect of pp final state interactions but with conflicting results. *Ferreira* (1961) found that when these interactions were included only s-waves in the $T = 0$ state were needed to fit the data, whereas *Levy-Leblond* and *Gourdin* (1962), who also included pp final-state interactions, required s, p and d-waves.

Later, in 1964, as a continuation of the experiments of *Slater et al.*, measurements were made of the differential cross-section for the non-charge-exchange breakup reaction at 377, and 530 MeV/c, and the charge-exchange cross-section at 377 MeV/c (*Stenger et al.*, 1964). These data were included with the earlier data in an analysis again based on the simple impulse approximation, using $T = 0$ and 1 KN phase shifts, and deuteron form factors. The groundstate wave function was taken to be of the Hulthèn form, and the final-state wave functions were found to be adequately represented by plane waves. Energy independent analyses were carried out at the four available momenta assuming the pure s-wave phase shift solution for the $T = 1$ state. The best fit is given by using s, p, and d-waves in the $T = 0$ state, and the resulting phase shifts are shown in Table 3.3. The double entry corresponds to the Fermi-Yang ambiguity.

Table 3.3. *The $T = 0$ $KN(S = +1)$ phase shifts from the analysis of Stenger et al., (1964)*

$k(\text{MeV}/c)$	δ_s^0	δ_{p1}^0	δ_{p3}^0	δ_{d3}^0	δ_{d5}^0
350	2 ± 2	1 ± 2	7 ± 2	1 ± 1	-3 ± 1
		10 ± 4	2 ± 1	-4 ± 2	0 ± 1
530	8 ± 4	-8 ± 6	16 ± 4	-3 ± 4	-9 ± 3
		24 ± 6	0 ± 3	-11 ± 4	-4 ± 3
642	6 ± 6	-7 ± 19	26 ± 4	-1 ± 7	-6 ± 11
		40 ± 10	3 ± 9	-5 ± 14	-2 ± 4
812	12 ± 6	-7 ± 44	30 ± 4	-1 ± 15	-4 ± 24
		43 ± 17	8 ± 12	-8 ± 25	-1 ± 9

The polarization of the non-spectator proton in the charge-exchange reaction is predicted by the solutions of *Stenger et al.* to be large and of opposite sign for the Fermi-Yang ambiguous solutions. Recently, a measurement of this quantity has been made at 600 MeV/c (*Ray et al.*. 1969) which clearly favours the Yang set of phases. *Ray et al.* have analysed their polarization and differential cross-section data for the charge-exchange reaction in a simple impulse approximation by two different methods. The first is a standard phase shift analysis where the spin-flip and non-spin-flip amplitudes are expanded into a fixed number of partial waves. The second method is that of *Cutkosky* and *Deo* which was discussed in Section 3.1.1. The input $T = 1$ phase shifts were taken to be either (a) the Goldhaber pure s-wave, (Set I) or (b) the Cutkosky-Deo solution for $G^2 = 14$, (Sets IIF and IIY). Using the first method of analysis with set I input phases, two solutions were found giving an spd fit, both having a large $T = 0$ $p\frac{1}{2}$ phase shift (Yang type). Using the Cutkosky-Deo analysis with Set I imput phases, two similar solutions were found, plus an additional one with a large $T = 0$ phase shift. Using set IIF input phases, two sp solutions were found in the conventional-type analysis, and one with the Cutkosky-Deo method, which are qualitatively similar to the solutions found previously. The inclusion of $T = 0$ d-waves marginally improves the fit. Finally, using set IIY as input, both methods produce solutions definitely requiring d-waves.

Since the Fermi $T = 1$ solution is preferred from $K^+ p$ analyses, the results of *Ray et al.* imply that the simplest interpretation of $T = 0$ KN

scattering at 600 MeV/c is by s and p-waves alone with a large $p\frac{1}{2}$ phase-shift of the Yang type. Finally, these authors allowed the $T=1$ KN phase-shifts to be free also and fitted, in addition to their own data, the elastic K^+p cross-section and forward real part. Their final best solution at 600 MeV/c is shown in Table 3.4.

Table 3.4. *The s and p-wave KN phase shifts at 600 MeV/c from the analysis of Ray et al. (1969)*

	δ_s	δ_{p1}	δ_{p3}
$T=0$	6.0 ± 7.6	30.3 ± 6.4	-2.7 ± 3.2
$T=1$	-32.7 ± 3.0	-8.6 ± 3.0	4.2 ± 3.0

3.3 $S=-1$ Interaction

The majority of $S=-1$ data that exists below ~ 1 GeV/c comes from studies of K^-p interactions, although a small amount of $K_2^0 p$ data also exists. Because the initial K^-p state contains both isospin 0 and 1 parts the role of the deuteron in understanding the $S=-1$ $\bar{K}N$ interaction is much less important than in the corresponding $S=-1$ situation. We will therefore concentrate entirely on reactions initiated from pure $\bar{K}N$ states and not discuss the deuteron further.

At threshold, initial K^-p states can give rise to elastic scattering, and, in addition, the exothermic production of hyperons, i.e. $\Lambda\pi, \Sigma\pi$ and $\Lambda\pi\pi$. Just above threshold two addition channels are open, $\bar{K}^0 n$ and $\Sigma\pi\pi$. A large body of data exists for the reactions

$$K^-p \to K^-p;\ \bar{K}^0 n,$$

$$\to \pi^-\Sigma^+;\ \pi^+\Sigma^-;\ \pi^0\Sigma^0,$$

$$\to \pi^0\Lambda,$$

in the form of total cross-section, differential cross-sections, and differential polarizations of the final-state baryon (*Sakitt et al.*, 1965; *Kim*, 1966; *Humphrey* and *Ross*, 1962; *Kittel et al.*, 1966; *Kadyk et al.*, 1968; *Kadyk et al.*, 1966; *Berley et al.*, 1970; *Watson et al.*, 1963; *Abrams et al.*, 1965; *Donald et al.*, 1966; *Bertanza et al.*, 1969; *Armenteros et al.*, 1968a; *Armenteros et al.*, 1968c; *Cool et al.*, 1966; *Holley et al.*, 1967; *Wang*, 1965; *Bugg et al.*, 1968; *Gelfand et al.*, 1966; *Armenteros et al.*, 1969), although most of the data come from bubble-chamber studies and are not of very high accuracy. There is also a very limited amount of data on the reac-

7*

tions

$$K_2^0 p \to K_1^0 p\,,$$
$$\to \pi^+ \Sigma^0\,,$$
$$\to \pi^+ \Lambda\,,$$

(*Kadyk et al.*, 1966; *Kadyk et al.*, 1968; *Sayer et al.*, 1968; *Donald et al.*, 1966).

The multichannel nature of the $\bar{K}N$ interaction, even at threshold, greatly complicates analyses, but below ~ 300 MeV/c two important pieces of evidence exist which simplify matters considerably. Firstly, no three-body final states are observed, and secondly, the angular distributions are consistent with the interaction being purely s-wave, although there is some evidence for the need for p-waves in the data on $K_2^0 p \to \Lambda \pi^+$ above about 250 MeV/c. These facts have enabled analyses of the very low energy region to be carried out in some detail, and we will therefore consider analyses below ~ 1 GeV/c in two parts, (a) those below ~ 500 MeV/c and (b) those above this momentum, although some will overlap both regions.

3.3.1 Analyses below ~ 500 MeV/c

a) Scattering Length Models

Early analyses of low-momentum $K^- p$ reactions used a model in which the $\bar{K}N$ isospin amplitudes were expanded in powers of the centre-of-mass momentum. For the s-waves this gives

$$f^T(s) = \frac{A_T(s)}{1 - iq\,A_T(s)}\,, \tag{3.3.1}$$

where

$$A_T(s) = a_T(s) + i\,b_T(s)\,.$$

In the s-wave region below ~ 300 MeV/c the approximation was made of keeping $a_T(s)$ and $b_T(s)$ constant throughout the momentum range. Apart from the four parameters a_0, b_0, a_1, and b_1, two others are needed for a complete description of the $\bar{K}N$ interaction. These are usually taken to be (*i*) a phase angle Φ, where

$$\Phi \equiv \Phi_0 - \Phi_1\,, \tag{3.3.2}$$

and $\Phi_T(T = 0, 1)$ is the phase of the $\pi\Sigma$ production amplitude in an isospin state T, and (ii) the ratio ε of the cross-section for $\pi\Lambda$ production to that for total hyperon production in a $T = 1$ state, i.e.

$$\varepsilon = \frac{\sigma(\pi\Lambda)}{[\sigma(\pi\Sigma) + \sigma(\pi\Lambda)]_{T=1}}\,, \tag{3.3.3}$$

where e.g. $\sigma(\pi\Sigma)$ means $\sigma(\bar{K}N\rightarrow\pi\Sigma)$. In terms of these quantities the hyperon production cross-sections are given by

$$\sigma(K^-p\rightarrow\pi^{\pm}\Sigma^{\mp})= \frac{1}{6}\sigma_0+\frac{1}{4}(1-\varepsilon)\,\sigma_1\pm\left[\frac{1}{b}\,\sigma_0\sigma_1(1-\varepsilon)\right]^{\frac{1}{2}}\cos\Phi\,,$$

$$\sigma(K^-p\rightarrow\pi^0\Sigma^0)=\frac{1}{6}\sigma_0\,, \tag{3.3.4}$$

$$\sigma(K^-p\rightarrow\pi^0\Lambda)=\frac{1}{2}\varepsilon\sigma_1\,,$$

where σ_T is the total absorptive cross-section for $\bar{K}N\rightarrow\pi Y$ in an isospin state T.

Corrections due to the violation of charge indepencence are more important here than in K^+p scattering because of the considerable $K^-\bar{K}^0$ mass difference. In practice all analyses have taken account of electromagnetic mass differences between $K^-\bar{K}^0$ and np, and included modifications due to the Coulomb interaction, in a method due to *Dalitz* and *Tuan* (1960). This method gives rise to the following expressions for K^-p cross-sections;

$$\left(\frac{d\sigma_{el}(\vartheta)}{d\Omega}\right)=\left|\frac{\mathrm{cosec}^2(\vartheta/2)}{2Bq^2}\exp\left(\frac{2i}{qB}\ln[\sin(\vartheta/2)]\right)\right.$$

$$\left.+\frac{C}{2D}(A_0+A_1-2iq_0A_0A_1)\right|^2\,, \tag{3.3.5}$$

$$\sigma(K^-p\rightarrow\bar{K}^0n)=\pi C\,\frac{q_0}{q}\left|\frac{A_1-A_0}{D}\right|^2\,, \tag{3.3.6}$$

$$\sigma_{0,1}=4\pi C\,\frac{\mathrm{Im}\,A_{0,1}}{q}\left|\frac{1-iq_0A_{1,0}}{D}\right|^2\,, \tag{3.3.7}$$

where B is the Bohr radius of the initial K^-p system; q_0 is the centre-of-mass momentum in the $\bar{K}^0n$ channel;

$$D=1-\frac{i}{2}(A_0+A_1)\,[q_0+qC(1-i\lambda)]-qq_0C(1-i\lambda)\,A_0A_1\,,$$

C is the Coulomb correction factor for s-wave scattering, given by

$$C=\frac{2\pi}{qB}\left[1-\exp\left(\frac{-2\pi}{qB}\right)\right]^{-1}\,,$$

and λ is a function of the interaction radius, which is usually taken to be ~ 0.5 F. The Φ may then be written

$$\Phi = \Phi_0 + \text{Arg}\left[\frac{1 - iq_0 A_1}{1 - iq_0 A_0}\right], \tag{3.3.8}$$

where Φ_0 is the value at the $\bar{K}^0 n$ threshold.

Analyses of $K^- p$ data based on the above formulation have been carried out by several groups for $k \lesssim 300$ MeV/c (*Humphrey* and *Ross*, 1962; *Sakitt et al.*, 1965; *Kim*, 1965; *Kim*, 1966; *Kittel et al.*, 1966), and established the existence of two solutions, one of which was somewhat preferred on statistical grounds. Later measurements of $K_2^0 p$ interactions (*Kadyk et al.*, 1966; *Donald et al.*, 1966) definitely confirmed the preferred solution as the correct one. This latter solution has the feature of predicting a resonant $T = 0$ state just below the $\bar{K} N$ threshold whose position (~ 1410 MeV) and width (~ 30 MeV) make it identifiable with the Y_0^* (1405) observed in production experiments.

The prediction of the $Y_0^*(1405)$ as an $s\frac{1}{2}$ virtual bound state of the $\bar{K} N$ system was a considerable success of the "constant scattering length" (CSL) model. A further success was the determination of the spin-parity of the $Y_0^*(1520)$ resonance from an analysis of $K^- p$ reactions from $250 - 520$ MeV/c using s, p and d-waves (*Watson et al.*, 1963). An analysis has also been made of data up to 400 MeV/c using effective range terms in the expansion of $A_T(s)$ (*Kittel* and *Otter*, 1966). However, the CSL model has defficiencies which hardly justify this extension.

Although the CSL model is simple and can explain the data below 300 MeV/c in an economical way it is theoretically unsatisfactory because it ignores the multichannel nature of the problem in a potentially serious manner. To see this we can continue the s-wave $T = 0$ amplitude below the $\bar{K} N$ threshold (by the prescription $q \rightarrow i|q|$), and then take its imaginary part. This gives, from Eq. (3.3.1)

$$\text{Im} f^0(s) = \frac{b_0}{(1 + a_0|q|)^2 + (b_0|q|)^2}, \tag{3.3.9}$$

which does not vanish at the $\pi \Sigma$ threshold, as it clearly should. Objections of this kind may be irrelevant if all we require is an interpolation formula for the physical region, but if we use the resulting amplitude outside this region, as is done in dispersion relations, then more attention must be paid to this problem. Because the unphysical region gives important contributions to forward dispersion relations the use of the CSL model has been largely replaced by models base on the use of the K-matrix.

b) Zero-Range K-Matrix Models

Unitary amplitudes with better analytic properties than those of the CSL model may be most simply constructed by use of the K-matrix (*Dalitz* and *Tuan*, 1960; *Jackson* and *Wyld*, 1959 a; 1959 b) which is defined, for each partial wave, by

$$K^{-1} = F^{-1} - iQ,\qquad\qquad(3.3.10)$$

where, for two-body channels, Q is a diagonal matrix of channel momenta, and F is a matrix of partial-wave amplitudes. The matrix

$$Q^l K^{-1} Q^l\qquad\qquad(3.3.11)$$

is an analytic function of energy which does not contain branch points at the channel thresholds. Time-reversal invariance and Hermiticity ensure that K is a real, symmetric matrix and so for the $\bar{K}N$ problem the isospin components may be written

$$K^0 = \begin{pmatrix} \alpha_0 & \beta_0 \\ \beta_0 & \gamma_0 \end{pmatrix} \begin{matrix} \bar{K}N \\ \pi\Sigma \end{matrix}\;,\qquad\qquad(3.3.12)$$
$$\begin{matrix} \bar{K}N & \pi\Sigma \end{matrix}$$

$$K^1 = \begin{pmatrix} \alpha_1 & \beta_\Sigma & \beta_\Lambda \\ \beta_\Sigma & \gamma_{\Sigma\Sigma} & \gamma_{\Sigma\Lambda} \\ \beta_\Lambda & \gamma_{\Sigma\Lambda} & \gamma_{\Lambda\Lambda} \end{pmatrix} \begin{matrix} \bar{K}N \\ \pi\Sigma \\ \pi\Lambda \end{matrix}\;,\qquad(3.3.13)$$
$$\begin{matrix} \bar{K}N & \pi\Sigma & \pi\Lambda \end{matrix}$$

where the elements are functions of energy.

If we now use Eq. (3.3.12) in (3.3.10) and express the result for the elastic channel in the form of Eq. (3.3.1) we find

$$\text{Im}\,A_0(s) = \frac{q_\Sigma \beta_0^2}{1 + (q_\Sigma \gamma_0)^2}\;,$$

where q_Σ is the centre-of-mass momentum in the $\pi\Sigma$ channel. This expression now vanishes at the $\pi\Sigma$ threshold, as required. In the K-matrix formulation Eq. (3.3.8) still hold but now the six parameters $a_0, b_0, a_1, b_1, \varepsilon$, and Φ are all given in terms of the K-matrix elements.

Analyses have been made recently taking the elements of the K-matrix to be energy-independent. This is the zero-range approximation (ZRA). In the work of *Martin* and *Sakitt* (1969a) K^-p data below ~ 300 MeV/c were analyzed using a pure s-wave interaction. A unique solution was found giving a good fit to the data in this region in terms of the nine real K-matrix parameters. The resulting K-matrix elements are shown in Table 3.5. In this table, and in all other tables of K-matrix elements, the entries are such that only the last quoted figure is significant. The $T=0$ part of this solution, when continued into the unphysical region, shows a resonance at ~ 1415 MeV with a width ~ 30 MeV,

Table 3.5. *The s-wave K-matrix parameters (in Fermis) from the analysis of Martin and Sakitt (1969a)*

K^0	$\bar{K}N$	$\pi\Sigma$	K^1	$\bar{K}N$	$\pi\Sigma$	$\pi\Lambda$
$\bar{K}N$	-1.88	-0.92	$\bar{K}N$	0.22	-0.78	-0.38
$\pi\Sigma$	-0.92	-0.36	$\pi\Sigma$	-0.78	0.92	-0.17
			$\pi\Lambda$	-0.38	-0.17	0.46

values which are compatible with those obtained using the CSL model. Another recent analysis using the ZRA (*Martin* and *Ross*, 1970) also fits data below 300 MeV/c, but includes data on $K_2^0 p$ interactions, specifically

$$\sigma(K_2^0 p \to K_1^0 p) = \pi \left| \frac{1}{2} \sum_{T=0,1} \left(\frac{\alpha_T}{1-iq'\alpha_T} \right) - \frac{A_1}{1-iq'A_1} \right|^2, \quad (3.3.14)$$

$$\sigma(K_2^0 p \to \pi Y) = \frac{2\pi}{q'} \frac{\text{Im}\,A_1}{|1-iq'A_1|^2}, \quad (3.3.15)$$

and

$$\sigma_{\text{tot}}(K_2^0 p) = \sum_{T=0,1} \left[\frac{\pi\alpha_T^2}{1+(q'\alpha_T)^2} \right] + 2\pi \frac{|A_1|^2 + \text{Im}\,A_1/q'}{|1-iq'A_1|^2}, \quad (3.3.16)$$

where q' is the $K_2^0 p$ centre-of-mass momentum, and α_T are the s-wave KN ($S=+1$) scattering lengths.

The advantage of using the $K_2^0 p \to \pi^+ \Lambda$ cross-section is that it is pure $T=1$. The disadvantage of using the other $K_2^0 p$ data is that it involves $S=+1$ amplitudes, and although the $T=1$ part of the KN interaction is reasonably well-known at these momenta, the $T=0$ part is very poorly known (see Section 3.2).

These authors also find a unique fit to the data below 300 MeV/c, and the resulting K-matrix parameters are shown in Table 3.6. The pure s-wave solution has a sign ambiguity in some of the K-matrix elements since one can easily verify that the $K^- p$ and $K_2^0 p$ cross-sections remain invariant under the transformation

$$\beta_0 \leftrightarrow -\beta_0, \quad (3.3.17)$$
$$\beta_\Sigma \leftrightarrow -\beta_\Sigma$$

Table 3.6. *The s-wave K-matrix parameters (in Fermis) from the analysis of Martin and Ross (1970)*

K^0	$\bar{K}N$	$\pi\Sigma$	K^1	$\bar{K}N$	$\pi\Sigma$	$\pi\Lambda$
$\bar{K}N$	-2.4	-1.2	$\bar{K}N$	0.0	-0.71	-0.38
$\pi\Sigma$	-1.2	-1.1	$\pi\Sigma$	-0.71	0.3	-0.2
			$\pi\Lambda$	-0.38	-0.2	0.2

and *either*

$$\beta_A \leftrightarrow -\beta_A \quad \text{or} \quad \gamma_{\Sigma A} \leftrightarrow -\gamma_{\Sigma A}.$$

The agreement between the two solutions cannot be judged by the diagonal errors of the K-matrix parameters alone because the associated variance matrix has large off-diagonal terms. A better criterion is to compare calculated values for the scattering amplitudes themselves, and, in particular, for the threshold parameters. We will see in Sec. 4.1 below that both analyses give good agreement for the $\bar{K}N$ scattering lengths.

An attempt was made by *Martin* and *Ross* to determine the p-wave $T=1$ $\bar{K}N$ amplitude by extending their analysis to include all data below 400 MeV/c which involved either the $K_2^0 p$ or $\pi\Lambda$ channels. The s-wave ZRA K-matrix was still used but the $T=1$ p-waves were parameterized by a CSL model. This now required p-waves in Eq. (3.3.14)–(3.3.16), and, in particular for the $S=+1$ part of the interaction. Although the $T=$ KN p-waves are very small below 400 MeV/c, the values of the $T=0$ phase shifts as given by *Stenger et al.* (see Table 3.3) are large, and interfere with the $\bar{K}N$ p-waves in Eq. (3.3.14). The s-wave K-matrix parameters do not change much from those obtained in their pure s-wave fit.

The $\bar{K}N$ scattering lengths obtained in these ZRA K-matrix analyses will be discussed in Section 4.1 below.

c) Effective-Range K-Matrix Models

A multichannel effective-range parameterization was first given by *Ross* and *Shaw* (1961) based on the expansion of the inverse K-matrix (with suitable threshold factors removed). Some time ago this parameterization was used by *Kim* (1967a) for the matrix M, where

$$K = Q^l M^{-1} Q^l. \tag{3.3.18}$$

Thus

$$M(s) = M_0 + \tfrac{1}{2} R^{1-2l} C_l (Q^2 - Q_0^2), \tag{3.3.19}$$

where R is an effective-range matrix; $C_0 = 1$; $C_1 = -3$; and the subscript zero means that the quantities are to be evaluated at the $\bar{K}N$ threshold.

Table 3.7. *The s-wave parameters of the matrix M_0 (in F^{-1}), and effective-ranges (in F), from the analysis of Kim (1967a)*

M_0^0	$\bar{K}N$	$\pi\Sigma$	M_0^1	$\bar{K}N$	$\pi\Sigma$	$\pi\Lambda$
$\bar{K}N$	0.00	-1.11	$\bar{K}N$	-3.60	-2.86	2.08
$\pi\Sigma$	-1.11	2.0	$\pi\Sigma$	-2.86	-1.40	1.81
R^0	0.54	-0.9	$\pi\Lambda$	2.08	1.81	-2.3
			R^1	-0.13	-0.8	-1.2

Table 3.8. *The p-wave parameters of the matrix M_0 (in F^{-3}), and effective ranges (in F), from the analysis of Kim (1967a)*

$p\frac{1}{2}$ Scattering

M_0^0	$\bar{K}N$	$\pi\Sigma$	M_0^1	$\bar{K}N$	$\pi\Sigma$	$\pi\Lambda$
$\bar{K}N$	18	-11	$\bar{K}N$	-18.1	7	4.7
$\pi\Sigma$	-11	-11	$\pi\Sigma$	7	-0.3	-7.0
			$\pi\Lambda$	4.7	-7.0	-3

$p\frac{3}{2}$ Scattering

M_0^0	$\bar{K}N$	$\pi\Sigma$	M_0^1	$\bar{K}N$	$\pi\Sigma$	$\pi\Lambda$
$\bar{K}N$	11	-1.4	$\bar{K}N$	5	-14	-12
$\pi\Sigma$	-1.4	5.6	$\pi\Sigma$	-14	-15.8	-13.6
			$\pi\Lambda$	-12	-13.6	-16
			R^1	-0.7	0.27	0.3

Kim kept s, p and d-waves in his analysis, using the effective-range K-matrix for the s and p-waves, but with non-zero effective ranges only in the diagonal terms of the $s\frac{1}{2}$ and $p\frac{3}{2}$ amplitudes. No effective range terms were used for the $p\frac{1}{2}$ wave. The d_{13} wave was parameterized in the simple CSL approximation, and to take account of the $Y_0^*(1520)$ a Breit-Wigner resonance formula was used for the d_{03} wave. A total of 44 parameters was used. *Kim* analysed all two-body K^-p data, as they existed at the time, below 550 MeV/c. However, no attempt was made (except in the d_{03} wave), to allow for three-body final states, which are non-negligible at these momenta. A unique fit was obtained for the s-wave parameters but the published p-waves, although giving the lowest χ^2 value, are not well separated from other possible solutions. The s and p-wave parameters of the matrix M^0 are shown in Tables 3.7 and 3.8. The main features of the solution are (i) that when continued into the unphysical region the $Y_0^*(1405)$ is predicted with mass

~ 1403 MeV and width ~ 50 MeV; (ii) the Y_1^* (1385) is predicted to have a negligible coupling to the $\bar{K}N$ system, and (iii) the predicted value of the ratio

$$R = \frac{\sigma(K_2^0 p \to K_1^0 p)}{\sigma(K_2^0 p \to \pi^+ \Lambda) + 2\sigma(K_2^0 p \to \pi^+ \Sigma^0)}$$

is consistent with later measurements only if the Yang $I = 0$ KN phase shifts are used, and is inconsistent with the Fermi set. This agrees with the direct measurement of the polarization in $K^+ n$ scattering, (Ray et al., 1969.). However, Martin and Ross, in their fit to this ratio, can find solutions with either set of $T = 0$ KN phase shifts. Once again, we will defer until Section 4.1 discussion of the threshold parameters deduced from this analysis.

d) Other Analyses

Another K-matrix analysis has recently appeared of data in the region 350–430 MeV/c (Berley et al., 1970). These authors have made fits to new data, produced by them, on the differential cross-sections for $K^- p \to \Lambda \pi^0$; $\Sigma^0 \pi^0$ and $\bar{K}^0 n$, and the differential polarization of the Λ and Σ final-state hyperons, plus other published data on $K^- p \to K^- p$, and $\Sigma^\pm \pi^\mp$ in this region. Solutions were found, both assuming the standard Breit-Wigner amplitude for the Y_0^*(1520) resonance and a constant scattering length for the $d_{\frac{3}{2}} T = 1$ amplitude, but differing in their treatment of the s and p-wave backgrounds. One method uses the K-matrix in the zero-range approximation, and the other uses the effective-range approximation based on the M-matrix exactly as parametrized by Kim. The searches were performed using as starting values the solutions of either Kim or Martin and Sakitt. The parameters for the s and p-waves of each solution are shown in Tables 3.9, 3.10, 3.11, and 3.12. The signs of β_0, β_Σ, and β_Λ are now fixed because of the interference with the $d_{\frac{3}{2}}$ wave. Both of these solutions give equally good fits to the data, and so, in particular, effective range terms are not required, at least up to 430 MeV/c. However, even if the effective-range solution

Table 3.9. *The s-wave K-matrix parameters (in Fermis) from the analysis of Berley et al. (1970)*

K^0	$\bar{K}N$	$\pi\Sigma$	K^1	$\bar{K}N$	$\pi\Sigma$	$\pi\Lambda$
$\bar{K}N$	−2.04	−1.28	$\bar{K}N$	0.56	0.66	0.39
$\pi\Sigma$	−1.28	−0.40	$\pi\Sigma$	0.66	1.07	0.10
			$\pi\Lambda$	0.39	0.10	0.50

Table 3.10. *The p-wave K-matrix parameters (in* F^3*) from the analysis of Berley et al. (1970)*

$p\frac{1}{2}$ *scattering*

K^0	$\bar{K}N$	$\pi\Sigma$		K^1	$\bar{K}N$	$\pi\Sigma$	$\pi\Lambda$
$\bar{K}N$	0.03	−0.216		$\bar{K}N$	−0.077	−0.007	−0.065
$\pi\Sigma$	−0.216	0.13		$\pi\Sigma$	−0.007	0.01	0.02
				$\pi\Lambda$	−0.065	0.02	0.19

$p\frac{3}{2}$ *scattering*

K^0	$\bar{K}N$	$\pi\Sigma$		K^1	$\bar{K}N$	$\pi\Sigma$	$\pi\Lambda$
$\bar{K}N$	0.155	0.011		$\bar{K}N$	0.049	−0.029	−0.043
$\pi\Sigma$	0.011	0.1		$\pi\Sigma$	−0.029	−0.25	−0.05
				$\pi\Lambda$	−0.043	−0.05	0.05

Table 3.11. *The s-wave parameters of the matrix* M_0 *(in* F^{-1}*), and effective ranges (in F), from the analysis of Berley et al. (1970)*

M_0^0	$\bar{K}N$	$\pi\Sigma$		M^1	$\bar{K}N$	$\pi\Sigma$	$\pi\Lambda$
$\bar{K}N$	0.03	−1.50		$\bar{K}N$	−4.34	−4.88	0.91
$\pi\Sigma$	−1.50	2.6		$\pi\Sigma$	−4.88	−2.98	1.75
R^0	0.23	1.3		$\pi\Lambda$	0.91	1.75	0.22
				R^1	−3.3	0.9	−0.6

Table 3.12. *The p-wave parameters of the matrix* M_0 *(in* F^3*), and effective ranges (in F), from the analysis of Berley et al. (1970)*

$p\frac{1}{2}$ *scattering*

M_0^0	$\bar{K}N$	$\pi\Sigma$		M_0^1	$\bar{K}N$	$\pi\Sigma$	$\pi\Lambda$
$\bar{K}N$	−4.4	−8.3		$\bar{K}N$	−15	6.7	3
$\pi\Sigma$	−8.3	−4.2		$\pi\Sigma$	6.7	−2.8	−12
				$\pi\Lambda$	3	−12	−14

$p\frac{3}{2}$ *scattering*

M_0^0	$\bar{K}N$	$\pi\Sigma$		M_0^1	$\bar{K}N$	$\pi\Sigma$	$\pi\Lambda$
$\bar{K}N$	8.0	−3.0		$\bar{K}N$	−21.9	− 65	−13.0
$\pi\Sigma$	−3.0	16		$\pi\Sigma$	−65	−122	− 7
				$\pi\Lambda$	−13.0	− 7	− 2.8
				R^1	− 0.5	1	0.40

is accepted the values of these terms differ greatly from those of *Kim* in the important $T=0$ s-wave amplitude. We will return to this point in Section 4.3.

3.3.2 Analyses above ~500 MeV/c

No coupled-channel analyses have been published of data above ~500 MeV/c, although such calculations are currently in progress. Furthermore, with one exception, all analyses have been of the energy-dependent type. In these latter analyses the form of the partial-wave amplitudes are taken to be either (a) a smooth background, (b) a resonance or (c) a combination of both resonance and background.

A frequently used form for the background amplitude is pair of straight lines in the Argand diagram with a join somewhere in the middle of the range. For example, if we define

$$T(s) = q f_{l\pm}(s),$$

then a possible form for the background is

$$T^B = a + b_1(k - 0.8), \quad k > 0.8 \text{ GeV}/c,$$
$$= a + b_2(k - 0.8), \quad k < 0.8 \text{ GeV}/c, \tag{3.3.20}$$

with a, b_1 and b_2 complex numbers. If a pure resonance form is used it is usually represented by

$$T^R = \frac{t}{\varepsilon - i}, \tag{3.3.21}$$

where

$$t = \frac{(\Gamma_i \Gamma_f)^{\frac{1}{2}}}{\Gamma}; \quad \varepsilon = \frac{2(W_R - W)}{\Gamma}.$$

$\Gamma_i(\Gamma_f)$ is the partial width for the decay of the resonance, of full width Γ, into the initial (final) state. The full width is frequently given an energy-dependent form

$$\Gamma = \Gamma(q) = \Gamma_0 \left\{ \frac{q D_l(q)}{q_R D_l(q_R)} \right\},$$

where Γ_0 is a constant and D_l is the usual non-relativistic barrier penetration factor. However, it is not clear that the current data require such refinements. Finally, if both background and resonance are used the S-matrix is assumed to be factorizable into a product of background and resonance terms, leading to

$$T = T^B + T^R + 2i T^B T^R, \tag{3.3.22}$$

where T^B and T^R are given by Eq. (3.3.20) and (3.3.21).

The above formulation has been used by *Armenteros et al.* (1968b, 1968c, 1969) in a series of analyses of data obtained by them in bubble chamber experiments. The method consists of trying to fit the data by a combination of background and resonance terms in various partial waves, and by observing the quality of the fit as the combinations are changed deduce the existence of resonance states. Frequently, known resonances which lie just outside the energy range analysed are included in the parameterization, but the resonance parameters themselves are not varied.

This technique has been used by *Armenteros et al.* to analyse data for $K^- p \rightarrow K^- p$, $\bar{K}^0 n$ and $\pi \Sigma$ between 430–800 MeV/c and 600–1200 MeV/c. In addition, energy-independent analyses have been made for the reaction $K^- p \rightarrow \pi^0 \Lambda$ between 600–1200 MeV/c (*Armenteros et al.*, 1968a).

The results of these various analyses can be summarised briefly by saying that good fits to the data can only be achieved if certain resonances are included, and that the same resonances are required in all reactions, where invariance laws allow them to occur (the resonance parameters have been summarised by *Barbaro-Galtieri et al.*, 1970). However, the resonance parameters as deduced from the various channels are not always completely consistent and further confirmation must await more detailed multichannel analyses.

4. Low-Energy Parameters

In Section 3 we have discussed various direct analyses of kaon-nucleon scattering data below ~ 1 GeV/c. In this section we will consider how the results of those analyses can be used to deduce values for the low-energy parameters of the kaon-nucleon system, where by low-energy parameters we mean such quantities as scattering lengths, coupling constants etc.

4.1 $S = -1$ Scattering Lengths

Information about the $s = -1$ scattering lengths follows directly from the K-matrix analyses of very low-energy $\bar{K}N$ data described in Section 3.3.1. In Table 4.1 we show the predicted values of the $T = 0$ s-wave scattering length from the pure s-wave K-matrix fits (*Martin* and *Sakitt*, 1969b; *Martin* and *Ross*, 1970), and from the s, p, d effective-range analysis of *Kim* (1967a). Also shown in this table are the values predicted from two of the earlier analyses using the CSL formulation (*Kim*, 1965;

Table 4.1. *Values of the* $T=0$ *s-wave* $\bar{K}N$ *scattering length (in Fermis)*

$A_0 = a_0 + ib_0$	Reference	Method
$-(1.67 \pm 0.04) + i(0.72 \pm 0.04)$	*Kim* (1965)	CSL
$-(1.63 \pm 0.07) + i(0.51 \pm 0.05)$	*Sakitt et al.* (1965)	CSL
$-(1.65 \pm 0.04) + i(0.73 \pm 0.02)$	*Von Hippel* and *Kim* (1968)	KM
$-(1.66 \pm 0.02) + i(0.69 \pm 0.02)$	*Martin* and *Sakitt* (1969a)	KM
$-(1.74 \pm 0.04) + i(0.70 \pm 0.01)$	*Martin* and *Ross* (1970)	KM

Table 4.2. *Values of the* $T=1$ $\bar{K}N$ *scattering length (in Fermis)*

$A_1 = a_1 + ib_1$	Reference	Method
$(0.00 \pm 0.06) + i(0.69 \pm 0.03)$	*Kim* (1965)	CSL
$-(0.19 \pm 0.08) + i(0.44 \pm 0.04)$	*Sakitt et al.* (1965)	CSL
$-(0.13 \pm 0.02) + i(0.51 \pm 0.03)$	*Von Hippel* and *Kim* (1968)	KM
$-(0.09 \pm 0.03) + i(0.54 \pm 0.02)$	*Martin* and *Sakitt* (1969a)	KM
$-(0.05 \pm 0.04) + i(0.63 \pm 0.06)$	*Martin* and *Ross* (1970)	KM

Sakitt et al., 1965). Table 4.2 shows the $T=1$ s-wave scattering length from the same analyses. The agreement between the various calculations is very satisfactory, but it would be desirable to have confirmation of these values from analyses based on alternative parameterizations.

Reliable theoretical calculations of these quantities are not yet available. For example, if we evaluate the sign-subtracted forward dispersion relation, Eq. (2.2.16), at threshold then we have a simple sum rule for the difference of the $K^{\pm}p$ scattering lengths. However, due to the errors on the K^+p s-wave scattering length, and uncertainties concerning the precise values of the KN coupling constants and integrals over the unphysical region, it is not yet possible to deduce an accurate value for the K^-p s-wave scattering length.

The p-wave scattering lengths are, of course, more difficult to determine. The p-wave solution of *Kim* (1967a) is not unique and so little significance can be attached to the precise values predicted by using his K-matrix elements. *Martin* and *Ross* (1970) give the following values for the $T=1$ p-wave scattering lengths, based on their CSL analysis of $K_2^0 p$ and $\pi\Lambda$ data below 400 MeV/*c*,

$$A_{p1}^1 = -(0.067 \pm 0.037)\,\mathrm{F}^3 + i(0.0024 \pm 0.0010)\,\mathrm{F}^3 ,$$
$$A_{p3}^1 = -(0.060 \pm 0.037)\,\mathrm{F}^3 + i(0.0037 \pm 0.0012)\,\mathrm{F}^3 . \tag{4.1.1}$$

These values are essentially determined by data in the range 200–400 MeV/*c*. and in this region *Kim*'s *K*-matrix solution gives

$$A_{p1}^1 \simeq -0.07 + i\ 0.02\text{F}^3\ ,$$
$$A_{p3}^1 \simeq +0.06 + i\ 0.002\text{F}^3\ . \tag{4.1.2}$$

The scattering lengths deduced from the constant *K*-matrix analysis of *Berley et al.* (1970) are

$$A_{p1}^1 \simeq -0.08 + i\ 0.008\text{F}^3\ ,$$
$$A_{p3}^1 \simeq +0.05 + i\ 0.005\text{F}^3\ . \tag{4.1.3}$$

The difference in sign of Re A_{p3}^1 between Eq. (4.1.2) and (4.1.3), and Eq. (4.1.1) leads to a difference in the form of the angular distribution for the reaction $K_2^0 p \to K_1^0 p$. The solution with the negative sign approximates this distribution better, but the one with the positive sign differs from it by less than two standard deviations, and so it is clear that more data are required to resolve this point.

Finally, the *T*=0 *p*-wave scattering lengths from the analysis of *Berley et al.* are

$$A_{p1}^0 \simeq 0.03 + i\ 0.04\text{F}^3\ ,$$
$$A_{p3}^0 \simeq 0.16 + i\ 0.00\text{F}^3\ . \tag{4.1.4}$$

4.2 $S=+1$ Scattering Lengths

4.2.1 $T=0$ Scattering

In their experiment on $K^+ d$ scattering *Stenger et al.* (1964) took additional data at 230 MeV/*c*, and assumed that at this momentum the *T*=0 interaction was a pure *s*-wave. The *s*-wave phase shift was parameterized according to

$$\sin \delta_s^0 = a_s^0 q\ ,$$

and the data fitted, using the impulse approximation, in terms of the $T=1$ *s*-wave KN scattering length, which was taken to be (-0.31 ± 0.01) F (see below), and Coulomb interactions. The best value of a_s^0 was found to be

$$a_s^0 = (0.04 \pm 0.04)\ \text{F}\ , \tag{4.2.1}$$

which is consistent with the value which one obtains by extrapolating to threshold the *s*-wave phase shift of Table 3.3 using the usual scattering length expansion

$$q \cot \delta_s^0 = \frac{1}{a_s^0}\ .$$

Chand (1967) has used a modified form of the static multiple-scattering model of *Chand* and *Dalitz* (1962) to calculate K^+d cross-sections at 230 MeV/c. He also finds a_s^0 to be small, and by fitting the total charge-exchange cross-section of *Slater et al.* (1961) finds

$$a_s^0 = -0.11 \begin{matrix} +0.06 \\ -0.04 \end{matrix} \text{F}, \qquad (4.2.2)$$

which is consistent with the value given in Eq. (4.2.1). These two calculations thus agree that a_s^0 is small, but its precise value (or even its sign) is as yet unkown.

If we take the Yang set of $T=0$ phase shifts, as confirmed by the K^+n polarization measurement at 600 MeV/c (*Ray et al.*, 1969), then the $p\frac{1}{2}$ phase shift is reasonably well-determined. Moreover the phase shift is consistent with the expected threshold behaviour, and can be fitted by the expansion

$$q^3 \cot \delta_{p1}^0 = \frac{1}{a_{p1}^0}.$$

Using the values given in Table 3.3 we find

$$a_{p1}^0 = (0.11 \pm 0.04)\,\text{F}^3. \qquad (4.2.3)$$

The errors on the $p\frac{3}{2}$ phase-shift are too large to enable the $p\frac{3}{2}$ scattering length to be estimated in this way, and the same applies to the d-waves.

4.2.2 $T=1$ Scattering

a) s-Wave

We mentioned in Section 3.1.1 that although low-energy phase-shift analyses favour an s-wave $T=1$ phase shift which is negative, solutions can also be found with $\delta_s^1 > 0$ which statistically are not greatly inferior. However, such solutions are difficult to reconcile with low-energy K^-p scattering and forward dispersion relations. This can be seen very simply by evaluating the sign-subtracted forward dispersion relation for protons at threshold. From Eq. (2.2.16) we have

$$\left(\frac{M+m}{2M}\right)(a_s^- - a_s^+) = m \sum_{Y=A,\Sigma^0} \frac{R_Y}{\omega_Y^2 - m^2} + \frac{m}{\pi}\int_{\omega_{\pi A}}^{m} d\omega' \frac{\text{Im} f_-(\omega')}{\omega'^2 - m^2}$$

$$+ \frac{m}{4\pi^2} \text{P} \int_m^\infty d\omega' k' \frac{[\sigma_-(\omega') - \sigma_+(\omega')]}{\omega'^2 - m^2}, \qquad (4.2.4)$$

where $a_s^\pm$ are the s-wave $K^\pm$-p scattering lengths. The physical integrals for $k' \gtrsim 300$ MeV/c can be accurately evaluated using measured total cross-section (see the compilation by *Price et al.*, 1969), and a Regge-pole parameterization for momenta above 25 GeV/c (e. g. that of *Dass et al.*, 1969). Furthermore, we have seen in Section 4.1 that the value of a_s^- is rather well-known from analyses of low-energy K^-p interactions. Evaluating the relevant terms in Eq. (4.2.4) gives (in Fermis)

$$a_s^+ \simeq -1.8\,U - 0.032\,G^2 - (1.40 \pm 0.02)\,,$$

where

$$U = \frac{m}{\pi} \int\limits_{\omega_{\pi A}}^{\bar{\omega}} d\omega' \, \frac{\mathrm{Im}\,f_-(\omega')}{\omega'^2 - m^2}\,,$$

$\bar{\omega}$ corresponds to $k' = 300$ MeV/c, and

$$G^2 \simeq (g_{KAp}^2 + g_{K\Sigma p}^2)/4\pi\,.$$

The integral U will be dominated by s and p-waves, and the p-wave contribution is positive. Thus, if we denote by U_0 the s-wave part of U we have,

$$a_s^+ \lesssim -1.8\,U_0 - 0.032\,G^2 - (1.40 \pm 0.02)\,.$$

Now U_0 may be evaluated using one of the K-matrix analyses described in Section 3.3.1. If we use the results of *Martin* and *Sakitt*, as an example of the zero-range models, we find

$$a_s^+ \lesssim -0.032\,G^2 - (0.11 \pm 0.13),$$

and so, for $G^2 > 0$, $a_s^+ \lesssim -(0.11 \pm 0.13)$, and for $G^2 \simeq 5$, which we shall see in Section 4.3 is one of the lowest estimate for G^2,

$$a_s^+ \lesssim -(0.27 \pm 0.13)\ \mathrm{F}\,.$$

If we use the effective-range solution of *Kim* we have

$$a_s^+ \lesssim -0.032\,G^2 + (0.11 \pm 0.11)\,,$$

and for $G^2 \simeq 5$,

$$a_s^+ \lesssim (0.0 \pm 0.1)\ \mathrm{F}\,.$$

Thus, regardless of the set of low-energy parameters we use, the $T = 1\ KN$ phase-shift solution with a dominant positive s-wave (which requires $a_s^+ \sim 0.3\,F$) is incompatible with forward dispersion relations. We shall therefore consider only the negative s-wave solution in what follows.

Given the s-wave $T = 1$ phase shifts shown in Table 3.1 there are a number of different ways of extrapolating them to threshold. *Goldhaber et al.* used the effective range formula

$$q\cot\delta_s^1 = \frac{1}{a_s^1} + \frac{r_s^1}{2}\,q^2\,,$$

Table 4.3. *Values of the s-wave $T = 1$ KN scattering length a_s^1 (in F), and effective range r_s^1 (in F)*

a_s^1	r_s^1	Reference
-0.29 ± 0.015	0.5 ± 0.15	*Goldhaber et al.* (1962)
-0.28 ± 0.01		*Lea et al.* (1968)
-0.286 ± 0.004	$0.54 \pm 0.09\ (G^2 = 0)$	
-0.283 ± 0.006	$0.54 \pm 0.09\ (G^2 = 7)$	*Cutkosky* and *Deo* (1970)
-0.292 ± 0.006	$0.35 \pm 0.09\ (G^2 = 14)$	
-0.31 ± 0.01	$0.40 \pm 0.14\ (k < 0.7\ \text{GeV}/c)$	*Martin* and *Perrin* (1969)
-0.32 ± 0.01	$0.22 \pm 0.23\ (k < 0.86\ \text{GeV}/c)$	
-0.305 ± 0.012	$C_s^1 = -0.023 \pm 0.004\ \text{F}^3$	*Martin* and *Thompson* (1970)

and their best-fit values are shown in Table 4.3. All subsequent estimates of a_s^1 are consistent with the value found by *Goldhaber et al.* Nevertheless, differences are likely to occur in the various methods as data improve in the low-energy region, and so we shall briefly discuss some other methods that have been used.

Firstly, values of a_s^1 are obtainable from the phase-shift analyses discussed in Section 3.1.1. In the analysis of *Lea et al.*(1968) the polynomial expansion (3.1.8) was used to fit data up to 1.45 GeV/c. This expansion has absolutely nothing to recommend it theoretically, and fitting data up to 1.45 GeV/c is not a good way to determine threshold parameters. Nevertheless, the value of a_s^1 shown in Table 4.3 is also in agreement with the value of *Goldhaber et al.* This fact alone is probably an indication that with present data almost any extrapolation to threshold will be suitable. Also shown in Table 4.3 are the estimates of *Cutkosky* and *Deo*, as a function of $G^2 \equiv (g_A^2 + g_\Sigma^2)/4\pi$ obtained from their analysis of data below 650 MeV/c, using the modified effective-range expansion Eq. (3.1.9).

An alternative method of deducing a_s^1 which attempts to include theoretical constraints from forward dispersion relations is due to *Martin* and *Perrin* (1969). These authors express the elastic $K^+ p$ differential cross-section, and total cross-section data below 860 MeV/c by the expressions

$$\frac{d\sigma}{d\Omega} = \sum_{n=0}^{N} W_n(q^2)\, P_n(\cos \vartheta), \qquad (4.2.5)$$

and

$$\sigma_T = 4\pi W_0(q^2),$$

with

$$W_n(q^2) = q^{2n}(A_n + B_n q^2 + C_n q^4 + \cdots). \qquad (4.2.6)$$

8*

If, furthermore, the real part of the forward laboratory amplitude is written

$$\operatorname{Re} f_+(\omega) = \frac{W}{M}(A_R + B_R q^2 + C_R q^4 + \cdots),\tag{4.2.7}$$

then, in the elastic region, the following consistency conditions hold between the two sets of parameters

$$A_0 = A_R^2,$$
$$B_0 + A_1 = A_0^2 + 2A_R B_R,$$
$$C_0 + B_1 + C_2 = 2A_0 B_0 + 2A_R C_R + B_R^2,$$

and so on. The calculation prodeeds as follows. A fit to the data is made using Eq. (4.2.6) for a fixed value of N. Since the value of $A_0 = (a_s^2)^2$, from Eq. (4.2.7) the value of $\operatorname{Re} f_+(m)$ is also determined. Using an input value for $\operatorname{Re} f_-(m)$, the effective KN coupling constant is calculated from the sign-subtracted forward disperson relation, Eq. (2.2.16), evaluated at threshold. This value is then used in the dispersion relation for the derivative of $\operatorname{Re} f_+(\omega)$ at $\omega = m$ i. e.

$$\frac{\partial \operatorname{Re} f_+(\omega)}{\partial \omega} = \frac{1}{\pi}\frac{\partial}{\partial \omega}\left[P\int_m^\infty d\omega' \frac{\operatorname{Im} f_+(\omega')}{\omega' - \omega}\right]$$
$$-\frac{1}{\pi}\int_{\omega_{\pi\Lambda}}^\infty d\omega' \frac{\operatorname{Im} f_-(\omega')}{(\omega' + \omega)^2} - \frac{R_\Lambda G}{(\omega_\Lambda + \omega)^2},\tag{4.2.8}$$

where $G = (g_{K\Lambda p}^2 + 0.84 g_{K\Sigma^0 p}^2)/4\pi$.

Since, from Eq. (4.2.7),

$$(M + m)\frac{\partial \operatorname{Re} f_+(\omega)}{\partial \omega}\bigg|_{\omega = m} = 2mM B_R + A_R,$$

the evaluation of (4.2.8) gives an estimate for B_R. Similarly, an estimate of C_R may be obtained by evaluating a dispersion relation for $\partial^2 \operatorname{Re} f_+(\omega)/\partial \omega^2$ at $\omega = m$, and hence the consistency conditions can be checked. The low-energy $K^+ p$ contribution to the derivative dispersion relations is calculated for each fit using the values of the parameters $A_0, B_0, \ldots$ etc. predicted by that fit.

Two fits were made, one using data below 700 MeV/c with $N = 1$, and the second using data up to 860 MeV/c, with $N = 2$. Since

$$A_0 = (a_s^1)^2, \quad \text{and} \quad B_0 = -(a_s^1)^3 (a_s^1 + r_s^1),$$

the best values of A_0 and B_0 may be used to calculate a_s^1 and r_s^1 (assuming a_s^1 to be negative). The values obtained for the two fits are shown in Table 4.3.

The final method is due to *Hamilton* and *Woolcock* (1963), and also uses forward dispersion relations. It has been used successfully by these authors to obtain accurate values for the s-wave πN scattering lengths. If we expand the forward $K^+ p$ amplitude into partial-wave amplitudes we have

$$f_+(\omega) = \frac{W}{M} \left[f_s^1(\omega) + f_{p1}^1(\omega) + 2 f_{p3}^1(\omega) + \cdots \right]. \tag{4.2.9}$$

Using the expression on the left-hand side of the doubly-subtracted forward dispersion relation (2.2.21) gives

$$\frac{W}{M} \operatorname{Re} f_s^1(\omega) = \left(\frac{M+m}{2M} \right) \left[\left(1 + \frac{\omega}{m} \right) a_s^+ + \left(1 - \frac{\omega}{m} \right) a_s^- \right] + k^2 C_s^1(\omega),$$

$$\tag{4.2.10}$$

where the s-wave curvature coefficient C_s^1 is given by

$$C_s^1(\omega) = \sum_{Y=A,\,\Sigma^0} \frac{R_Y}{(\omega_Y + \omega)(\omega_Y^2 - m^2)} + \frac{1}{\pi} \int_{\omega_{\pi A}}^{m} d\omega' \frac{\operatorname{Im} f_-(\omega')}{k'^2(\omega' + \omega)}$$

$$+ \frac{P}{4\pi^2} \int_{m}^{\infty} d\omega' \frac{1}{k'} \left[\frac{\sigma_+(\omega')}{\omega' - \omega} + \frac{\sigma_-(\omega')}{\omega' + \omega} \right] \tag{4.2.11}$$

$$- \frac{1}{q^2} \frac{M}{W} \left[\operatorname{Re} f_{p1}^1(\omega) + 2 \operatorname{Re} f_{p3}^1(\omega) \right]$$

$$+ \text{ higher waves.}$$

A fit to the s-wave phase-shifts has recently been made (*Martin* and *Thompson*, 1970) using the parametric form (4.2.10). In practice, if $C_s^1(\omega)$ is expanded about threshold it is found necessary to keep only the first term $C_s^1(m) = C_s^1$ in the fit. Since this calculation was part of an iterative attempt to calculate both s and p-waves scattering lengths it will be discussed more fully below, but the final value of a_s^1, given in Table 4.3, is, once again, consistent with previous estimates.

b) p-Waves

The p-wave scattering lengths are more difficult to determine because of the dominance of s-waves in $K^+ p$ scattering below 1 GeV/c. Recently, however, an attempt has been made to estimate these quantities from forward dispersion relation sum rules (*Martin* and *Thompson*, 1970).

The first sum rule is for the combination $(a_{p1}^1 + 2a_{p3}^1)$, and follows directly from Eq. (4.2.11) by taking the limit $\omega \to m$. Thus,

$$\left(\frac{M}{M+m}\right)(a_{p1}^1 + 2a_{p3}^1)$$

$$= \sum_Y \frac{R_Y}{(\omega_Y + m)(\omega_Y^2 - m^2)} + \lim_{\omega \to m} \frac{P}{\pi} \int_{\omega_{\pi A}}^m d\omega' \frac{\mathrm{Im} f_-(\omega')}{(\omega' + \omega) k'^2} \quad (4.2.12)$$

$$+ \lim_{\omega \to m} \frac{P}{4\pi^2} \int_m^\infty d\omega' \frac{1}{k'}\left[\frac{\sigma_+(\omega')}{\omega' - \omega} + \frac{\sigma_-(\omega')}{\omega' + \omega}\right] - C_s^1 .$$

The integrals in this equation are all finite but require some care when evaluated numerically.

The second sum rule is for the combination $(a_{p1}^1 - a_{p3}^1)$, and is found by evaluating an unsubtracted forward dispersion relation for $B_+(\omega, 0)$ at threshold. Thus,

$$8\pi M(a_{p1}^1 - a_{p3}^1)$$

$$= -\frac{2\pi}{M} a_s^1 + \frac{1}{\pi} \int_m^\infty d\omega'\left[\frac{\mathrm{Im} B_+(\omega', 0)}{\omega' - m} - \frac{\mathrm{Im} B_-(\omega', 0)}{\omega' + m}\right] \quad (4.2.13)$$

$$- \frac{1}{\pi} \int_{\omega_{\pi A}}^m d\omega' \frac{\mathrm{Im} B_-(\omega', 0)}{\omega' + m} - \frac{1}{2M} \sum_{Y = A, \Sigma^0} \frac{g_{KY\bar{p}}^2}{\omega_Y + m} .$$

To evaluate these two sum rules the following input data were used.

The s-wave unphysical integrals and low-energy $K^- p$ physical integrals were evaluated using the K-matrix solution of *Martin* and *Sakitt* (1969a), and for the KN coupling constants their values were used (see Section 4.3.1) i. e.

$$g_{KAp}^2/4\pi = 5.0 \pm 1.9 ,$$

$$g_{K\Sigma^0 p}^2/4\pi = 1.0 \pm 1.5 . \quad (4.2.14)$$

Because of the absorptive nature of the $K^- p$ interaction both the real *and* imaginary parts of $B_-(\omega, 0)$ have the same threshold behaviour, and hence will have important p-wave contributions in the low-energy region. These have been estimated by using the p-wave zero-range K-matrix solution of *Berley et al.* (1970) in the region below 430 MeV/c.

The p-wave unphysical regions were assumed to be dominated by the $Y_1^*(1385)$ resonance, and here the narrow width approximation was used, i. e.

$$\operatorname{Im} f_{Y*}(\omega) = \frac{\pi}{2} \alpha \left(\frac{M_R}{M} \right) \delta(\omega - \omega_R),$$

where

$$\alpha = \left(\frac{g^2}{4\pi} \right) \left[\frac{(M_R + M)^2 - m^2}{3 M_R^2} \right] \left(\frac{q_R}{M} \right)^2,$$

(4.2.15)

and M_R is the mass of the resonance. Unitary symmetry (given the width of the $N^*(1236)$) predicts $g^2/4\pi \sim 2.4$. However, since we are using values for the KN coupling constants which are reduced from their $SU(3)$ values it is more reasonable to use a value for g^2 which is also reduced from its symmetry values. (Such a view would follow naturally from the quark model, for example.) There is some evidence from high-energy photoproduction of $Y_1^*(1385)$ that $g^2/4\pi$ is reduced by a factor 2–3 from its $SU(3)$ value, which would be consistent with using the values of the KN coupling constants given in Eq. (4.2.14) (*Harari*, 1969). Therefore $g^2/4\pi = 1.2$ was used and an arbitrary 50% error assigned to this term.

The remaining terms in Eq. (4.2.12) can be evaluated in terms of measured values of total cross-sections (*Price et al.*, 1969). Above 2 GeV/c the Regge-pole model of *Dass et al.* (1969) was used.

Evaluation of the sum rule (4.2.13) requires more detailed input information.

For $K^+ p$ scattering below 2 GeV/c the best solution of a recent phase shift analysis using s, p, d and f-waves was used (*Lea et al.*, 1970a). It might be objected that since the p-wave $K^+ p$ phase-shifts are used to evaluate a term in the sum rule it is somewhat circular to use the sum rule to calculate the p-wave scattering lengths. However, this objection is invalid because the integral is over the *absorptive* part of B_+ and the p-waves in the low-energy region contribute very little to this integral, partly due to their intrinsic smallness and partly due to their threshold behaviour. Moreover, the total contribution of the $K^+ p$ integral below 2 GeV/c is very small and the values of the p-wave scattering lengths would be little altered even if the p-wave phase shifts were in considerable error.

For the $K^- p$ integral from 430–1200 MeV/c we used the partial-wave amplitudes of *Armenteros et al.* (1968b, 1969). Above 1.2 GeV/c no partial-wave analyses exist for $K^- p$ scattering and we have thus been forced to use the extrapolated Regge-pole parameters of *Dass et al.*, although they do not produce values of $\operatorname{Im} B_-(\omega, 0)$ which are continuous with the values obtained using the parameterizations of *Armenteros et al.*

S and p-wave scattering lengths may now be obtained in an iterative manner as follows. A first estimate for a_s^1 was obtained by using the

Table 4.4. *Values of the p-wave* $T = 1$ *KN scattering lengths,* a_{p1}^1, *and* a_{p3}^1 *(in* F^3*)*

a_{p1}^1	a_{p3}^1	Reference
-0.035 ± 0.006	0.007 ± 0.008	*Martin* and *Thompson* (1970)
-0.009 ± 0.004	$0.006 \pm 0.002 \ (G^2 = 0)$	
-0.018 ± 0.005	$0.011 \pm 0.002 \ (G^2 = 7)$	*Cutkosky* and *Deo* (1970)
-0.028 ± 0.003	$0.016 \pm 0.001 \ (G^2 = 14)$	
$(a_{p1}^1 + 2a_{p3}^1) = -0.01 \pm 0.01$		*Martin* and *Perrin* (1969)

s-wave phase shifts of *Goldhaber et al.* to construct values for Ref_s^1. A least-squares fit to these values was made in terms of the two parameters a_s^1 and C_s^1, using Eq. (4.2.10) with a_s^- having the value given by *Martin* and *Sakitt*. Next, using the fitted values of a_s^1 and C_s^1, together with the input data described above, the *p*-wave scattering lengths, were calculated from the sum rules (4.2.12) and (4.2.13). We may now preserve the total nuclear cross-section of *Goldhaber et al.* by subtracting off the *p*-wave contribution, thereby forming new *s*-wave cross-sections. To do this we assumed that the *p*-wave phase shifts were adequately represented by a scattering length expansion. The new *s*-wave cross-sections may now be used, via the *s*-wave fitting procedure, to predict new *s*-wave parameters and hence, via the sum rules, new *p*-wave scattering lengths. Only one iteration of this procedure was necessary to produce a solution for the *s* and *p*-wave scattering lengths which did not change within their errors from one iteration to the next.

The final values for the *s*-wave parameters are shown in Table 4.3, and the final results from the sum rules were

$$a_{p1}^1 + 2a_{p3}^1 = -(0.020 \pm 0.018) \ \text{F}^3 \ ,$$

and

$$a_{p1}^1 - a_{p3}^1 = -(0.042 \pm 0.010) \ \text{F}^3 \ . \tag{4.2.16}$$

Although the errors in Eq. (4.2.16) are correlated it seems, nevertheless, worthwhile to give estimates for the individual *p*-wave scattering lengths, and solving these equations gives the results shown in Table 4.4.

Very little is known about the *p*-wave scattering lengths. Conventional phase-shift analyses suggest that at somewhat higher momenta the $p\frac{1}{2}$ phase-shift is negative and the $p\frac{3}{2}$ phase-shift is positive, with $|\delta_{p1}^1| > |\delta_{p3}^1|$, which would agree with the results obtained here, but the poor quality of existing data makes it very difficult to deduce reliable values for the scattering lengths.

Cutkosky and *Deo* have presented a series of solutions for the *p*-wave scattering lengths as a function of the parameter $G^2 = (g_A^2 + g_\Sigma^2)/4\pi$

obtained from their phase-shift analysis of low-energy K^+p data (see Section 3.1.1) and their results are shown in Table 4.4. The reasonable agreement between the results of *Cutkosky* and *Deo*, and those of *Martin* and *Thompson* is very encouraging, and may suggest that the degree of self-consistency within the input data for the p-wave sum rules is greater than one might suspect. Also shown in Table 4.4 is the value of $(a_{p1}^1 + 2a_{p3}^1)$ found by *Martin* and *Perrin* in their partial-wave analysis of K^+p differential cross-sections.

4.3 Kaon-Nucleon Coupling Constants

4.3.1 Forward Dispersion Relations

Forward dispersion relations have been used more than any other technique to calculate values for the kaon-nucleon coupling constants, and in this section we will discuss the results that have been obtained from various forms of these dispersion relations.

a) Sign-subtracted Sum Rule

The simplest and most widely studied expression for the KN coupling constants is the sign-subtracted forward dispersion relation evaluated at threshold. Since this sum rule exhibits several features common to nearly all KN forward dispersion relations we will discuss it in some detail.

For convenience we will consider the amplitude,

$$T(\omega) \equiv T_n^-(\omega) - 2T_p^-(\omega) \,,$$

which is pure $T = 0$ in the $S = -1$ channel, and therefore isolates the Λ-pole term. It also has the advantage of removing the unkown $Y_1^*(1385)$ contribution. Evaluating at threshold an unsubtracted forward dispersion relation for $T(\omega)$ gives,

$$\frac{[(M_\Lambda - M)^2 - m^2]}{\omega_\Lambda^2 - m^2} \left(\frac{m}{2M^2}\right) \frac{g_{K\Lambda p}^2}{4\pi}$$

$$= -\left(\frac{M+m}{4M}\right)(3a_s^1 - a_s^0 - 2\bar{a}_s^0) - \frac{m}{\pi} \int_{\omega_{\pi\Sigma}}^m d\omega' \frac{\mathrm{Im}\,\bar{f}^0(\omega')}{\omega'^2 - m^2} \quad (4.3.1)$$

$$+ \frac{m}{4\pi^2} \mathrm{P} \int_m^\infty d\omega' \frac{1}{k'} [\sigma_-^n(\omega') - \sigma_+^n(\omega') - 2\sigma_-^p(\omega') + 2\sigma_+^p(\omega')] \,,$$

where a_s^T and $\bar{a}_s^T$ are the s-wave KN and $\bar{K}N$ scattering lengths for $T = 0,1$; $\bar{f}^0(\omega)$ is the $T = 0$ $\bar{K}N$ forward scattering amplitude; and $\sigma_\pm^N(N = n, p)$ is the $K^\pm N$ total cross-section.

The input data used to evaluate this sum rule are: the s-wave scattering lengths; total cross-sections; a model for the high-energy part of the physical integral; and a model for $\text{Im}\,\bar{f}^0$ in the low-energy physical, and unphysical regions.

The values of the s-wave scattering lengths were discussed in Sec. 4. The $K^\pm$-p total cross-section are accurately known from just above threshold to $\sim 25\ \text{GeV}/c$. The $K^\pm n$ cross-sections are less well-known, but are fortunately weighted by a factor of $\frac{1}{2}$. At high momenta the model used is usually based on Regge poles, and for the unphysical region one of the models discussed in Section 3.3.1 can be used.

Eq. (4.3.1) has been evaluated by many authors (*Zovko*, 1966a; *Lusignoli et al.*, 1966; *Carter*, 1967; *Davies et al.*, 1967; *Kim*, 1967b; *Martin* and *Poole*, 1967, 1968; *Queen et al.*, 1969b; *Granovski* and *Starikov*, 1968; *Rood*, 1967; *Martin* and *Sakitt*, 1969b; *Martin* and *Ross*, 1970), and as an illustration of the results obtained we show in Table 4.5

Table 4.5. *Contributions to the sum rule, Eq. (4.3.1), for $g_{K\Lambda p}^2/4\pi$, from the analysis of Martin and Sakitt (1969b)*

Contribution	Value	Error
Scattering lengths	-18.2	0.6
s-wave $T = 0$ $\bar{K}N$ scattering below 300 MeV/c and the unphysical region	35.1	1.7
$K^- N\ 0.3 - 5.0\ \text{GeV}/c$	-14.9	0.4
$K^+ N\ 0 - 5.0\ \text{GeV}/c$	5.7	0.3
Asymptotic region $k > 5\ \text{GeV}/c$	-2.7	0.1
Total	5.0	1.9

the results of *Martin* and *Sakitt* (1969b) who have evaluated the unphysical region using the result of their zero-range K-matrix analysis. Two points are clear from the results of Table 4.5:

(i) there are strong cancellations between the various terms, and

(ii) the $T = 0$ unphysical region gives a very important contribution to the sum rule.

The value of $g_{K\Lambda p}^2/4\pi$ obtained in the above calculation, i. e. 5.0 ± 1.9, is compatible with all previous calculations using either the CSL or ZRA K-matrix to evaluate the unphysical integral. However, if one uses the effective-range K-matrix solution of *Kim* to evaluate this integral then the unphysical region is larger by some 20–25%, and because of the cancellations in Eq. (4.3.1) this means that the value of $g^2_{K\Lambda p}$ is greatly enhanced. *Kim*, in his original analysis (*Kim*, 1967b) found $g^2_{K\Lambda p}/4\pi$ = 13.5 ± 2.1.

It has been pointed out, however, that the width of the $Y^*_0(1405)$ is particularly sensitive the effective-range parameters in the K-matrix (*Martin et al.*, 1969a). For example, if the parameter R^0 in Table 3.7 is changed from 0.54F to 0.44F then $g^2_{K\Lambda p}/4\pi$ falls from 13.5 to 11.0, and the same effect can be produced by keeping the diagonal terms unchanged but including an off-diagonal term of -0.2F. It is not, of course, clear that such changes do not destroy the fit to the low-energy $\bar{K}N$ data, but in view of the result of *Berley et al.* (1970) that effective range terms are not required up to 430 MeV/c it is somewhat surprising that such large terms are found in *Kim's* analysis.

By considering the $T = 1$, $S = -1$ forward amplitude a sum rule for $g^2_{K\Sigma^- n}$ may be constructed similar to Eq. (4.3.1) but involving only $K^{\pm}n$ cross-sections in the physical region. It is usual to quote the result for $g^2_{K\Sigma^0 p}/4\pi = \frac{1}{2}g^2_{K\Sigma^- n}/4\pi$, and all analyses agree that this coupling constant is small and probably $\lesssim 2$ where the inequality is due to the neglect of the $Y^*_1(1385)$ contribution. Since *Kim's* parameterization includes this term his result can be stated as an equality. He finds $g^2_{K\Sigma^0 p}/4\pi = 0.2 \pm 0.4$.

The value of $g^2_{K\Lambda p}$ found by *Kim* is far larger than all other estimates and is, moreover, compatible with $SU(3)$ symmetry if we assume the usual value for the F/D ratio. It is therefore important to decide between the various solutions of low-energy $\bar{K}N$ scattering, or alternatively, to use dispersion relations in which the unphysical region plays a smaller role. We will consider in more detail, in Section 4.3.2., the question of consistency of models of low-energy $\bar{K}N$ scattering, but for the present we will discuss results obtained by using other forward dispersion relations.

b) Energy-subtracted Relations

The sum rule (4.3.1) uses values of the forward real parts at only one energy, in practice at threshold, and produces results that are sensitive to the input values of the s-wave scattering lengths. The energy-subtracted relations enable information at many energies to be used simultaneously, but in order for this to be possible values of $\mathrm{Re} f_{\pm}(\omega)$ must be available.

In practice values of $\mathrm{Re} f_{\pm}(\omega)$ are obtained by fitting the elastic differential cross-section by a Legendre polynomial expansion

$$\left(\frac{d\sigma}{d\Omega}\right)_{\mathrm{CM}} = \sum_{l=0}^{L} a_l P_l(\cos\vartheta),$$

and then using this expansion to extrapolate the cross-section to the forward point. The real parts of $f_\pm(\omega)$ are then obtained by use of the optical theorem i. e.

$$[\operatorname{Re} f_\pm(\omega)]^2 = \left(\frac{W}{M}\right)^2 \sum_{l=0}^{L} a_l - \left(\frac{k\sigma_\pm}{4\pi}\right)^2.$$

This method does not give the sign of the real part, and in practice either a definite sign is assumed (e. g. for K^+p scattering), or that sign is chosen which produces a better statistical fit, when used in the dispersion relation. It is very difficult to give a meaningful error on $\operatorname{Re} f_\pm$ when the above procedure is used, but a common method is to estimate a rough error by comparing fits obtained with the optimum value of $l = L_0$ with fits obtained using $l = L_0 \pm 1$.

Perrin and *Woolcock* have used Eq. (2.2.19) in a linear least-squares fit to values of $\operatorname{Re} f_\pm(\omega)$ in terms of the two parameters K and the effective KN coupling constant

$$G^2 = (g^2_{K\Lambda p} + 0.84 g^2_{K\Sigma^0 p})/4\pi ,$$

which is obtained by transferring the Λ and Σ poles to a common mass. In their original calculation (*Perrin* and *Woolcock*, 1968) they used about 60 values of $\operatorname{Re} f_\pm$ in the fit, and evaluated the unphysical region integrals using the CSL model of *Kim* (1965). They found $G^2 = 12.3 \pm 2.8$. However, in a later paper (*Perrin* and *Woolcock*, 1969) using a revised set of 66 values of $\operatorname{Re} f_\pm(\omega)$ they found $G^2 = 8.0 \pm 1.7$, if the CSL parameterization was used, and $G^2 = 10.9 \pm 1.7$ using *Kim's* effective-range K-matrix solution. As a check on their method these authors then evaluated $\operatorname{Re} f_\pm(\omega)$ at $\omega = m$ using the best values of K and G^2. Values for the s-wave scattering lengths were found that were consistent with the input values, but with rather large errors. It is clear from the above results that using values of $\operatorname{Re} f_\pm(\omega)$ at several energies away from threshold produces values for the KN coupling constants which are somewhat less sensitive to the particular model used for the unphysical region, but *are* sensitive to the input values of $\operatorname{Re} f_\pm(\omega)$.

The calculations of *Perrin* and *Woolcock* are examples of linear extrapolation techniques, and other such methods exist but have not been widely used.

In the method due to *Haber-Schaim* (1956) the identity

$$\frac{1}{\omega'^2 - \omega^2} \equiv \frac{\omega^2}{\omega'^2(\omega'^2 - \omega^2)} + \frac{1}{\omega'^2} ,$$

is used in the unsubstracted dispersion relation for $T^-(\omega)$ giving

$$
\begin{aligned}
L(\omega) \equiv \omega\, \mathrm{Re}\, T^-(\omega) &- \frac{\omega^4}{\pi} \int_{\tilde{\omega}}^{m} d\omega' \frac{\mathrm{Im}\, f_-(\omega')}{\omega'^2(\omega'^2 - \omega^2)} \\
&- \frac{\omega^4}{4\pi^2} P \int_{m}^{\infty} d\omega' \frac{k'[\sigma_-(\omega') - \sigma_+(\omega')]}{\omega'^2(\omega'^2 - \omega^2)} \\
&= \frac{\omega^2}{\pi} \int_{\tilde{\omega}}^{m} d\omega' \frac{\mathrm{Im}\, f_-(\omega')}{\omega'^2} + \omega^2 \sum_Y \frac{R_Y}{\omega_Y^2 - \omega^2} \\
&+ \frac{\omega^2}{4\pi^2} \int_{m}^{\infty} d\omega' \frac{k'[\sigma_-(\omega') - \sigma_+(\omega')]}{\omega'^2} \, .
\end{aligned}
$$

(4.3.2)

For $\omega^2 \gg \omega_Y^2$, $L(\omega)$ versus ω^2 is a straight line with intercept ΣR_Y at $\omega^2 = 0$. This method has been used to fit a very limited set of values of $\mathrm{Re}\, f_\pm$ using the CSL model for the unphysical region (*Zovko*, 1966a; *Davies et al.*, 1967; *Ara* and *Rashid*, 1967) and the results obtained are compatible with other calculations using the CSL model, but with larger errors.

The Haber-Schaim method requires values of $\mathrm{Re}\, f_+$ and $\mathrm{Re}\, f_-$ at equal energies and these are rarely available. Another method (*Schnitzer* and *Salzmann*, 1959) is based on the twice-subtracted dispersion relation (2.2.21). This equation can be written in the form

$$
J_\pm(\omega) = \frac{1}{2}\,\mathrm{Re}\, f_-(m)\left(1 \mp \frac{\omega}{m}\right) + \frac{1}{2}\,\mathrm{Re}\, f_+(m)\left(1 \pm \frac{\omega}{m}\right),
$$

where $J_\pm(\omega)$ can be found from (2.2.21). In practice one calculates from data the function $P(\omega)$, which equals $J_+(\omega)$ for $\omega > m$ and equals $J_-(\omega)$ for $\omega < m$. This quantity must be linear function of ω, whose slope and intercept at $\omega = 0$, determine the s-wave $K^\pm p$ scattering lengths. *Zovko* (1966a) has used this method, in conjuction with the CSL parameterization, to calculate the KN couplings, by varying G^2 to minimize the least-squares deviation of $P(\omega)$ from a straight line. The least value of G^2 was ~ 5, but again the number of values of $P(\omega)$ used was rather small.

c) Other Subtractions

Calculations have been carried out using the broad-area subtracted dispersion relation Eq. (2.2.23). These relations reduce the contribution of the $Y_0^*(1405)$ by converting part of the unphysical integral to an integral over $\mathrm{Re}\, f_-(\omega)$. Taking $\beta = \frac{1}{2}$ and using the CSL extrapolation, *Martin* and *Poole* (1967, 1968) find $g_{KAp}^2/4\pi = 4.6 \pm 1.3$, where the error is estimated by making reasonable variations in the K-matrix parameters. This value is stable when ω_0 was varied in the range such the 1340 MeV $< W <$ 1395 MeV. This result is confirmed by *Chan* and *Meiere* (1968) who set ω_0 equal to the lowest pion-hyperon threshold. These authors also considered the effective-range K-matrix model for the unphysical region and in this latter case find $g^2_{K \Lambda p}/4\pi = 13.5 \pm 1.8$. *Chan* and *Meiere* investigated the behaviour of the coupling constant as β was varied in the range $0 < \beta < 1$. For *Kim*'s parameterization the value of $g_{K \Lambda p}$ was stable, whereas for $\beta \gtrsim 0.6$ the value obtained using the CSL extrapolation changed considerably. However, this erratic behaviour is not surprising since ω_0 was chosen to be $\omega_{\pi Y}$, and the CSL model is known to have incorrect analytic properties at this point.

d) Finite Contour Relations

Another method which also uses values of $\mathrm{Re} f_+(\omega)$ at several values of ω is based on the finite contour relations Eq. (2.2.33). In the first application of these relations (*Martin* and *Ross*, 1968) the amplitude for $\omega > \omega_c$ was assumed to be given by the Regge-pole model. In that case the integral round the circle may be expanded in a rapidly convergent power series in the variable $x = \omega + \omega_\Lambda$. If we now transfer the Σ-pole to the position of the Λ-mass, then Eq. (2.2.33) may be written

$$F(x) - \sum_{n=3}^{\infty} b_n x^n = \frac{-R}{x} + b_0 + b_1 x + b_2 x^2 , \qquad (4.3.3)$$

where R is proportional to the effective KN coupling constant, and $F(x)$ may be found from Eq. (2.2.33). In practice $F(x)$ was calculated at about 20 values of x and the values of b_n for $n \geq 3$ were fixed at the values predicted by the Regge-pole model of *Rarita* and *Phillips* (1965). A least-squares fit was then made to the left-hand side in terms of the parameters R, b_0, b_1 and b_2, with the latter three parameters constrained to lie close to those of the Regge-pole model. Using the CSL model for the unphysical region gave

$$(g^2_{K \Lambda p} + 0.84 g^2_{K \Sigma^0 p})/4\pi = 4.5 \pm 3.8 , \qquad (4.3.4)$$

while using *Kim*'s parameterization leads to the result

$$(g^2_{K \Lambda p} + 0.84 g^2_{K \Sigma^0 p})/4\pi = 5.7 \pm 3.8 . \qquad (4.3.5)$$

In a later application of the same method (*Martin* and *Perrin* 1970), the Regge-pole model was not assumed, but instead the denominator of the integral round the circle was expanded in a power series, leading to the form

$$H(\omega) = \frac{R}{\omega + \omega_A} + \sum_{n=0}^{\infty} b_n \frac{\omega^n}{\omega_c^{n-1}},$$

where

$$H(\omega) = -\operatorname{Re} f_+(\omega) + \frac{\mathrm{P}}{\pi} \int_{\omega_{\pi A}}^{\omega_c} \mathrm{d}\omega' \left[\frac{\operatorname{Im} f_+(\omega')}{\omega' - \omega} + \frac{\operatorname{Im} f_-(\omega')}{\omega' + \omega} \right],$$

$$R = 0.061 (g_{K\Lambda p}^2 + 0.84 g_{K\Sigma^0 p}^2)/4\pi,$$

and

$$b_n = -\operatorname{Re} \frac{\omega_c^{n-1}}{2\pi i} \int_{\text{Circle}} \mathrm{d}\omega' \frac{f_+(\omega')}{\omega'^{n+1}}.$$

Approximately 70 values of $\operatorname{Re} f_\pm$ below 3 GeV/c were fitted in terms of G and the parameters b_n, with ω_c chosen to be ~ 10 GeV. Values of $\operatorname{Re} f_-$ were not used below ~ 600 MeV/c because they were too dependent on the model used for the low-energy $\bar{K}N$ region. Typical fits required terms up to $n = 2$ in the expansion. All three types of low-energy $\bar{K}N$ models were used and the following results were obtained,

$$(g_{K\Lambda p}^2 + 0.84 g_{K\Sigma p}^2)/4\pi = \begin{cases} 8.8 \pm 2.0 & \text{CSL model}, \\ 10.5 \pm 2.0 & \text{ERA model}, \\ 7.2 \pm 2.0 & \text{ZRA model}. \end{cases}$$

These results are very similar to those obtained by *Perrin* and *Woolcock* (1969) using energy-subtracted dispersion relations.

4.3.2 Consistency of Models for $\bar{K}N$ Scattering

We have seen above that the $T = 0$ unphysical region gives large contributions to forward dispersion relations, and that several models exist, which although fitting the low-energy $\bar{K}N$ data, differ somewhat in their behaviour when continued below the physical threshold. For this reason it is important to try and decide which of the various models is a better one to use.

One self-consistency check follows directly from determinations of the KN coupling constants using either energy-subtracted or finite-contour dispersion relations. At the end of these calculations the best-fit values of the free parameters can be used to predict values for the s-wave $K^\pm p$ scattering lengths, and these can then be compared with the input values.

Table 4.6. *Input and output values of the s-wave K^+p scattering lengths, $a_s^\pm$ (in F), from fits to* $\mathrm{Re} f_\pm(\omega)$

a_+	a_-		Model	Reference
-0.29 ± 0.02	-0.83 ± 0.07	Input	CSL	*Perrin* and *Woolcock* (1969)
-0.35 ± 0.21	-0.86 ± 0.26	Output		
-0.29 ± 0.02	-0.90 ± 0.03	Input	ERA	
-0.33 ± 0.20	-1.00 ± 0.26	Output	K-matrix	
	-0.87 ± 0.04	Input	CSL	*Martin* and *Perrin* (1970)
	-0.86	Output		
	-0.89 ± 0.04	Input	ERA	
	-1.07	Output	K-matrix	
	-0.90 ± 0.04	Input	ZRA	
	-0.85	Output	K-matrix	

Table 4.7. *Values of the $K\Lambda p$ coupling constant and its derivative, as a function of ω, from the analysis of Queen et al. (1969b)*

ω (MeV)	CSL		ERA K-Matrix	
	g^2	$\partial g^2/\partial\omega$ (GeV^{-1})	g^2	$\partial g^2/\partial\omega$ (GeV^{-1})
498	3.8 ± 0.6	2.4 ± 4.3	12.1 ± 2.7	90.5 ± 7.4
520	3.2 ± 0.7	-32.1 ± 5.5	13.4 ± 2.7	49.7 ± 6.0
540	2.3 ± 0.8	-60.8 ± 8.6	14.2 ± 2.7	26.9 ± 9.2

The input and output values obtained by *Perrin* and *Woolcock* (1969) are shown in Table (4.6) for calculations using the CSL and ERA K-matrix models. There is self-consistancy between input and output scattering lengths, but the output values have rather large errors. The finite-contour relation can be used in a similar manner, and the results obtained (*Martin* and *Perrin*, 1970) are also shown in Table 4.6 for a_s^-. There is some apparent discrepancy when *Kim*'s parameterization is used but the two values could still be consistent within their errors.

Other consistency tests have been devised based mainly on the observation that because the KN couplings are *constants* the value predicted by a forward dispersion relation should be independent of the energy at which the dispersion relation is evaluated.

Queen et al. (1969b) have derived consistency tests by writing dispersion relations for the derivative of the coupling constant with respect to ω and checking whether this quantity is consistent with zero. The dispersion relations are obtained by differentiating the sign-subtracted relations (2.2.16), and ω was taken to be just above the $\bar{K}N$ elastic threshold.

The values of $g^2_{K\Lambda p}/4\pi$ and its derivative obtained by these authors using the CSL and ERA K-matrix solutions are shown in Table 4.7. Both parameterizations appear to fail the test. However, there are difficulties associated with derivative dispersion relations which make the reliability of such calculation uncertain at this time.

For example, differentiating the dispersion relation increases the importance of higher partial waves, which are essentially unknown, and moreover the effect of errors on these terms is further enhanced by the strong cancellations that exist between the various terms in the dispersion relations. Thus, at $\omega = 540$ MeV, the positive and negative contributions to the derivative are ~ 200 if *Kim's* parameterization is used.

The calculations of *Queen et al.* have been extended by considering the values of g^2 obtained when ω is varied in the unphysical region as well as the low-energy physical region (*Martin, Queen,* and *Violini,* 1969a, 1969b; *Rogers,* 1969). *Martin et al.* consider the dispersion relation for the quantity

$$T(\omega) \equiv f_-(\omega) - f_+(\omega_0),$$

with $\omega_0 = 498$ MeV and ω in the range $400 \text{ MeV} \lesssim \omega \lesssim 520$ MeV. By varying ω in the unphysical region the contribution of the $Y_0^*(1405)$ to the sum rule for $g^2_{K\Lambda p}$ is changed considerably. All three low-energy models were studied, and the results obtained are shown in Table 4.8.

Table 4.8. *Values of $g^2_{K\Lambda p}/4\pi$ as a function of ω, from the analysis of Martin et al. (1969a)*

ω(MeV)	CSL	ERA K-Matrix	ZRA K-Matrix
400	2.5	9.2	4.1
425	3.2	10.6	4.5
450	3.3	11.6	4.5
475	3.2	12.5	4.3
498	3.1	13.1	4.1
520	2.8	13.7	3.6

Since the typical error on the coupling constant is ~ 2.7 the results of Table 4.8 appear to show consistency. However, these errors are highly correlated, and so the derivative should be calculable with a smaller error that might be thought initially. From Table 4.8 the derivative at $\omega = 498$ MeV is ~ 31 GeV^{-1} for *Kim's* parameterziation, and since the error on this quantity from Table 4.7 is ~ 7 these authors reject this solution as not being self-consistent. For the CSL model the derivative is ~ -8, and the error from Table 4.7 is ~ 4. Since the deviation from zero is only twice this error *Martin et al.* accept the CSL solution as being self-consistent. Also, because the ZRA K-matrix parameterization gives rise to a similar gradient this solution is also accepted. It is worth bearing in mind, however, that the errors obtained from derivative sum rules could well be in error. A final point to note about these calculations is the different values obtained for the gradient. For example, using *Kim's* parameterization, *Martin et al.* find that at $\omega = 498$ MeV, the derivative is ~ 31 GeV^{-1}, while from Table 4.7 *Queen et al.* find ~ 91 GeV^{-1}, the difference being apparently attributable to the fact that the latter authors include mass differences in their calculations whereas *Martin et al.* do not.

Rogers (1969) has carried out similar calculations to those above, using *Kim's* parameterization only, and working with $K^{\pm}p$ relations, but varying ω over the whole unphysical region as well as the low-energy physical region. The results of the calculation are in good qualitative agreement with those of *Martin et al.*

4.3.3 Other Determinations

Although forward dispersion relations have been widely applied to obtain values for the KN coupling constants, other methods have also been used. For example, the couplings occur as parameters in dynamical calculations of associated production and associated photoproduction, but the results obtained, while often suggesting rather small values, have considerable uncertainties associated with the models used. In this section we will consider briefly some other methods which, while not producing definitive results at present, could well be important for future work.

a) Work of *Cutkosky* and *Deo*

In the phase-shift analysis of *Cutkosky* and *Deo*, described in Section 3.1, the contributions from the hyperon pole terms were explicitly separated out, and the KN coupling constants were among the parameters varied to fit the K^+p data. In a previous analysis (*Cutkosky* and *Deo*, 1968 b) these authors had fitted the K^+p elastic differential cross-section at a few selected momenta below 2 GeV/c, and from the results obtained had concluded that

$$G^2 \equiv g^2_{K\Lambda p}/4\pi + g^2_{K\Sigma^0 p}/4\pi = 15^{+6}_{-4}.$$

For their full phase-shift analysis (*Cutkosky* and *Deo*, 1970), it seems most reasonable to find that value of G^2 which gives a minimum in χ^2 when summed over all data fitted. From the tables given by *Cutkosky* and *Deo* one deduces that for the preferred (Fermi) set of phases this value is ~ 9, but the minimum obtained as a function of G^2 is broad and changing χ^2 by one produces a change in G^2 of ~ 3. The best value obtained from this analysis is therefore $G^2 = 9 \pm 3$. Thus, although the central value agrees with determinations from energy-subtracted forward dispersion relations and finite-energy relations, more precise data are needed for an accurate estimate.

b) Phenomenological Bounds

An interesting attempt to obtain phenomenological bounds for the coupling constants has recently been made by *Rogers* (1969) using a method which is independent of the unphysical region contribution. The method is based on exploiting the properties of the integral

$$\psi(\lambda) = \int_{-1}^{+1} d\alpha \, \frac{\sigma(\alpha)}{1 - \alpha\lambda}, \tag{4.3.6}$$

where $\sigma(\alpha) \geq 0$ for $-1 \leq \alpha \leq 1$ and λ lies in the range $-1 < \lambda < 1$. If the integral is evaluated at a set of points $\lambda_i (\lambda = 1, 2, ..., N)$ and

$$\psi_i \equiv \psi(\lambda_i),$$

then it can be shown (*Tiktopoulos* and *Treiman*, 1968) that the set ψ_i must satisfy certain relations among themselves. The full set of such conditions has been given by *Rogers*, but the most useful are those obtained for every pair ψ_i and ψ_k. If $-1 < \lambda_j < \lambda_k < 1$, then

$$\begin{gathered}
(1 + \lambda_j) \, \psi_j \leq (1 + \lambda_k) \, \psi_k, \\
(1 - \lambda_j) \, \psi_j \geq (1 - \lambda_k) \, \psi_k.
\end{gathered} \tag{4.3.7}$$

To use these relations *Rogers* starts from the $K^+ p$ forward dispersion relation subtracted once at $\omega = \omega_s < 0$ i. e.

$$\frac{\operatorname{Re} f_+(\omega) - \operatorname{Re} f_-(-\omega_s)}{\omega - \omega_s}$$

$$= I(\omega) - \sum_{Y = \Lambda, \, \Sigma^0} \frac{R_Y}{(\omega + \omega_Y)(\omega_Y + \omega_s)} \tag{4.3.8}$$

$$+ \frac{1}{\pi} \int_m^\infty d\omega' \left[\frac{\operatorname{Im} f_+(\omega')}{(\omega' - \omega)(\omega' - \omega_s)} - \frac{\operatorname{Im} f_-(\omega')}{(\omega' + \omega)(\omega' + \omega_s)} \right].$$

The unphysical integral is

$$I(\omega) = \frac{-1}{\pi} \int\limits_{\omega_{\pi A}}^{m} d\omega' \frac{\operatorname{Im} f_-(\omega')}{(\omega' + \omega)(\omega' + \omega_s)}$$

$$\equiv \frac{1}{\pi} \int\limits_{\omega_{\pi A}}^{m} d\omega' \frac{\chi(\omega')}{\omega' + \omega}$$

where

$$\chi(\omega) = \frac{-\operatorname{Im} f_-(\omega)}{\omega + \omega_s}.$$

Now if $\operatorname{Im} f_-(\omega) \geqq 0$ and $\omega_s < -m$ then $\chi(\omega) \geqq 0$ and the integral $I(\omega)$ can be reduced to the required form (4.3.1) by the transformations

$$\alpha = \frac{2\omega' - m - \omega_{\pi A}}{m - \omega_{\pi A}}$$

and

$$\lambda = \frac{-(m - \omega_{\pi A})}{2\omega + m + \omega_{\pi A}}.$$

Applying (4.3.7) to the transformed integral evaluated at a pair of points ω_1 and $\omega_2 > \omega_1 > m$ leads to the relations

$$(\omega_1 + \omega_{\pi A}) I(\omega_1) \leqq (\omega_2 + \omega_{\pi A}) I(\omega_2),$$

$$(\omega_1 + m) I(\omega_1) \geqq (\omega_2 + m) I(\omega_2).$$

Finally, if the dispersion relation (4.3.8) is written in the form

$$C(\omega) G^2 = H(\omega) + I(\omega),$$

where G^2 is an effective KN coupling constant we have the results

$$G^2 [C(\omega_1)(\omega_1 + \omega_{\pi A}) - C(\omega_2)(\omega_2 + \omega_{\pi A})]$$
$$\leqq (\omega_1 + \omega_{\pi A}) G(\omega_1) - (\omega_2 + \omega_{\pi A}) G(\omega_2), \tag{4.3.9}$$

and

$$G^2 [C(\omega_1)(\omega_1 + m) - C(\omega_2)(\omega_2 + m)]$$
$$\geqq (\omega_1 + m) G(\omega_1) - (\omega_2 + m) G(\omega_2). \tag{4.3.10}$$

Eq. (4.3.9) sets an upper bound on G^2 and (4.3.10) sets a lower bound, and in principle these bounds are obtainable from experimental data.

In practice *Rogers* set $\omega_s = -0.614$ GeV; $\omega_2 = 1.62$ GeV and considered ω_1 in the range 0.55 GeV $\leqq \omega_1 \leqq 1.5$ GeV, taking values of the $K^+ p$ forward real parts in this region from the phase-shift analysis of *Lea et al.* (1968). The upper limit obtained was very large, but the best estimate of the lower limit was found to be

$$G^2 = (g^2_{K A p} + 1.05 g^2_{K \Sigma^0 p})/4\pi > 14 \pm 4.$$

The major uncertainty in this calculation lies in the validity of the particular phase-shift solutions used.

c) Partial-Wave Dispersion Relations

Early work on the dynamics of the low-energy KN and $\bar{K}N$ systems preceeded on the assumption that the dominant forces arise from the exchange of a small number of stable, or quasistable, states in the crossed channels. Examples of such exchanges are $\Lambda, \Sigma, \varrho, \omega$ etc., and various combinations have been considered by several authors (an incomplete list is: *Lee*, 1960; *Ferrari et al.*, 1961a; *Islam*, 1961; *Ramakrishan et al.*, 1962; *Ebata et al.*, 1962; *Singh*, 1963; *Ino*, 1967; *Cho*, 1966, 1967; *Roy*, 1964; *Kumar*, 1965; *Pundari et al.*, 1968).

A commonly used procedure is to fix some of the coupling constants governing the exchange processes, either from experiment of symmetry schemes, and to leave others as free parameters. The N/D method is then used to produce unitary partial-wave amplitudes, and the free coupling constants varied to fit the low-energy data.

Little definitive information can be drawn from these calculations, however, because of a variety of reasons. For example, sometimes only a limited set of particle-exchanges was considered, and sometimes the number of free parameters involved exceeded the constraining capacity of the data as they existed at the time. A more serious objection is the neglect of other terms. In particular, it was pointed out by several authors (*Lee*, 1960; *Yamaguchi*, 1959; *Barshay*, 1959; *Ferrari et al.*, 1961b) that the exchange of a non-resonant pair of pions produces a singularity in the KN partial-wave amplitudes which approaches very close to the physical region and would *a priori* be expected to be of some importance. Because of the length and nearness of the cut in the complex energy plane resulting from this process [see Figs. (2.2) and (2.3)], it is clear that the exchange of such a state must be considered rather carefully. There is also the persistent neglect of short-range forces, which cannot be adequately represented by single-particle exchanges and, moreover, would naturally be expected to be important for the low partial waves. Finally, the N/D method has not led to reliable results for πN scattering, and there is no reason to suppose that it will be better for the less well-known KN system.

Attempts to remedy some of the above defects have been made in a calculation of KN scattering by *Warnock* and *Frye* (1965). These authors considered the exchange of a large number of stable and quasistable states as well as background terms represented by low-order polynomials in the manner of *Cini* and *Fubini* (1960), i. e. the invariant amplitudes

were parameterized by
$$A^T(s, t) = C^T + C_s^T s + C_t^T t + \cdots ,$$

and similarly for $B^T(s, t)$. The influence of the exchange of a low-energy pair of s-wave pions was represented by the exchange of a scalar resonance, although the effect of this term was found to be small and was omitted in the final fits. Partial-wave amplitudes were obtained by projection from fixed-variable dispersion relations, and rescattering corrections estimated by the use of KN phase-shift information. The data available at the time of their calculation only allowed a very rough estimate of these corrections but they appeared to be small in this model. Five coupling constants and eight background parameters were varied to fit the $spd\ T = 0$ phase shifts of *Stenger et al.* (see Table 3.3), and the s-wave $T = 1$ phase shifts of *Goldhaber et al.* (see Table 3.1). For the Fermi $T = 0$ set an acceptable fit could only be achieved for a negative $K\Lambda N$ coupling constant, and if this is constrained to be zero then the best solution gave $g_{K\Sigma N}^2/4\pi = 58 \pm 9$, a value which is in total disagreement with all forward dispersion relation calculations. For the Yang $T = 0$ phases the best fit corresponded to the $K\Sigma N$ coupling being negative, and again if this is constrained to be zero then the best value of $g_{K\Lambda N}^2/4\pi$ is 7 ± 16. Thus, although the central value agrees with forward dispersion relation calculations the error is very large indeed.

More accurate data would undoubtedly improve the above calculation but then it would be important to give a better treatment of the nearby cut. An attempt to do this has been made by *Martin* and *Spearman* (1964) who concentrated on calculating the effect of the exchange of a non-resonant $\pi\pi$ pair. This was done using as input, s-wave $T = 0\ \pi\pi$ phase shifts similar to those of *Chew* and *Mandelstam* (1960), the parameters of known K^* resonances, and a parameterized form for the s-wave πK interaction. All other singularities were represented by two poles in each partial-wave amplitude. Unitary amplitudes were produced by the N/D method.

By varying the residues of the poles, together with the s-wave πK scattering length combination $a_s^+ = \frac{1}{3}(a_s^{\frac{1}{2}} + 2a_s^{\frac{3}{2}})$ a fit was obtained to the $K^+ p$ elastic differential cross-section data of *Goldhaber et al.* The contribution of the $\pi\pi$ cut was found to be small, in agreement with the work of *Warnock* and *Frye*. However, the $\pi\pi$ phase-shifts used by *Martin* and *Spearman* are no longer adequate, and more detailed calculations (*Martin*, 1968b) show that this term could be far larger, and still be compatible with the data of *Goldhaber et al.* It is clear that before any definitive determination of the KN couplings can be obtained from partial-wave dispersion relations more accurate KN data are needed, and more information about low-energy $\pi\pi$ and πK interactions will be required.

4.4 Forward Scattering Amplitude

The imaginary part of the forward amplitude may be obtained simply from the total cross-section by use of the optical theorem. The real part may then be obtained using any of the dispersion relations discussed in Section 2. These calculations have been made using the unsubtracted relations (2.2.16) (*Zovko*, 1966b), the once-subtracted relation (2.2.17) with $\omega_0 = m$ (*Zovko*, 1966b; *Queen*, 1967), the twice-subtracted relation (2.2.21) (*Carter*, 1968; *Lusignoli et al.* 1966a, 1967), the relation (2.2.23) with $\beta = \frac{1}{2}$ (*Martin* and *Poole*, 1968), and the finite contour relation (2.2.33) (*Martin* and *Perrin*, 1970).

The various calculations, are, in general, in fairly good agreement with each other, although there are some differences due to variations in the input data used. These variations are not, however, as large as one might suspect by examining contributions to the real parts from various terms in the dispersion relations. This is because individual terms are highly correlated and, in particular, the model used to evaluate the unphysical region is closely related to the values used for the KN coupling constants.

Table 4.9. *Predicted values of* $\mathrm{Re}\, f_{\pm}(\omega)$ *(in F) from the calculation of Carter (1968) compared with the best-fit values of Martin and Perrin (1970)*

k_{Lab}	$\mathrm{Re}\, f_{+}^{p}(\omega)$		$\mathrm{Re}\, f_{-}^{p}(\omega)$	
(GeV/c)	Fit	Prediction	Fit	Prediction
0.625	−0.46	−0.51	0.21	0.14
0.725	−0.45	−0.50	0.43	0.41
0.825	−0.43	−0.48	0.46	0.44
0.925	−0.40	−0.45	0.53	0.53
1.025	−0.40	−0.45	0.26	0.27
1.125	−0.44	−0.50	−0.25	−0.22

As examples of these calculations we show in Table 4.9 values of $\mathrm{Re}\, f_{\pm}^{p}(\omega)$ from the analysis of *Carter*. In these calculations the CSL model was used for the unphysical region and the KN coupling constant found from the sign-subtracted sum rule (2.2.16) at $\omega = m$. *Predictions* of the real parts were then made using the twice-subtracted relation (2.2.21). Also shown in this table are the best-fit values of $\mathrm{Re}\, f_{\pm}^{p}(\omega)$ found by using the finite contour relations (2.2.33) to fit the experimental quantities deduced from Legendre polynomial fits to the elastic differential cross-section data (*Martin* and *Perrin*, 1970). The agreement between the two is within experimental errors.

5. Summary and Outlook

This final section will summarize the information that has been obtained about kaon-nucleon scattering and the low-energy parameters, and we will comment on those analyses and experiments that would be most useful to understand the interaction further.

All existing analysis of K^+p scattering agree on the value of the s-wave scattering length, once the s-wave is accepted as being repulsive, (see Table 4.3). Evidence from forward dispersion relations supports this choice of sign, but a definitive test would be to measure the differential cross-section in the Coulomb interference region below $500 \text{ MeV}/c$. Several experiments to measure the differential cross-section below $1 \text{ GeV}/c$ are currently in progress and should help greatly in reducing the uncertainties in the s-wave phase shift.

The p-wave scattering lengths are less well known (see Table 4.4), although two analyses agree that $a^1_{p1} < 0$; $a^1_{p3} > 0$ with $|a^1_{p1}| > |a^1_{p3}|$. These results agree with phase-shift analyses at higher momenta (Fermi-type solution). A definitive test would be a measurement of the recoil proton polarization below $\sim 800 \text{ MeV}/c$, but this is a very difficult experiment. Another poorly known quantity is the total inelastic cross-section, which rises rapidly above $\sim 850 \text{ MeV}/c$, and provides a strong constraint on phase-shift analyses. Some accurate measurements of this quantity would therefore be very useful.

As K^+p data improve, different methods of phase-shift analysis can be explored, and, in particular, models exploiting more theoretical requirements can be used. The work of Cutkosky and Deo has already gone some way to encorporating momentum-transfer analyticity properties, but the results are rather insensitive to the strengths of the hyperon poles. Another method is to use parametric forms based on partial-wave dispersion relations, and work along these lines is currently in progress (*Martin*, 1970).

With the new data from current low-energy experiments we can reasonably expect that in the not too distant future fairly accurate s and p-wave $T = 1$ amplitudes below $1 \text{ GeV}/c$ will be available, with perhaps a lesser knowledge of the d-waves. These amplitudes would then enable $T = 0$ information to be extracted from either K^0p or K^+d data. The former would be theoretically preferable, but in practice K^+d scattering is more accessible experimentally. We have seen in Section 3.2 that low-energy K^+d data are very sparse and of poor quality, and *any* new measurements would be useful. However, more data of a higher accuracy would mean that models for deuteron breakup would have to be developed which were reliable over a larger energy range than is presently the case.

Table 5.1. *Values of the KN coupling constants*; $g_A^2 \equiv g_{K\Lambda p}^2/4\pi$, $g_\Sigma^2 \equiv g_{K\Sigma^0 p}^2/4\pi$ *and* $G^2 \simeq g_A^2 + g_\Sigma^2$

Value	Source
	Threshold forward dispersion relation sum rules using
$g_A^2 \sim 5 \pm 2$; $g_\Sigma^2 \lesssim 2$	ZRA K-matrix Unphysical region
$g_A^2 \sim 14 \pm 2$; $g_\Sigma^2 \sim 0 \pm 1$	ERA K-matrix Unphysical region
	Fits to Re$f_\pm(\omega)$ using either energy-subtracted dispersion
	relations or finite contour relations, with
$G^2 \sim 8 \pm 2$	ZRA K-matrix Unphysical region
$G^2 \sim 11 \pm 2$	ERA K-matrix Unphysical region
$G^2 \sim 9 \pm 3$	$K^+ p$ phase-shift analysis
$g_A^2 \sim 7 \pm 16$; $g_\Sigma^2 \sim 0$	KN partial-wave dispersion relations
$G^2 \gtrsim 14 \pm 4$	Dispersion relation bound

At present all we know is that the Yang set of $T = 0$ phases is preferred, i. e. a large positive $p\frac{1}{2}$ amplitude and a small negative $p\frac{3}{2}$ (see Tables 3.3 and 3.4). The scattering lengths are not known with any confidence, although a_s^0 appears to be very small.

All analyses of $\bar{K}N$ data agree on the values of the s-wave scattering lengths, (see Tables 4.1 and 4.2). However, all these analyses use basically the same model, and it would be desirable to have confirmation of these values using different parameterizations, e. g. based on the multichannel N/D equations. In view of the importance of the unphysical region in forward dispersion relation calculations it would be highly desirable to repeat the spd analysis of the region below ~ 500 MeV/c to reliably determine the s-wave effective ranges and p-wave scattering lengths. This will require measurements of differential cross-sections for hyperon production as well as the polarization of the final state baryon. Although most of the data would undoubtedly come from $K^- p$ initial states, once the $S = +1$ KN amplitudes are better known then $K_2^0 p$ data could also be effectively used. At momenta above ~ 500 MeV/c no coupled-channel analyses exist, although such calculations are in progress.

The situation for the KN coupling constants is summarized in Table 5.1. Values obtained from forward dispersion relation threshold sum rules are more sensitive to the low-energy $\bar{K}N$ model used than methods using the dispersion relations away from threshold, as is only to be expected. Determinations using energy-subtracted dispersion relations or finite-contour relations produce similar results which are not so sensitive to the unphysical region, but are sensitive to the input values of the forward real parts. It would be interesting to see whether the same values result from other energy dependent methods, e. g. that of *Schnitzer* and *Salzman*. The values obtained from $K^+ p$ phase-shift analysis, and KN partial-wave dispersion relations are consistant with the forward dispersion relation values but the latter has very large errors. The lower bound obtained by Rogers, however, is larger, although the results depend on the validity of the particular forward amplitudes used. It would be interesting to see how sensitive this method is to variations in the input data.

Although consistency tests of low-energy $\bar{K}N$ models favour those using either the CSL or ZRA K-matrix, it is clear that a significant improvement on existing determinations of the KN coupling constants will only come with more reliable phenomenological analyses, and these, in turn, must await more accurate data.

Finally, the real parts of the forward amplitude, as calculated from various dispersion relations, are in reasonable agreement with each other, and with the experimental values, despite considerable uncertainties in the input data.

References

Abrams, G. S., Sechi-Zorn, B.: Phys. Rev. **139**, B 454 (1965).
Adler, S. L.: Phys. Rev. **137** B, 1022 (1965).
de Alfaro, V., Fubini, S., Furlan, G., Rossetti, G.: Phys. Letters **21**, 576 (1966).
Andersson, S., Daum, C., Erne, F. C., Lagnaux, J. P., Sens, J. C., Udo, F., Wagner, F.: Phys. Letters **30** B, 56 (1969a).
— — — *Lagnaux, J. P., Sens, J. C., Udo, F.:* Phys. Letters **28** B, 611 (1969b).
Ara, G., Rashid, A. M. H.: Prog. Theor. Phys. **38**, 1338 (1967).
Armenteros, R., Baillon, P., Bricman, C., Ferro-Luzzi, M., Nguyen, H. K., Pelosi, V., Plane, D. E., Schmitz, N., Burkhardt, E., Filthuth, H., Kluge, E., Oberlack, H., Ross, R. R., Barloutaud, R., Granet, P., Meyer, J., Narjoux, J. L., Pierre, F., Porte, J. P., Prevost, J.: Nucl. Phys. B **8**, 183 (1968a); B **8**, 195 (1968b); B **8**, 223 (1968c).
— — — *Ferro-Luzzi, M., Plane, D. E., Schmitz, N., Burkhardt, E., Fitthuth, H., Kluge, E., Oberlack, H., Ross, R. R., Barloutaud, R., Granet, P., Meyer, J., Porte, J. P., Prevost, J.:* Nucl. Phys. B **14**, 91 (1969).
Asbury, J. G., Dowell, J. D., Kato, S., Lundquist, D., Novey, T. B., Yokosawa, A., Barnett, B., Koehler, P. F. M., Steinberg, P.: Phys. Rev. Letters **23**, 194 (1969).
Barbaro-Galtieri, A., Derenzo, S. E., Price, L. R., Rittenberg, A., Rosenfeld, A. H., Barash-Schmidt, N., Bricman, C., Roos, M., Söding, P., Wohl, C. G.: Rev. Mod. Phys. **42**, 1 (1970).
Barger, V., Phillips, R. J. N.: Phys. Letters **31** B, 643 (1970).
Barrelet, E.: private communication, (1968).
Barshay, S.: Phys. Rev. **110**, 743 (1958).
Berley, D., Yamin, P., Kofler, R., Mann, A., Meisner, G., Yamanoto, S., Thompson, J., Willis, W.: Phys. Rev. **1**, D 1996 (1970).
Bertanza, L., Bigi, A., Carrara, R., Casali, R., Pazzi, R., Deiley, D., Hart, E. L., Ruhm, D. C., Willis, W. J., Yamanoto, S. S., Wong, N. S.: Phys. Rev. **177**, 2036 (1969).
Bland, R. W.: UCRL Report No. 18131, (1968).
— *Goldhaber, G., Trilling, G. H.:* Phys. Letters **29** B, 618 (1969).
Bransden, B. H.: In: High Energy Physics Ed. E. H. S. Burhop, New York: Academic Press 1969, Vol. 3.
Bugg, D. V., Gilmore, R. S., Knight, K. M., Salter, D. C., Stafford, G. H., Wilson, E. J. N., Davies, J. D., Dowell, J. D., Hattersley, P. M., Horner, R. J., O'Dell, A. W., Carter, A. A., Tapper, R. J., Riley, K. F.: Phys. Rev. **168**, 1466 (1968).

Burrowes, H. C., Caldwell, D. O., Frisch, D. H., Hill, D. A., Ritson, D. M., Schluter, R. A.: Phys. Rev. Letters **2**, 117 (1959).
Carreras, B., Donnachie, A., Kirsopp, R.: DNPL/P31, (1970a).
— — Nucl. Phys. B **19**, 349 (1970b).
Carter, A. A.: Phys. Rev. Letters **18**, 801 (1967).
— Cambridge Univ. Report HEP 68–10, unpublished.
Chan, C. H., Meiere, F. T.: Phys. Rev. Letters **20**, 568 (1968).
Chand, R., Dalitz, R. H.: Ann. Phys. (N. Y.) **20**, 1 (1962).
— Ann. Phys. (N. Y.) **42**, 81 (1967).
Chew, G. F.: Phys. Rev. **80**, 196 (1950).
— *Wick, G. C.:* Phys. Rev. **85**, 636 (1952).
— *Goldberger, M. L.:* Phys. Rev. **87**, 778 (1952).
— — *Low, F. E., Nambu, Y.:* Phys. Rev. **106**, 1337 (1957).
— *Mandelstam, S.:* Phys. Rev. **119**, 467 (1960).
Cho, K. S.: Nuovo Cimento **43** A, 840 (1966); **47** A, 707 (1967).
Cini, M., Fubini, S.: Ann. Phys. **10**, 352 (1960).
Ciulli, S.: Nuovo Cimento, **61** A, 787 (1969).
Cook, V., Keefe, D., Kerth, L. T., Murphy, P. G., Wenzel, W. A., Zipf, T. F.: Phys. Rev. Letters **7**, 182 (1961).
— — — *Murphy, P. G., Wenzel, W. A., Zipf, T. F.:* Phys. Rev. **129**, 2743 (1963).
Cool, R. L., Giacomelli, G., Kycia, T. F., Leontic, B. A., Li, K. K., Lundby, A., Teiger, J.: Phys. Rev. Letters **17**, 102 (1966).
Cutkosky, R. E., Deo, B. B.: Phys. Rev. **174**, 1854 (1968a).
— — Phys. Rev. Letters **20**, 1272 (1968b).
— — Phys. Rev. **1**, D 2547 (1970).
Dalitz, R. H., Tuan, S. F.: Ann. Phys. (N. Y.) **10**, 307 (1960).
Dass, G. V., Michael, C., Phillips, R. J. N.: Nucl. Physics B **9**, 549 (1969).
Davies, G. H., Queen, N. M., Lusignoli, M., Restignoli, M., Violini, G.: Nucl. Phys. B **3**, 616 (1967).
Dolen, R., Horn, D., Schmid, C.: Phys. Rev. **166**, 1768 (1968).
Donald, R. A., Edwards, D. N., Lys, J., Nisar, T., Moore, R. S.: Phys. Letters, **22**, 711 (1966).
Ebata, T., Takahashi, A.: Prog. Theor. Phys. **27**, 223 (1962).
Faddeev, L. D.: JEPT **12**, 1014 (1961a);
— Soviet Phys. Doklady **6**, 384 (1961b).
— Soviet Phys. Doklady **7**, 600 (1963).
Femino, S., Jannelli, S., Mezzanares, F.: Nuovo Cimento **50** A, 371 (1967).
Ferrari, F., Frye, G., Pusterla, M.: Phys. Rev. **123**, 315 (1961a); **123**, 308 (1961b).
Ferreira, E.: Not. Fisica **8**, 4 (1961).
Focardi, S., Minguzzo-Ranzi, A., Monari, L., Saltini, G., Serra, P.: Phys. Letters **24** B, 314 (1967).
Gelfand, N. M., Harmsen, D., Levi-setti, R., Predazzi, E., Raymund, M., Doede, J., Manner, W.: Phys. Rev. Letters. **17**, 1224 (1966).
Giacomelli, G., Lugaresi-Serra, P., Mandrioli, G., Rossi, A. M., Griffiths, F., Hughes, I. S., Jacobs, D. A., Jennings, R., Wilson, B. C., Ciapetti, G., Costantini, V., Martellotti, G., Zanello, D., Castelli, E., Sessa, M.: Nucl. Phys. B **20**, 301 (1970).
Glauber, R. J.: Phys. Rev. **100**, 242 (1955).
Goldhaber, S., Chinowsky, W., Goldhaber, G., Lee, W., O'Halloran, T., Stubbs, T. F., Pjerrou, G. M., Stork, D. H., Ticho, H. K.: Phys. Rev. Letters **9**, 135 (1962).
Granovskii, Ya. I., Starikov, V. N.: Soviet J. Nucl. Phys. **6**, 444 (1968).
Haber-Schaim, U.: Phys. Rev. **104**, 1113 (1956).
Hamilton, J., Woolcock, W. S.: Rev. Mod. Phys. **35**, 737 (1963).
Harari, H.: Proc. of the Daresbury Conf. on Electron and Photons, (1969).

Von Hippel, F., Kim, J. K.: Phys. Rev. Letters, **20**, 1303 (1968).

Hirata, A. A., Wohl, G. G., Goldhaber, G., Trilling, G. H.: Phys. Rev. Letters **21**, 1485 (1968).

Hirsch, W., Gidal, G.: Phys. Rev. **135** B, 191 (1964).

Holley, W., Beall, E. F., Keefe, D., Kerth, L. T., Thresher, J. J., Wang, C. L., Wenzel, W. A.: Phys. Rev. **154**, 1273 (1967).

Humphrey, W. E., Ross, R. R.: Phys. Rev. **127**, 1305 (1962).

Ino, T.: Prog. Theor. Phys. **37**, 398 (1967).

Islam, M. M.: Nuovo Cimento **20**, 546 (1961).

Jackson, J. D., Wyld, H. W.: Phys. Rev. Letters **2**, 355 (1959a).

— — Nuovo Cimento **13**, 84 (1959b).

Kadyk, J. A., Oren, Y., Goldhaber, G., Goldhaber, S., Trilling, G. H.: Phys. Letters **17**, 599 (1966).

— *Chan, J. H., Goldhaber, G., Trilling, G. H.:* UCRL-18325 (1968).

Kato, S., et al.: Phys. Rev. Letters **24**, 615 (1970).

Kim, J. K.: Phys. Rev. Letters **14**, 29 (1965).

— Columbia University Report, Nevis 149, (1966).

— Phys. Rev. Letters **19**, 1074 (1967a).

— Phys. Rev. Letters **19**, 1079 (1967b).

Kittel, W., Otter, G.: Phys. Letters **22**, 115 (1966).

Kittel, W., Otter, G., Wacek, I.: Phys. Letters **21**, 349 (1966).

Kumar, A.: Phys. Rev. **139**, B 486 (1965).

Kycia, T. F., Kerth, L. T., Baender, R. G.: Phys. Rev. **118**, 553 (1960).

Lea, A. T., Martin, B. R., Oades, G. C.: Phys. Letters **23**, 380 (1966).

— — — Phys. Rev. **165**, 1770 (1968).

— — and UCL Spark Chamber Group, Phys. Letters **32** B, 214 (1970a).

Lee, B. W.: thesis University of Pennsylvania, (1960), unpublished.

Levy-Leblond, J. M., Gourdin, M.: Nuovo Cimento **23**, 1163 (1962).

Lovelace, C.: In: Pion-Nucleon Scattering, Ed. G. L. Shaw and D. Y. Wong, Wiley-Interscience (1969).

Lusignoli, M., Restignoli, M., Snow, G. A., Violini, G.: Phys. Letters **21**, 229 (1966).

— — *Violini, G., Snow, G. A.:* Nuovo Cimento **45** A, 1257 (1966a); **49**A, 705 (1967).

MacDowell, S. W.: Phys. Rev. **116**, 774 (1959).

Martin, A. D., Spearman, T. D.: Phys. Rev. **136**, B 1480 (1964).

— *Poole, F.:* Phys. Letters **25** B, 343 (1967); Nucl. Physics. B **4**, 467 (1968).

— *Ross, G. G.:* Phys. Letters **26** B, 527 (1968).

— *Queen, N. M., Violini, G.:* Nucl. Phys. B **10**, 481 (1969a); Phys. Letters **29** B, 311 (1969b).

— *Perrin, R.:* Nucl. Phys. B **10**, 125 (1969).

— *Ross, G. G.:* Nucl. Phys. B **16**, 479 (1970).

— *Perrin, R.:* Nucl. Phys. B **20**, 287 (1970).

Martin, B. R.: Phys. Rev. Letters, **21**, 1286 (1968a).

— Phys. Rev. **175**, 2034 (1968b).

— *Sakitt, M.:* Phys. Rev. **183**, 1345 (1969a).

— — Phys. Rev. **183**, 1352 (1969b).

— *Thompson, G. D.:* Nucl. Phys. B **22**, 285 (1970).

— unpublished (1970).

Moorhouse, R. G.: Ann. Rev. Nucl. Sci. **19**, 301 (1969).

Perrin, R., Woolcock, W. S.: Nucl. Phys. B **4**, 671 (1968).

— — Nucl. Phys. B **12**, 26 (1969).

Pomeranchuk, I. Ia.: JEPT **3**, 306 (1956).

Price, L. R., Barash-Schmidt, N., Benary, O., Bland, R. W., Rosenfeld, A. H., Wohl, C. G.: URCL-20000 (1969).

Pundari, S. B., Dutta-Roy, B.: Phys. Rev. **165**, 1663 (1968).

Queen, N. M.: Nucl. Physics B1, 207 (1967).
— *Restignoli, M., Violini, G.:* Fortschr. Physik **17**, 467 (1969a).
— *Leeman, S., Yeomans, F. E.:* Nucl. Phys. B **11**, 115 (1969b).
Ramakrishnan, A., Balachandran, A. P., Raman, K.: Nuovo Cimento **24**, 369 (1962).
Rarita, W., Phllips, R. J. N.: Phys. Rev. **139** B, 1336 (1965).
Ray, A. K., Burris, R. W., Fisk, H. E., Kraemer, R. W., Hill, D. G., Sakitt, M.: Phys. Rev. **183**, 1183 (1969).
Restignoli, M., Sertorio, L., Toller, M.: Phys. Rev. **150**, 1389 (1966).
Rogers, T. W.: Phys. Rev. **178**, 2478 (1969).
Rood, H. P. C.: Nuovo Cimento **50** A, 493 (1967).
Roper, L. D., Bailey, D. S.: Phys. Rev. **155**, 1744 (1967).
Ross, M., Shaw, G.: Ann. Phys. (N. Y.) **13**, 147 (1961).
Roy, D. P.: Phys. Rev. **136**, B 804 (1964).
Sakitt, M., Day, T. B., Glasser, R. G., Seeman, N., Friedman, J., Humphrey, W. E., Ross, R. R.: Phys. Rev. **139**, B 719 (1965).
Sayer, G. A., Beall, E. F., Devlin, T. J., Shepard, P., Soloman, J.: Phys. Rev. **169**, 1045 (1968).
Schnitzer, H. J., Salzman, G.: Phys. Rev. **113**, 1153 (1959).
Singh, G. P.: Prog. Theor. Phys. **30**, 327 (1963).
Slater, W., Stork, D. H., Ticho, H. K., Lee, W., Chinowsky, W., Goldhaber, G., Goldhaber, S., O'Hallovan, T.: Phys. Rev. Letters **7**, 378 (1961).
Stenger, V. J., Slater, W. E., Stork, D. H., Ticho, H. K., Goldhaber, G., Goldhaber, S.: Phys. Rev. **134** B, 1111 (1964).
Stubbs, T. F., Bradner, H., Chinowsky, W., Goldhaber, G., Goldhaber, S., Slater, W., Stork, D., Ticho, H.: Phys. Rev. Letters **7**, 188 (1961).
Tiktopoulos, G., Treiman, S.: Phys. Rev. **167**, 1437 (1968).
Wang, C. L.: UCRL-11881 (1965).
Warnock, R. L., Frye, G.: Phys. Rev. **138**, B 947 (1965).
Watson, M. B., Ferro-Luzzi, M., Tripp, R. D.: Phys. Rev. **131**, 2248 (1963).
Yamaguchi, Y.: Prog. Theor. Phys. Suppl. **11**, 37 (1959).
Zovko, N.: Z. Phys. **192**, 346 (1966a).
Zovko, N.: Z. Phys. **196**, 16 (1966b).

Dr. *B. R. Martin*
Department of Physics
University College London
London/England

The ΛKN Coupling and Extrapolation below the $\bar{K}N$ Threshold

A. D. MARTIN

Contents

1. The Sum-Rule Value for the ΛKN Coupling and the Extrapolation over the Y_0^* (1405)
   Resonance 142
2. Finite-Contour Dispersion Relation 145
3. Other Methods for Determining the ΛKN Coupling 148
4. K^- Absorption on Nuclei and Extrapolation below the K^-p Threshold 148
Acknowledgement 151
References 151

First we review, briefly, the problems associated with the determination of the ΛKN coupling constant from forward dispersion relations, in particular, stressing the sensitivity of the sum-rule prediction to the extrapolation of the K^-p amplitudes below threshold. We then describe an attempt to overcome this problem which is based on finite-contour dispersion relations. Finally, we discuss the possibility of using the data for K^- absorption on nuclei to check the extrapolation of the K^-p solutions below threshold.

1. The Sum-Rule Value for the ΛKN Coupling and the Extrapolation over the Y_0^* (1405) Resonance

With the available data for $K^{\pm}N$ total cross sections, $\sigma_{\pm}$, and the knowledge of the S-wave kaon nucleon scattering lengths it might seem that the conventional forward dispersion relations (or sum rules as they are often called) would provide a reliable means of calculating the ΛKN and ΣKN coupling constants. These relations are of the form

$$
\frac{cg_\Lambda^2}{m^2 - \omega_\Lambda^2} + \frac{dg_{\Sigma^0}^2}{m^2 - \omega_\Sigma^2} = \frac{\operatorname{Re} f_-(m) - \operatorname{Re} f_+(m)}{2m} - \frac{1}{4\pi^2} \operatorname{P} \int_m^\infty \frac{\sigma_- - \sigma_+}{k_L} \, d\omega
$$

$$
- \frac{1}{\pi} \operatorname{P} \int_{\omega_{\Lambda\pi}}^m \frac{\operatorname{Im} f_-}{k_L^2} \, d\omega \tag{1}
$$

where $f_{\pm}$ are the forward $K^{\pm}N$ scattering amplitudes in the lab. frame and where m, ω, k_L are the kaon mass, lab. energy and momentum respectively. For the $K^{\pm}p$ and $K^{\pm}n$ relations the coefficients c and d are

$$c_p = 0.061, c_n = 0; \quad d_p = \tfrac{1}{2}d_n = 0.051 . \tag{2}$$

To determine g_Λ, the (unrationalized) $\Lambda K^- p$ coupling constant, we evaluate the $\bar{K}N$ isospin $I=0$ combination of these relations. Unfortunately, however, the sum of the terms on the right hand side turns out to be small difference between the relatively large and poorly known contribution from the integral over the unphysical region below the $\bar{K}N$ threshold (which is dominated by the $Y_0^*(1405)$ resonance) and the remaining contributions. As a result of this cancellation the prediction for g_Λ^2 is very sensitive to the form of the extrapolation used to calculate the unphysical region integral. Extrapolating with either constant scattering length (CSL) or constant K-matrix (ZRA) parameters yields values consistent with $g_\Lambda^2 = 5 \pm 2$, whereas if the multichannel effective range solution (ERA) of *Kim* (1967) is used the prediction is $g_\Lambda^2 = 13.5 \pm 2$. We refer to the preceding review article for a discussion and a list of references of these calculations and also of the various $K^- p$ solutions.

Each of the above-mentioned extrapolations is based on a parametric form which has been used to analyse the low energy $K^- p$ data. In the CSL model the vital $I = 0$ $\bar{K}N$ S-wave amplitude is described in terms of a constant complex scattering length $A_0 \equiv a_0 + i b_0$ (with $a_0 \approx -1.7$ fm and $b_0 \approx 0.7$ fm). For the other models the amplitude is parametrized in terms of an effective-range expansion of the inverse K-matrix

$$K^{-1} \equiv M = \begin{pmatrix} \alpha & \beta \\ \beta & \gamma \end{pmatrix} + \frac{1}{2}(k^2 - k_0^2)^{\frac{1}{2}} \begin{pmatrix} r_\alpha & r_\beta \\ r_\beta & r_\gamma \end{pmatrix} (k^2 - k_0^2)^{\frac{1}{2}} \tag{3}$$

where k is the diagonal matrix of the channel c. m. momenta, and k_0 is the value of k at the $K^- p$ threshold. The matrix elements left-to-right and top-to-bottom refer to the $\bar{K}N$ and $\Sigma\pi$ channels respectively. The constant K-matrix or zero-range (ZRA) model retains only the three real parameters α, β, γ for the $I = 0$ S-wave, whereas in *Kim's* parameterization (ERA) the diagonal effective-ranges r_α and r_γ were also taken as parameters.

The crucial observation is that the difference between the value of g_Λ^2 obtained using either the CSL or the ZRA extrapolations and the much larger value obtained from the ERA extrapolation is attributable to the introduction of the sizeable S-wave effective-ranges, in particular to the parameter r_α. An example of the sensitivity of the value of g_Λ^2 to variations of the effective-ranges is shown in Fig. 1. The curves for g_Λ^2 are obtained using *Kim's* ERA parameters but within each case

one of the S-wave effective-ranges varied about the value obtained by
Kim. Clearly, both the parameters r_α and r_β will need to be well-determined
if the sum-rule prediction for g_A^2 is to be reliable.

One possible check of the extrapolations of the $K^- p$ solutions is
to compare their predictions for the mass and the width of the virtual
bound-state Y_0^* resonance with the values observed in production experiments. The problem here is that since there are no $\pi Y \to \pi Y$ data
($Y = \Lambda$ or Σ) the available low energy $K^- p$ data are completely described

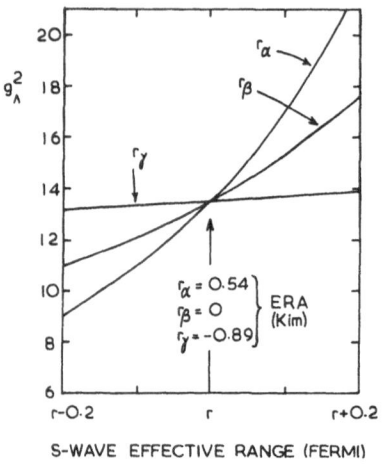

Fig. 1. The sensitivity of the sum rule value for g_A^2, as calculated using the ERA extrapolation,
to variation of each of the S-wave effective ranges in turn

at each energy by six S-wave parameters $(a_0, b_0, a_1, b_1, \varepsilon, \phi)$ and so only
these combinations of the K-matrix elements are well-determined. As
a consequence the extrapolation of the $I = 0\ \pi\Sigma \to \pi\Sigma$ amplitude is
expected to be much less reliable than the extrapolation of the $\bar{K}N \to \bar{K}N$
amplitude which is used in the evaluation of the sum rule. However
even if the observed width cannot be used as a reliable check the same
is not true for the Y_0^* mass.

Below the $K^- p$ threshold all the coupled $I = 0$ S-wave amplitudes
contain the factor $(1 + \kappa A_0)^{-1}$ where $\kappa \equiv -ik$ and where the scattering
length $A_0 \equiv a_0 + ib_0$. Thus for large negative a_0 we obtain a virtual
bound-state resonance in the neighbourhood of $\kappa_R = -1/a_0$. Now the
$K^- p$ data require a_0 to be between -1.6 and -1.7 fm just above threshold
and assuming that it is energy independent the prescription $\kappa_R = -a_0^{-1}$
would correspond to a resonant mass between 1408 and 1411 MeV.
Actually since the resonance is near threshold it turns out that the

resonance peak occurs in the imaginary part of the amplitudes some 5 MeV higher. The energy dependence of the scattering length in the ZRA parametrizations,

$$A_0^{-1} = \alpha - \beta^2/(\gamma - ik_{\Sigma\pi}), \qquad (4)$$

is found not to help and in fact shifts the resonance to slightly higher mass still. The extrapolation with ZRA and CSL parameters therefore yields a Y_0^* resonance in the region of 1415 MeV upwards as compared to the observed value of 1405 MeV. Now from eqs. (3) and (4) we find that the resonance mass can be decreased by the introduction of a positive r_α and/or a positive r_β (taking $\beta < 0, \gamma > 0$). The sign for r_β depends on the signs of β and γ. By inspection of Fig. 1, for example, we see from the slopes dg_Λ^2/dr that the introduction of such effective-ranges will increase the sum-rule value from $g_\Lambda^2 \approx 5$.

Kim's ERA fit of the $K^- p$ data is an example of such a parametrization and extrapolating with his parameters (in particular $r_\alpha = 0.54$ fm) leads to $g_\Lambda^2 \approx 13.5$ and a Y_0^* mass of 1400 MeV. However *Kim* did not include the parameter r_β (which is vital to the extrapolation) in his fit to the data. Moreover since r_α and r_γ are essentially determined by the $S - D$ interference in the region of the Y_0^* (1520) the results depend on the validity of the effective-range approximation over the full range $1400 < E < 1520$ MeV, the lower end of which is already penetrated by the left-hand two-pion exchange cut.

We conclude that the present knowledge of the $I = 0$ S-wave $\bar{K} N$ amplitude is inadequate to allow a reliable sum-rule prediction for g_Λ^2. We have argued that the value, $g_\Lambda^2 \approx 5$, using the ZRA extrapolation will be a lower limit; since the introduction of r_α and r_β, which is necessary to extrapolate to the observed Y_0^* mass, lead to an increased sum-rule value for g_Λ^2.

2. Finite-Contour Dispersion Relation

To improve on the above estimate of g_Λ^2 we seek a relation which reduces the importance of the unphysical region contribution. To do this *Martin* and *Perrin* (1970) applied *Cauchy's* theorem to $f_+(\omega')/(\omega' - \omega)$ around the contour C shown as the dashed line in Fig. 2. For ω inside the circle and just above the real axis the following finite-contour relation is obtained

$$\frac{c_p g_\Lambda^2}{\omega_\Lambda + \omega} + \frac{d_p g_{\Sigma^0}^2}{\omega_\Sigma + \omega} = H(\omega) + \text{Re} \frac{1}{2\pi i} \int_{\text{circle}} \frac{f_+(\omega')}{\omega' - \omega} d\omega' \qquad (5)$$

with

$$H(\omega) = -\operatorname{Re} f_+(\omega) + \frac{1}{\pi} \mathrm{P} \int_{\omega_{A\pi}}^{\omega_c} \left(\frac{\operatorname{Im} f_+}{\omega' - \omega} + \frac{\operatorname{Im} f_-}{\omega' + \omega} \right) d\omega'. \qquad (6)$$

$f_\pm$ are the $K^\pm p$ forward amplitudes and the last integral in eq. (5) is around the circular contour of radius ω_c and centre $\omega' = 0$. If this relation is evaluated for values of ω away from the $K^- p$ threshold then not only is the unphysical region contribution suppressed in comparison with eq. (1), but more important the Λ pole contribution is enhanced by approximately a factor 6 relative to the unphysical region contribution. This can easily be verified by approximating the unphysical region integral by a $Y_0^*(1405)$ pole: $\operatorname{Im} f_-(\omega) = X \delta(\omega - \omega_{Y*})$.

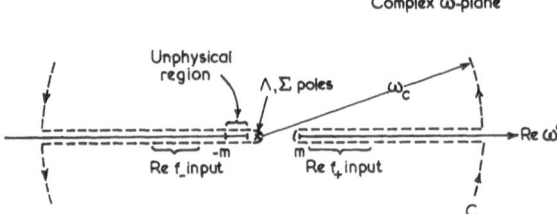

Fig. 2. The contour C used to obtain eq. (5). The real part data are used only in the indicated regions (not to scale)

In a typical application ω_C was chosen to be 10 GeV and $H(\omega)$ was calculated at those values of ω, satisfying $|\omega| < 3$ GeV, for which either $\operatorname{Re} f_+$ or $\operatorname{Re} f_-$ can be estimated by extrapolating the observed $K^\pm p$ angular distributions to the forward direction. For this range of ω the integral around the circle is well approximated by the first few terms of a power series expansion in ω and so Eq. (5) can be written in the form

$$H(\omega) = \frac{0.061 \overline{g^2}}{\omega + \omega_A} + b_0 + b_1 \omega + \ldots + b_n \omega^n.$$

The correction arising from the transference of the Σ-contribution to the Λ-pole position is negligible for physical values of ω. The parameters $\overline{g^2} (\equiv g_\Lambda^2 + 0.84 g_{\Sigma 0}^2)$, $b_0, \ldots b_n$ were then determined by a least-squares fit to 70 known values of $H(\omega)$. The sensitivity of these predictions were tested to variation of ω_C and n. In practice the optimum value of n was 1 or 2. The predictions for $\overline{g^2}$ are shown in Table 1. As expected they are much less sensitive to the extrapolation model which is used to calculate the unphysical region contribution.

Table 1. *The values of* $\overline{g^2} \equiv g_\Lambda^2 + 0.84 g_{\Sigma^0}^2$ *obtained by the least-squares fit. Also, the values of the real part of the* $K^- p$ *S-wave scattering length,* a_-, *as determined, first, from the finite-contour relation and second directly from the* $K^- p$ *parameters. The particular CSL and ZRA solutions that are used are due to Kim (1965) and Martin and Ross (1970) respectively.*

	$\overline{g^2}$	a_- (fermi)	
		From eq. (5)	From $K^- p$ parameters
CSL	8.8 ± 2	-0.86 ± 0.03	-0.87 ± 0.04
ERA	10.5 ± 2	-1.07 ± 0.03	-0.89 ± 0.04
ZRA	7.2 ± 2	-0.85 ± 0.03	-0.90 ± 0.04

The $K^- p$ parameters are only used in the least-squares fit to calculate $\mathrm{Im}\, f_-$, even though they also contain information on the real part, $\mathrm{Re}\, f_-(m)$. The relation is not fitted at the $K^- p$ threshold because the value of $H(-m)$ is too dependent on the particular $K^- p$ parameters that are used. However, on the other hand this dependence leads to a useful self-consistency test for the $K^- p$ solutions. With the parameter values $\overline{g^2}, b_0, \ldots b_n$ that have been determined by the fit we can evaluate eq. (5) at $\omega = -m$ and so predict $\mathrm{Re}\, f_-(m)$. A comparison of this value with that calculated directly from the $K^- p$ parameters is shown in Table 1 for each of the $K^- p$ solutions. We see that *Kim's* ERA solution fails this consistency test and that in fact the CSL solution turns out to be the most self-consistent parametrization. We conclude that the best value of $\overline{g^2}$ is that obtained using the CSL parameters and so the overall prediction from the finite-contour relation is

$$g_\Lambda^2 + 0.84 g_{\Sigma^0}^2 = 8.8 \pm 3 \,.$$

This does not necessarily mean that the CSL extrapolation is correct since the consistency test is essentially a check of the unphysical *integral* over $\mathrm{Im}\, f_-$ and not of $\mathrm{Im}\, f_-$ itself.

Besides being less sensitive to the extrapolation, another advantage of the finite contour method is that it is independent of the high energy behaviour of the amplitudes, whereas the validity of the sum rule relies on the *Pomeranchuk* hypothesis.

Somewhat similar results for $\overline{g^2}$ have been obtained (*Perrin* and *Woolcock*, 1969) by fitting the real parts using a dispersion relation subtracted at $\omega = 0$. However in this case the consistency test does not appear so stringent, probably because the unphysical region is not suppressed as much as it is in the finite-contour relation.

3. Other Methods for Determining the ΛKN Coupling

We briefly mention other techniques which have recently been used to estimate g_Λ^2. *Cutkosky* and *Deo* (1970) have exploited the analyticity properties in $\cos\theta$ to extrapolate from the K^+p angular data to the Λ-pole position. By conformally mapping the cut $\cos\theta$-plane to the inside of an ellipse and extrapolating with an optimally convergent polynomial expansion for the amplitude they obtain $7 \lesssim \overline{g^2} \lesssim 21$. Another technique is to compare the high energy $K^-n \to \Lambda\pi^-$ and $\pi N \to N\pi$ backward scattering data. Using the available data around 4 GeV/c *Martin* and *Michael* (1970) determined the ratio of the coupling of the N_α Regge trajectory to ΛK and πN states. Extrapolating this result to the nucleon pole yields $g_\Lambda^2 = 14.5 \pm 5$. Finally *Rogers* (1969), using the positivity of $\mathrm{Im}\, f_-$ in the unphysical region, has obtained bounds for g_Λ^2. A discussion of these methods can be found in the accompanying review articles by *B. R. Martin* and *C. Michael*.

4. K^- Absorption on Nuclei and Extrapolation below the K^-p Threshold

Although all the K^-p solutions give similar behaviour for the $\bar{K}N \to \bar{K}N$ and $\bar{K}N \to \pi Y$ amplitudes above the K^-p threshold we have seen that they may differ significantly when extrapolated into the unphysical region. One way of obtaining information below threshold is to consider the data for K^- absorption by nuclei. When a K^- in a mesic atom undergoes capture by the virtual nucleons bound in a nucleus the basic process is one of the following

$$K^-p \to \Sigma^-\pi^+, \Sigma^+\pi^-, \Sigma^0\pi^0 \quad \text{or} \quad \Lambda\pi^0,$$

$$K^-n \to \Sigma^-\pi^0, \Sigma^0\pi^- \quad \text{or} \quad \Lambda\pi^-,$$

$$K^-NN \to \Sigma N \quad \text{or} \quad \Lambda N.$$

The K^- may be taken to be at rest. Thus the observed emission ratios resulting from single nucleon K^- capture within the nucleus, besides depending on various nuclear properties, depend on the behaviour of the $K\bar{N} \to \pi Y$ amplitudes some 20 MeV below the K^-p threshold energy, E_t. The distance below threshold, which we denote by $E_t - E$, is the typical energy required to extract the nucleon (on which capture occurs) from the nucleus. It will be different for different nuclei.

In fact one of the original arguments (*Schult* and *Capps*, 1961, 1962) for the existence of the Y_0^* resonance just below threshold was based on the variation of the branching ratio $R_1 \equiv \sigma(\Sigma^+\pi^-)/\sigma(\Sigma^-\pi^+)$ observed for K^- capture in deuterium and helium, as compared to the value for hydrogen.

More recently the emission ratios R_1, together with

$$R_2 = \sigma(\Sigma^+ \pi^- + \Sigma^- \pi^+)/\sigma(\Sigma^- \text{ no } \pi^+),$$

for K^- capture by different nuclei have been used to discuss the proton-to-neutron density ratio, γ, in the distant nuclear surface, the region where most of the mesic absorption occurs (*Burhop*, 1967; *Bloom, Johnson* and *Teller*, 1969; *Bethe* and *Siemens*, 1970). The emission of a Σ^- without a charged pion is taken to signify a primary K^- interaction on a neutron. If we assume the validity of the impulse approximation and neglect final state interactions, then

$$R_1 = \frac{\frac{1}{6}|M|^2 + \frac{1}{4} - |M|\cos\phi/\sqrt{6}}{\frac{1}{6}|M|^2 + \frac{1}{4} + |M|\cos\phi/\sqrt{6}}, \qquad (7)$$

$$R_2 \approx \gamma(\tfrac{2}{3}|M|^2 + 1), \qquad (8)$$

where $M \equiv M_0/M_1$ is the ratio of the $I=0$ and $I=1$ $\bar{K}N \to \pi\Sigma$ amplitudes and where ϕ denotes their relative phase. The observed ratios R_1 and R_2 have been used to estimate $|M|$ and ϕ (*Burhop*, 1967). The results for different nuclei show the rapid variation of ϕ with binding energy that is expected from the proximity of the Y_0^* resonance.

Here we wish to point out the possibility of using the observed emission ratio R_1 as a check of the extrapolations of the K^-p solutions below threshold. We do not consider the R_2 data as they depend on the proton-to-neutron density ratio in the nuclear surface. An additional uncertainty in the calculation of the ratio R_2 is that the K^-p interaction rate is modified by the Coulomb penetration factor whereas the K^-n interaction is not. The inclusion of this factor enhances the estimate of R_2 as obtained from eq. (8) by about 15%. It is therefore fortunate that R_1 turns out to be the most sensitive to the extrapolation.

The curves in Fig. 3 show the variation of the ratio R_1 as predicted by extrapolating the various K^-p solutions below threshold. The values of the emission ratio R_1 observed from K^- capture in various nuclei are listed in Table 2 and have been taken from a compilation of the data by *Burhop* (1967). The experimental branching ratio at rest in hydrogen is due to *Kim* (1966). In order to plot this data on Fig. 3 we have to estimate the average energy that would be needed to extract the proton (on which the capture occurs) from the nucleus. We can regard this as the sum of three parts: the binding energy of the last proton, the excitation energy of the residual nucleus and the *Coulomb* potential Ze^2/R felt by the bound proton. The binding energies of the last proton are given in Table 2 and the points shown on Fig. 3 correspond to setting the excitation (+ Coulomb) energies equal to 10 ± 5 MeV.

Table 2. *The experimental values of R_1 are taken from Burhop (1967). The energy, $E_t - E$, required to extract the proton from the nucleus is the sum of the binding energy of the last proton and an assumed excitation of the residual nucleus of 10 MeV.*

Material	R_1	B. E. of last proton	$E - E_t$
Hydrogen	0.485 ± 0.013	0	0
Deuterium	$0.88 \ \pm 0.1$	2.2	$- 12.2$
Helium	$3.4 \ \ \pm 0.4$	20	$- 30$
Light emulsion nuclei (C, N, O)	$1.73 \ \pm 0.17$	14	$- 24$
Heavy emulsion nuclei (Ag, Br)	$1.19 \ \pm 0.13$	6	$- 16$

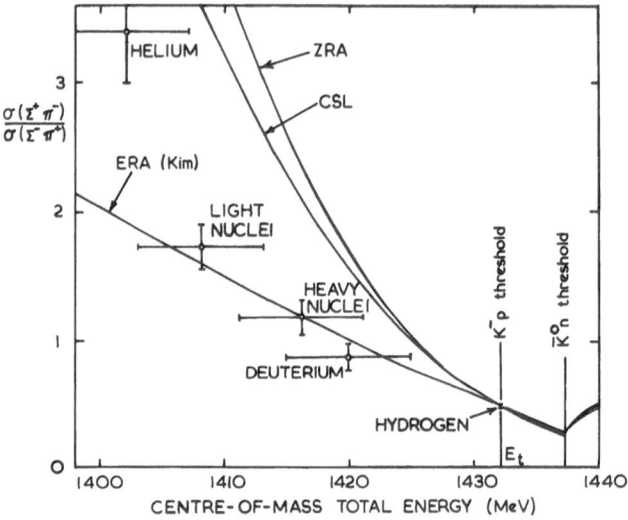

Fig. 3. A comparison of the ratio $R_1 \equiv \sigma(\Sigma^+ \pi^-)/\sigma(\Sigma^- \pi^+)$ calculated by extrapolating the $K^- p$ solutions below threshold with the values obtained from the K^- nuclear capture data (see Table 2)

In view of the crude treatment of the nuclear capture process Fig. 3 should probably only be regarded as an indication of the potential usefulness of a comparison of this data with the $K^- p$ extrapolations. In particular we have neglected (i) the effect of multiple scattering of the K^- within the nucleus, (ii) the possibility that one of the reaction products may undergo a secondary interaction in the same nucleus, such as

$$\Sigma^- p \to \Lambda n \quad \text{or} \quad \Sigma^0 n$$

$$\Sigma^+ n \to \Lambda p \quad \text{or} \quad \Sigma^0 p \tag{9}$$

and (iii) the possibility of real Y^* production. In the heavier nuclei the final state interactions (9) will be more symmetrical in Σ^- and Σ^+ and so should not affect R_1 appreciably, and hopefully approximation (i) will also have a negligible effect on R_1. Finally, although there is some evidence (*Bethe* and *Siemens*, 1970) that the residual nuclear excitation is about 10 MeV this assumption also requires further study. One way of avoiding the assumption would be to consider the ratio R_1 for a given final state, for example $\Sigma^+ \pi^- H^3 / \Sigma^- \pi^+ H^3$ for K^- capture in helium.

Another recent application of K^- capture data concerns K^- absorption by deuterium. *Bunnell et al.* (1970) have analysed the $K^- d \to \Lambda \pi^- p$ data using the impulse approximation. Adopting a model of the $K\bar{N} \to \Lambda \pi$ amplitude below the $\bar{K}N$ threshold they were in this way able to estimate the $\bar{K}N Y_1^*$ (1385) coupling constant.

Acknowledgement

It is a pleasure to thank Dr. *T. E. O. Ericson* for many interesting discussions concerning K^- absorption by nuclei.

References

Bethe, H. A., Siemens, P. J.: Nucl. Phys., B **21**, 589 (1970).
Bloom, S. D., Johnson, M. H., Teller, E.: Phys. Rev. Letters **23**, 28 (1969).
Bunnell, K. O., Cline, D., Laumann, R., Mapp, J., Uretsky, J. L.: Nuovo Cimento Letters **3**, 224 (1970).
Burhop, E. H. S.: Nucl. Phys. B **1**, 438 (1967).
Cutkosky, R. E., Deo, B. B.: Phys. Rev. D **1**, 2547 (1970).
Kim, J. K.: Phys. Rev. Letters **14**, 29 (1965).
— Columbia University Report, Nevis-149 (1966).
— Phys. Rev. Letters **19**, 1074 (1967).
Martin, A. D., Michael, C.: Phys. Letters **32** B, 297 (1970).
— *Perrin, R.:* Nucl. Phys., B **20**, 287 (1970).
— *Ross, G. G.:* Nucl. Phys. B **16**, 479 (1970).
Perrin, R., Woolcock, W. S.: Nucl. Phys. B **12**, 26 (1969).
Rogers, T. W.: Phys. Rev. **178**, 2478 (1969).
Schult, R. L., Capps, R. H.: Phys. Rev. **122**, 1659 (1961).
— — Nuovo Cimento **23**, 416 (1962).

Prof. Dr. *A. D. Martin*
Department of Physics
University of Durham
Durham/England

Nucleon-Nucleon Interactions below 1 GeV/c

G. KRAMER

Contents

I. Introduction 152
II. On-Shell-Methods 154
III. Off-Shell-Methods 161
IV. Comparison and Outlook 164
Acknowledgements 165
References 165

I. Introduction

The purpose of this meeting as I understand it is the discussion of the
theoretical analysis of the experimentally best known reactions with
a view towards common information about masses and coupling
constants of meson and nucleon resonances. Nucleon-nucleon scattering
is particularly suited for this enterprise since the exchange of mesons
gives the singularities in the unphysical region of the u- and t-channel.
These contributions, together with the elastic unitarity-cut, determine
the $N-N$ scattering amplitude for low energies, say below 400 MeV.
It is hoped that these contributions are still dominant for energies in
the laboratory system up to 750 MeV with only small modifications
produced from inelastic channels. In this review we shall limit ourselves
to a discussion of the theory of elastic pp and np scattering in this energy
range [1]. We shall not consider inelastic two-nucleon reactions such
as, for example

$$\gamma d \to np, \quad pp \to pp\gamma, \quad pp \to pn\pi^+ \quad \text{or} \quad pn \to pp\pi^-$$

where further complications arise because the basic interaction of the
extra particle − here a photon or pion − must be known. It is clear
that for energies above the pion production threshold some knowledge
about inelastic reactions, in particular pion production, is necessary to
make further progress in the understanding of the elastic channel. The
reason for the limitation to low energies is twofold: (1) the experimental
information is more accurate (2) inelastic channels should be unimportant.

Before discussing the content of the many papers about low energy NN scattering which might be of interest for us I shall describe very shortly the empirical information we have about this process (and also $\bar{N}N$ scattering which is needed, e. g., in fixed-t dispersion relations):

(1) total cross-sections for $pp, \bar{p}p$ and pn scattering
up to $E_{lab} = 30$ GeV, furthermore,

quite recently, for $\bar{p}n$ scattering
up to $E_{lab} = 60$ GeV [2].

(2) $d\sigma/dt$ for pp, pn and partly also for $\bar{p}p$ scattering for various $(-t)$ intervals and energies up to 30 GeV [2].

(3) Some polarization measurements in the GeV range and for smaller momentum transfer [2].

(4) Various cross-section and polarization measurements below 1 GeV which are used for pp and np phase shift analysis
  a) pp and np $(1 - 450$ MeV), $I = 1$ is unique,
                          $I = 0$ not quite unique for energies
                          below 50 MeV [3],
  b) pp $(450 - 750$ MeV) no unique set [3] available,
  c) np $(540 - 750$ MeV) no unique set [3] available.

It is clear that for the low energy region (below 400 MeV) only the information in (4a) is of particular interest to us. The information at higher energies (1) to (3) is needed if attempts are made to use such general frames as fixed t dispersion relations to gain information about the unphysical u-cut.

The plan for our review of the theoretical analysis is the following. We shall investigate all approaches from the point of view of obtaining information about meson resonances, in particular their couplings to the nucleon. We shall start in Section II with the discussion of the on-shell approaches based on fixed t dispersion relations and partial wave dispersion relations. In Section III we describe characteristic off-shell frameworks based on relativistic or nonrelativistic wave equations.

The work we shall discuss has not been done in the last two years. Therefore, the most complete phase shift information published in 1969 by the Livermore group [3] has been used for comparison or fitting purpose only in the most recent papers. Thus we can expect that some of the results concerning meson coupling constants are not comparable because older phase shift information has been used. On the other hand the information about $I = 1$ phase shifts has not changed appreciably in the last five years [3]. Therefore, most of the differences in coupling constants are rather due to the different methods of analysis than to different empirical information.

II. On-Shell-Methods

The problem of evaluation of meson resonance or multipion couplings
from NN or πN scattering lies in the fact that major portions (poles or
cuts) of these contributions lie in the unphysical region. For example,
the physical region for $N_1 + N_2 \rightarrow N_3 + N_4$ together with the crossed
reactions is $(s + t + u = 4m^2)$

   a) $N_1 + N_2 \rightarrow N_3 + N_4$: $s \geq 4m^2$, $t \leq 0$,
   b) $N_1 + N_3 \rightarrow N_2 + N_4$: $t \geq 4m^2$, $s \leq 0$,
   c) $N_1 + N_4 \rightarrow N_3 + N_2$: $u \geq 4m^2$, $t \leq 0$,

whereas the two pion exchange starts at $t \geq 4\mu^2$ and $u \geq 4\mu^2$ which is
much nearer to the physical region of channel (a) than are the physical
singularities from channel (b) and (c). Therefore, the main "forces" for
channel (a) are the exchanges of the π, the 2π, the 3π etc. systems. The
u- and t-channel exchanges cannot be uncorrelated pions only just
because resonances exist in these many-pion systems. On the other
hand it is an open question whether the description of the t- and u-
channel exchanges by meson resonances alone is sufficient.

In this section we shall describe the results obtained with forward
dispersion relations and with partial wave dispersion relations.

It is well known that the T matrix for nucleon-nucleon scattering
depends on five independent functions α, β, γ, δ, and ε. They depend
only on s and t and are defined by the GNO spin decomposition [4]

$$T = \frac{m}{E} (\alpha + \beta \ \sigma_1 \cdot n \sigma_2 \cdot n + i\gamma \ (\sigma_1 + \sigma_2) \cdot n + \delta \ \sigma_1 \cdot m \sigma_2 \cdot m + \varepsilon \ \sigma_1 \cdot l \varrho_2 \cdot l) \quad (2.1)$$

where $E = (p^2 + m^2)^{1/2}$, σ_1 and σ_2 are spin vectors of the incoming nucleons
and l, m and n are unit vectors in the directions $p + p'$, $p - p'$ and $p \times p'$
respectively; p and p' are the three momenta of the incoming and outgoing nucleons in the centre of mass system, m is the mass of the nucleon.
In the forward direction the amplitudes α, β, γ, δ, and ε obey the well-
known dispersion relations [4]. For example, in the case of np scattering,
the dispersion relation for the spin independent amplitude α, with a
subtraction at $s = 4m^2$ has the following form [4]

$$\mathrm{Re}\,\alpha(s) = \mathrm{Re}\,\alpha(4m^2) + \frac{2\Gamma_\alpha m(s - 4m^2)}{B(4m - B)(s - s_d)}$$

$$+ (s - 4m^2) \frac{1}{\pi} \int_{4m^2}^{\infty} ds' \frac{\mathrm{Im}\,\alpha(s')}{(s' - 4m^2)(s' - s)} \quad (2.2)$$

$$+ u \frac{1}{\pi} \int_{4\mu^2}^{\infty} du' \frac{\mathrm{Im}\,\alpha(u')}{u'(u' - u)}$$

where $s_d = (2m - B)^2$, B is the binding energy of the deuteron and Γ_α is the residue of the deuteron pole. μ is the pion mass. The forward dispersion relation for α has the advantage that except for the unphysical cut in u from $4\mu^2$ to $4m^2$ the integrand is known from total cross section measurements for np or pp and $\bar{p}n$ or $\bar{p}p$ scattering. On the other hand the important one- and multi-π-exchanges contribute to $\operatorname{Im}\alpha(u)$ for $\mu^2 \leq u \leq 4m^2$. For example, if the multipion exchanges are approximated by the pion resonances, the unphysical u-cut of α has contributions from the π, η, ϱ, ω, ϕ, f, f', A_1, A_2 and the $\varepsilon(\varepsilon$ is 0^+, $I = 0)$ [5]. These are only the more or less established meson resonances. There are some more at higher masses [5]. Writing M for the mass of the exchanged particle, the cut contribution becomes a sum of terms of the form

$$\frac{R_\alpha T}{M^2 + 2mT} \tag{2.3}$$

with residues R_α for every resonance α which can be expressed by the coupling constants defined in (2.4). Here T is the lab kinetic energy of the incident particle $2mT = s - 4m^2$. From (2.3) it is clear that more precise information about R_α can be obtained if the squared mass of the resonance is $M^2 \simeq 2mT$. Therefore, very little can be learned from fits to low energy data about resonances at higher masses. From the structure of (2.3) it is obvious that it is more or less impossible to obtain unique information about the couplings of all these resonances from the dispersion relation for α alone. This is clearly seen from a recent evaluation of the forward dispersion relation for α by *Pascual* and *Yndurain* [6]. They write only the π and η terms explicitly and expand contributions of the other mesons in a series of Chebyshev polynomials. If they fix the coupling of the pion by $f_\pi^2 = (\mu/2m)^2 \, g_{\pi n}^2 = 0.078$ which is the value derived from pion-nucleon scattering data [7], then f_η^2 is still undetermined inside the bounds $0.8 \geq f_\eta^2 \geq 0.3$. Also f_π^2 is not determined very well by this dispersion relation alone.

To improve the situation concerning the determination of the unphysical cut one can go to dispersion relations for all momentum transfers t or one can include the fixed-t dispersion relations for the other functions β, γ, δ, and ε. Taking also the information from the amplitudes β, γ, and ε has the advantage that at fixed-t contributions from the u-channel $N\bar{N}$ singlet and triplet processes can be separated from each other. That means that contributions from exchange of mesons like the π and η may be unambiguously separated from those due to ϱ, ω, ϕ, and ε. On the other hand one needs the imaginary parts of these functions in the physical s and u channels up to high energies. These are available only from phase shifts which at present are known only up to 750 MeV.

Dispersion relations for nonforward scattering have not been evaluated yet. Therefore, we can limit our discussion to the forward case. It turns out that only in the dispersion relation for the spin averaged amplitude α the physical s cut for large s is important. For β and ε the physical large s cut (above $E_{\mathrm{lab}} = 600$ MeV) and the physical u cut are not significant and are usually neglected or approximated by a constant of the value at the highest energy. Such an evaluation of the forward dispersion relations for α, β, γ, and ε has been made by *Bugg* two years ago [8]. He based his analysis on the phase shift analysis done by *Perring* [9] and an older one of the Livermore group [9]. His results are parameterized by the masses (which are the values found in other experiments except for the ε [5]) and a phenomenological 3π (P) cut and coupling constants for π^0, $\pi^\pm$, η, P, ϱ, ω, and ε. All mesons with higher masses such as the ϕ, f, f', A_1, and A_2 are neglected. The renormalized coupling constants are defined in terms of a Lagrangian for ε, η, and ω interactions as follows

$$L = \sqrt{4\pi}\,\bar{\psi}\left[g_\varepsilon\phi_\varepsilon + g_\eta\gamma_5\phi_\eta + ig_{1\omega}\gamma_\mu\phi_\omega^\mu + \frac{g_{2\omega}}{4m}\,\sigma_{\mu\nu}\left(\frac{\partial\phi_\omega^\mu}{\partial x_\nu} - \frac{\partial\phi_\omega^\nu}{\partial x_\mu}\right)\right]\psi\,. \quad (2.4)$$

The Lagrangians for the π and ϱ interactions are the same as for η and ω respectively, except for the replacement

$$\phi_\eta \to \tau\phi_\pi, \qquad \phi_\omega \to \frac{\tau}{2}\,\phi_\varrho\,.$$

Bugg finds that the dispersion integrals are significant only for s and p waves, in particular the dispersion integrals over the physical region receive appreciable contributions only from s and p waves and the dispersion integrals from the physical region of the crossed channel only in the s wave. This is in accord with the general argument that the phase shifts for the higher waves ($l \geq 2$) can be approximated for moderately high energies (say $E_{\mathrm{lab}} \leq 330$ MeV) by the Born terms of low mass exchange. Therefore, Bugg finds reasonable agreement of the $^3P_2$, $^3F_2$, ε_2, $^3F_3$, $^3F_4$, and $^1D_2$ phase shifts between 25 MeV $\leq E_{\mathrm{lab}} \leq 330$ MeV with values calculated from the Born amplitudes using the coupling constants derived from fitting the dispersion relations to the experimental data. This way one can get a good idea of the fit to the angular dependence of pp scattering. A similar point of view had been taken earlier by *Kopp* and *Söding* [10]. They took first the forward dispersion relation for α (pp scattering only) to include information contained in the low angular momentum phase shifts and second the partial wave amplitudes in the Born approximation for the states $^1D_2$, $^1G_4$, $^3F_3$, $^3F_4$, $^3H_4$. These two things were fitted to the earlier phase shift results of the Livermore group (Midpop III) [11].

Actually with this method the coupling strengths of most of the mesons are already fixed by comparing the Born amplitudes for the high partial waves to the empirical phase shifts. This is clearly seen by looking at the coupling constants obtained from an analysis based only on high partial wave Born amplitudes reported also in the paper of *Köpp* and *Söding*. Similar results had been obtained earlier by *Köpp* [12] and later also by *Arndt, Bryan*, and *MacGregor* [13]. Adding the dispersion relation for $\alpha(s)$ to the high partial wave Born terms has the effect that the coupling strengths of the mesons with larger masses as the ε, ω, and ϕ, could now also be determined with a reasonable accuracy since these mesons contribute mainly to the low angular momentum partial waves excluded from *Köpp*'s original analysis [12]. This is clearly seen from Table 1, where we report the couplings from the three analyses, *Bugg* [8], *Köpp* [12], *Köpp* and *Söding* [10]. Adding the forward dispersion relation for α improved considerably the accuracy of the ε, ω, and ϕ coupling. For comparison we also give the results obtained by *Arndt, Bryan*, and *MacGregor* [13]. They included np scattering in their analysis and all phases except $^1S_0$, $^3S_1$, $^3D_1$, and ε_1. The latter two were excluded since they clearly depend on unitarity because of the existence of the deuteron bound state. Furthermore *Arndt, Bryan*, and *MacGregor* even left the pion coupling and the σ-mass free. For the σ-mass they obtain the usual preferred low value around 400 MeV. Unfortunately the square of the pion-nucleon coupling constant came out 13% smaller than the usually preferred value from pion-nucleon scattering [7]. This indicates that unitarity is not completely negligible if all p-waves are included in the fit. By using an on-shell unitarization scheme such as for example the K-matrix or partial wave dispersion relations they find a better value for the pion-nucleon coupling constant. Surprisingly enough the other coupling constants are changed only very little by the unitarization procedure as chosen by these authors but not described further in their paper [13]. This is seen from line 5 in Table 1 where coupling constants from Born amplitudes with unitarity corrections are given [13]. Up to now we considered only such work where the unphysical region was represented by the pion and meson resonances (the finite widths of these resonances were neglected). We remark that all authors mentioned so far (except *Bugg* [8]) who start with meson resonances introduce a fictitious $\pi\pi$ resonance, the σ, with $I = 0$ and $J = 0^+$, and a rather low mass around 400 MeV. Sometimes this mass value is obtained from the fit, in other cases it is input. This σ is not established otherwise as a $\pi\pi$ resonance, whereas all the other particles, π, η, ϱ, ω, ϕ, and ε have been seen as meson resonances in their direct channel [5]. On the other hand a 0^{++} term with $I = 0$ with rather low mass seems to be necessary to produce additional attraction. It is the

Table 1. Meson-nucleon coupling constants obtained from different fits to low-energy pp and np scattering. The definition of the coupling constants is given by (2.4). The underlined numbers are input to the fits

Authors	Ref.	m_π (MeV)	$g_\pi^2/4\pi$	m_η (MeV)	$g_\eta^2/4\pi$	m_σ (MeV)	$g_\sigma^2/4\pi$	m_ε (MeV)	$g_\varepsilon^2/4\pi$	m_ϱ (MeV)	$\dfrac{g_{1\varrho}^2}{4\pi}$	$\dfrac{g_{2\varrho}}{g_{1\varrho}}$	m_ω (MeV)	$\dfrac{g_{1\omega}^2}{4\pi}$	$\dfrac{g_{2\omega}}{g_{1\omega}}$	m_ϕ (MeV)	$\dfrac{g_{1\phi}^2}{4\pi}$	$\dfrac{g_{2\phi}}{g_{1\phi}}$
Bugg[b]	[8]	136.5	13.6[a] ±0.7 / 14.7	550	0.2 ±2.3	—	—	613	14.1 ±1.5	750	7.52 ±4.0	2.5 ±1.0	750	13.5 ±1.7	0	—	—	—
Köpp, Söding	[10]	140	14.4	550	10.2 ±5.9	400	1.6 ±0.1	700	2.5 ±1.3	765	4.8 ±0.8	3.66	782	4.2 ±2.0	0	1020	0	0
Köpp	[10, 12]	140	14.4	550	0	400	1.6 ±0.3	700	8.0 ±3.6	765	5.2 ±1.2	3.66	782	8.5 ±9.0	0	1020	4.8 ±16.9	0
Arndt et al. (I)	[13]	136.5	12.74	—	—	421	2.33	—	—	763	3.7	4	783	3.06	0	—	—	—
Arndt et al. (II)	[13]	136.5	14.52	—	—	468	2.79	—	—	763	4.9	4	783	0.94	0	—	—	—
Scotti, Wong	[19]	135	14.0	548	12.1	437	3.05	—	—	591	5.1	3.0	780	2.77	0	1020	2.26	0
Ball et al.	[19]	140	13.0	548	0.3	545	5.15	—	—	750	5.4	3.66	780	1.36	0	1020	2.72	0
Schierholz (I)	[25]	140	14.4	550	9.9	400	1.4	700	5.4	711	3.0	4.47	781	8.5	-0.105	—	—	—
Schierholz (II)	[27]	140	14.4	550	8.0	400	1.4	700	6.8	711	2.4	4.78	781	9.05	-0.1	—	—	—
Bryan, Scott[c]	[28]	138.7	12.55	548.7	2.6	550	8.19	—	—	763	7.2	1.13	782.8	17.26	0	—	—	—
Green, Sawada[d]	[28]	138.7	14.7	548.7	2.0	416	2.35	782	14.7	763	2.6	3.74	782.8	23.0	0	—	—	—
Particle	[5]	134.9 / 139.6	—	548.8	—	—	—	700	—	765	—	—	783.7	—	—	1019.5	—	—

[a] The first number s for the neutral pion.

[b] In addition a particle $P(I=1, J=0^-)$ with mass m_P = 550 MeV, $g_P^2/4\pi$ = 6.6 ± 1.7.

[c] In addition a particle $\sigma_1(I=1, J=0^+)$ with mass m_{σ_1} = 600 MeV, $g_{\sigma_1}^2/4\pi$ = 1.65.

[d] In addition a particle $\sigma_1(I=1, J=0^+)$ with mass m_{σ_1} = 763 MeV, $g_{\sigma_1}^2/4\pi$ = 0.65.

general opinion that the σ term is an approximation of the uncorrelated $\pi\pi$ s-wave contribution. Actually with t-dispersion relations one can investigate this t-channel low partial wave $\pi\pi$ contribution. Such work based on the Cini-Fubini method [14] had been started by *Amati, Leader* and *Vitale* [15] in 1963 and developed further and corrected by *Durso* and *Signell* [16] and by *Durso* [16]. In particular the work of *Durso* [16] demonstrates that in the "2π basic" contribution the s-wave is by far the dominant one while the p-wave is small and the $d+$ higher wave contribution is even smaller. Furthermore, this s-wave $\pi\pi$ continuum as calculated by *Durso* has the same effect on NN scattering amplitudes as a σ with low mass. *Schierholz* fitted a σ with a mass of 400 MeV to *Durso*'s result and found the coupling constant to be $g_\sigma^2/4\pi = 1.4$ [17]. We mention that *Durso* in his calculation of the 2π continuum includes in the πN amplitudes besides the nucleon poles also the $\Delta(1236)$ contribution evaluated as in the wellknown paper by *Chew, Goldberger, Low,* and *Nambu* [18], with some corrections to account for the small s-wave πN scattering lengths. So it seems that it is physically justified to include the σ exchange in the unphysical cut. Nevertheless, some warning is in order. As *Durso* remarks himself, the model used to predict the t-channel s-wave is rather crude, and much more work starting from newer and better theoretical work on the πN scattering amplitudes is desirable to put *Durso*'s conclusions on safer grounds. On the other hand it seems to be well justified to approximate the t-channel p-wave by the ϱ resonance alone [16].

Up to now we did not consider in detail any framework which has unitarity built in from the beginning, with the exception of forward dispersion relations where unitarity is guaranteed by using experimental information in the physical s- and u-cut dispersion integrals.

An alternative method to enforce unitarity in the physical region $s \geq 4m^2$ (but not $u \geq 4m^2$ within the approximations made) is to use partial wave dispersion relations as derived from Mandelstam analyticity. This has been investigated in some detail since 1963 in papers by *Scotti* and *Wong* [19] and by *Ball, Scotti,* and *Wong* [19]. We shall explain this approach following the last publication of this group [20]. Let us consider, for example, the spin-singlet partial wave amplitude $t_j(p^2)$ normalized so that the elastic unitarity constraint is

$$\operatorname{Im} t_j(p^2) = \frac{mp}{E} |t_j(p^2)|^2 . \qquad (2.5)$$

The factor mp/E is the relativistic phase space factor. We denote the partial waves of the sum of the single-meson-exchange terms by $b_j(p^2)$. Then the approximate on-shell dispersion relation used by this group gives the following integral equation for $t_j(p^2)$

$$t_j(p^2) = b_j(p^2) + \frac{1}{\pi} \int_0^{p_c^2} d p'^2 \, \frac{mp'}{E'} \, \frac{|t_j(p'^2)|^2}{p'^2 - p^2 - i\varepsilon} \,. \tag{2.6}$$

Clearly the imaginary part of the amplitude satisfies the unitarity condition (2.5) because of the integral on the right side of (2.6). A cutoff p_c^2 is introduced because of the logarithmic divergence in $b_j(p^2)$ due to the exchange of vector mesons (ϱ, ω, and ϕ). In the earlier work of *Scotti* and *Wong* [19] the Born amplitudes were given a Regge-type behaviour at high energy in order to render the dispersion integrals convergent. In terms of analytic structures, one observes that the only singularities of the integral in (2.6) are two branch points at $p^2 = 0$ and $p^2 = p_c^2$ connected by a branch cut along the positive real axis. All other singularities of $t_j(p^2)$ are in $b_j(p^2)$. If we compare this with the singularities in nonrelativistic potential theory for a Yukawa potential corresponding to a meson mass μ we see the following. The amplitude $t_j(p^2)$ obtained from the potential has branch points at $p^2 = -\mu^2 n^2/4$ $(n = 1, 2, ...)$. The index n corresponds to the number of times the potential acts, i. e. to the number n of meson exchanges. On the other hand the $b_j(p^2)$ does not contain the exchange of any meson more than once. It can be argued that the exchange of two mesons is replaced by ϱ and σ, but there is no contribution of many exchanges of σ and ϱ in $b_j(p^2)$. This is usually justified with the remark that these contributions correspond to very short range which are effectively shielded by the repulsion due to ω and ϕ exchange. Furthermore a certain part of the rescattering process is included in the solution $t_j(p^2)$ through the nonlinearity of the Eq. (2.6). For a given Born term $b_j(p^2)$ the integral Eq. (2.6) can be solved by the N/D method. Unfortunately, there is a further problem in this approach besides the many approximations already discussed. The solution $t_j(p^2)$ from (2.6) does not have the threshold behaviour p^{2l}, where l is the orbital angular momentum. This behaviour is satisfied by the "potential" $b_j(p^2)$ but in general not satisfied by the integral term of the dispersion relations. The threshold behaviour in this dispersion relation must be the result of the cancellation between the dispersion integral and the rescattering terms in $b_j(p^2)$, originating from many-meson exchanges, which have been neglected here. Usually this is remedied by adding arbitrary an lth order pole in (2.6) with a residue which is adjusted to achieve the proper threshold behaviour. The location of the pole is an additional parameter of the model.

The results of the partial wave dispersion approach are given in the sixth and seventh lines of Table 1. *Scotti-Wong* [19] refers to the older work with Regge cut-off and most of the coupling constants left free, whereas *Ball-Scotti-Wong* [19] made a fit to the experimental cross section and polarization data directly with a cut-off p_c^2 corresponding to 600 MeV laboratory kinetic energy and constraints on coupling constants from approximative $SU(3)$ symmetry concerning ϱ, ω, and ϕ couplings and $SU(6)$ symmetry concerning the η coupling.

III. Off-Shell-Methods

To avoid the threshold difficulties and also for reasons of consistency it is desirable to include the multi-meson exchange terms in the left-hand cut singularity structure of the partial wave dispersion relations. But this would be rather complicated. A more direct way for summation of multi-meson exchanges is the Bethe-Salpeter equation in the ladder approximation. Unfortunately, the Bethe-Salpeter equation for nucleons has no solution because of divergence problems due to the spin of the nucleon and the vector mesons. Therefore, one must use some kind of an approximate scheme through the introduction of a cut-off or of form factors and Regge-like modifications of the vector meson exchanges. Since the Bethe-Salpeter equation is an essentially off-shell equation, the introduction of on-shell Regge behaviour is rather doubtful. Aside from the question of a cut-off, the Bethe-Salpeter equation for scattering of particles with spin is too complex so that no serious attempts have been made so far to apply it to nucleon-nucleon scattering.

In the last two years several authors have used some sort of a relativistic Lippman-Schwinger equation which is simpler than the Bethe-Salpeter equation for calculating the nucleon-nucleon scattering matrix with one-boson exchange potentials. This equation yields a relativistic invariant unitary scattering amplitude and is off-shell in the three-momentum (instead of four momentum in the case of the Bethe-Salpeter equation) of the center of mass system. In the literature there exist equations with different two-nucleon propagators depending on the approach taken to derive these equations. We shall discuss them shortly by limiting ourselves to the simpler case of scattering of particles with spin zero. For example. *Logunov* and *Tavkhelidze* [21] and also later *Blankenbecler* and *Sugar* [22] start from the Bethe-Salpeter equation and replace the explicitly invariant twoparticle propagator

$$G(k, P) = \frac{i}{2\pi} \frac{1}{(P+k)^2 - m^2 + i\varepsilon} \frac{1}{(P-k)^2 - m^2 + i\varepsilon} \qquad (2.7)$$

by an approximate two-particle propagator $g(k, P)$ which, considered as a function of $q^2 = p^2 - m^2$, has the singularity structure of G when both legs of G are on the mass-shell. G produces also contributions from inelastic states and from unphysical states with negative energy.

These contributions are eliminated in g. But g is supposed to have the cut structure of G in the physical region. Then one possible form for $g(k, P)$ is

$$g(k, P) = \int\limits_0^\infty \frac{dq'^2}{q^2 - q'^2 + i\varepsilon} \delta((P' + k)^2 - m^2)\, \delta((P' - k)^2 - m^2) \quad (2.8)$$

where $q'^2 = p'^2 - m^2$ and $P' = (P', 0, 0, 0)$. The integral on the right-hand side of (2.8) yields

$$g(k, P) = \delta(k_0) \frac{1}{4\sqrt{m^2 + k^2}} \frac{1}{q^2 - k^2 + i\varepsilon} \quad (2.9)$$

Because of the factor $\delta(k_0)$ we have an equation of the three dimensional (or instantaneous relative time) form. The second line of approach is rather old and more general. It starts from formulations of relativistic quantum theory for a direct particle interaction rather than by means of fields. Such formulations have been developed [23], for example, by *L. H. Thomas, Foldy, Sudarshan, Fong* and *Sucher* and *Coester* and others. In particular, *Coester* [23] emphasized this relativistic particle quantum mechanics as a possible alternative to quantum field theory. In addition he discussed the question of existence of generalized Møller wave operators and the cluster condition and went on to three-particle problems. Based on this *Schierholz* recently demonstrated the existence of Møller wave operators if the potential fulfills particular integral conditions [24]. From this the transition matrix is obtained which is reduced to partial waves. The off-shell partial wave amplitudes satisfy integral equations of the following form (here given for the singlet part)

$$T^j(k', k) = v^j(k', k) + \int\limits_0^\infty dq\, q^2\, v^j(k', q) \frac{1}{\sqrt{k^2 + m^2} - \sqrt{q^2 + m^2} + i\varepsilon} T^j(q, k).$$
$$(2.10)$$

This singular integral equation can be transformed into a Fredholm integral equation and *Schierholz* showed that solutions exist under the same assumptions for the potential $v^j(k', k)$ which guarantee the existence of the Møller operators [24]. Furthermore under these assumptions the solutions of the Fredholm equation are the only solutions of the original integral equation. $v^j(k', k)$ is the partial wave amplitude of the potential which is defined as the difference between the total and the free Hamiltonian in the center of mass system. It depends only on the relative momenta k' and k before and after scattering. Up to this point this theory does not offer physical arguments for the choice of the interaction potential. Of course one can formulate an equivalent theory for pion-nucleon interaction and try to derive from it the internucleon interaction. This has not been completely worked out so far. *Schierholz* [25] has chosen for v the off-energy-shell lowest order term of the field theoretical

S-matrix, as, for example, for one-pion exchange:

$$v(\boldsymbol{k}, \boldsymbol{k}) = - \frac{g_{\pi n}^2}{(2\pi)^3} \frac{4m^2}{\sqrt{(k'^2 + m^2)(k^2 + m^2)}}$$

$$\frac{1}{(\sqrt{k^2 + m^2} - \sqrt{k'^2 + m^2})^2 - (\boldsymbol{k} - \boldsymbol{k}')^2 - m_\pi^2}.$$

That way the theory agrees with field theory in lowest order [26]. Concerning the higher order terms one has the correspondence with the Bethe-Salpeter equation. *Schierholz* has also shown that his equation is completely equivalent to the Blankenbecler-Sugar equation [22] in the sense that both are approximations to the same Bethe-Salpeter equation with inelastic and negative energy states removed. They differ in the reduced propagator $g(k, P)$. Therefore, this equation and the Blankenbecler-Sugar equation do not lead to identical T matrices. This way all these off-energy-shell equations are connected with the ladder approximation of the Bethe-Salpeter equation, but not always in the same way. We remark that in these relativistic equations the potentials are strongly energy dependent. Only in the nonrelativistic approximation local potentials appear. However, this reduction can be justified only if both the momentum of the nucleon and the mass of the meson are small compared to the rest mass of the nucleon. This is certainly not the case for the vector mesons.

Schierholz [25] used his approach to describe pp scattering and recently also np scattering [27] and deuteron photodisintegration [27]. The coupling constants are adjusted to give a good fit to the Livermore phase shifts [3]. The results for the adjustment to pp scattering are seen in Table 1. To fulfill the condition of the theorem for the existence of unique solutions he must introduce Regge behaviour for the vector mesons with parameters which are also fitted to the phase shifts. Furthermore he introduces a form factor for the pion propagator with mass $\Lambda_\pi^2 = 18.5\,m_\pi^2$. He also obtains a fairly good description of the phase shifts in the energy range between 300 and 600 MeV [3]. Of course, if one considers the nonrelativistic limit of these equations one arrives at the Schrödinger equation with local potentials of central, tensor and spin-orbit type. Such one-boson-exchange potentials inserted into the Schrödinger equation have been used as a model for nucleon-nucleon phase shifts since 1962 by many authors [28]. Unfortunately, such potentials are highly singular and can be used for s and p waves only when a cut-off or form factors are introduced. In the recent work of *Bryan* and *Scott* [28] terms proportional to p^2/m^2 in the potential were retained eliminating the cut-off in the p-waves. *Green* and collaborators [28] solve this problem by introducing form factors for the vector-meson exchange which are fitted to the electromagnetic form factor of the proton and the neutron. Results on coupling constants of these groups are collected in the last two lines of Table 1. For fitting np scattering as well, both groups introduced a third scalar particle with $I = 1$ which can be associated with the δ resonance [5]. As already stated the nonrelativistic approximation of the vector-meson exchange potential is doubtful because of the large mass of the vector mesons. Therefore, we rather like to consider the results of the non-relativistic Schrödinger equation with caution and prefer to rely on the results obtained with the relativistic equation.

If we try to compare the results obtained with the on- and off-shell methods, relativistic or nonrelativistic, we run into the following problem. In the fixed-t dispersion relations the couplings of the mesons to the nucleon were defined as an approximation of the scattering matrix in the unphysical u-cut region. In the off-shell approaches this unphysical cut has contributions also from the multimeson exchanges, for example two pion exchange, which is the most important one. Presumably this does not affect too much the 1^- exchange but should have more effect on the 0^+ exchange. Therefore, coupling constants in the two approaches must not necessarily be comparable. It would be interesting to try to calculate the unphysical cut from the relativistic equation and compare it to the pure pole description. Before doing this it would be necessary to investigate the analytic structure of the scattering matrix following from the relativistic quantum mechanics with the usual choice of potentials. This structure may not be the same as the one which follows from field theory. For example it is known that in the equation with the propagator (2.9) unphysical cuts appear although with far away branch points [22]. Obviously locality is not built into this theory. Therefore, we should not be astonished to find such deviations from the usual analytic structure.

IV. Comparison and Outlook

Finally we come to a comparison of the coupling constants obtained with the different methods discussed in this review. The coupling parameters in the first three lines should be comparable. Unfortunately, different pp and np scattering data and also different mesons for exchange have been used in the analysis. Nevertheless, all agree that 0^+ exchange is needed with a coupling of the order $g^2/4\pi \sim 1$ with an exchanged mass around 400 MeV. Second the ϱ has a Pauli-coupling of order $g_{2\varrho}^2/4\pi \sim 60$. Depending on the assumed or fitted ratio of Pauli to Dirac coupling (of order ~ 3) different values for $g_{1\varrho}^2/4\pi$ are obtained. We remark that because of the greater strength the Pauli-coupling is determined more accurately than the Dirac-coupling. The coupling of the ω seems to be larger than the Dirac part of the ϱ coupling. In the off-shell approach the off-shell effects produce additional attraction which must be cancelled by a somewhat larger ω coupling. In the Schrödinger theory most of the repulsion coming from the vector mesons is reduced by going to the nonrelativistic limit. This leads to a large ω coupling constant around $g_{1\,\omega}^2/4\pi \sim 20$. Apparently, the change is small for the ϱ because the Pauli-coupling is the more important one and acts quite differently in the numerous partial waves compared to the Dirac coupling.

From the discussion it is so far very difficult to judge which coupling constants are more reliable. As already remarked one should omit the results obtained with the Schrödinger equation. This approach has its value for nuclear physics where one wants to start from a presentation of nuclear forces which is in some sense derived from elementary particle properties. Furthermore, only the on-shell methods (line $2-8$ in Table 1) should be compared with each other. From these the work of Bugg presumably is the most promising. It should be repeated with newer input data and supplemented with information from finite-energy sum rules [30]. Up to now presumably all results in lines $2-8$ of Table 1 are of the same quality. Concerning off-shell approaches we consider the work of *Schierholz* as the most reliable one. As already mentioned, the results from on-and off-shell approaches should not be compared, before one has not checked, which contributions in the unphysical u-cut come from multi-meson exchanges. It may be that a more realistic presentation of the unphysical cut in the fixed-t dispersion relations is given by the sum of ladder diagrams rather than by the pole terms alone. In this case the off-shell methods would yield the correct coupling constants. Of course, a more conclusive method to obtain the values of the coupling constants would be the extrapolation to the poles [29]. For this purpose presumably much more accurate experimental data are needed than are presently available.

Acknowledgements

I wish to thank *G. Schierholz* for helpful discussions on this report.

References

1. More recent reviews of the same subject are:
   Breit, G., Haracz, R. D.: High Energy Physics, Vol. I (Ed. by E. H. S. Burhop.). New York: Academic Press 1967.
   Breit, G.: The Potential Concept in Nucleon-Nucleon Interactions and Limitations of Existing Models, preprint from State University of New York at Buffalo.
   Green, A. E. S., MacGregor, M. H., Wilson, R.: Rev. Mod. Phys. **39** 498 (1967).

2. Systematic compilations of high energy pp scattering data can be found in:
  Kanada, H., et al.: Suppl. Prog. Theor. Phys. **41** and **42** (1967).
  Alexander, G., et al.: Nucl. Phys. B **5**, 1 (1968), B **7**, 282. (1968).
  A complete set of reference to $p\bar{p}$ scattering data can be found in:
  Owen, D. P., Peterson, F. C., Orear, I., Read, A. L., Ryan, D. G., White, D. H.: Phys. Rev. **181**, 1794 (1969).
3. *Mac Gregor, M. H., Arndt, R. A., Wright, R. M.:* Phys. Rev. **182**, 1714 (1969); Phys. Rev. **169**, 1149 (1968); Phys. Rev. **173**, 1272 (1968) and earlier papers of this group cited in these references.
  Seamon, R. E., Friedman, K. A., Breit, G., Haracz, R. D., Holt, I. M., Prakash, A.: Phys. Rev. **165**, 1579 (1968);
  and earlier papers of the Yale group which can be found in:
  Breit, G.: Rev. Mod. Phys. **39**, 560 (1967);
  Kazarinow, Yu. M., Lehar, F., Janout, Z.: Rev. Mod. Phys. **39**, 571 (1967).
4. *Goldberger, M. L., Nambu, Y., Oehme, R.:* Ann Phys. (N. Y.) **2**, 226 (1957).
5. For masses and quantum numbers of these meson resonances see: Particle Data Group. Rev. Mod. Phys. **41** (1969).
6. *Pascual, P., Yndurain, F. I.:* Nuovo Cimento **61** A, 225 (1969).
7. See the review of *Höhler, G.:* Pion-Nucleon Scattering (to be published) for more up to date values of f_π^2.
8. *Bugg, D. V.:* Nucl. Phys. B **5**, 29 (1968).
9. *Perring, I.:* Rev. Mod. Phys. **39**, 550 (1967);
  Arndt, R. A., MacGregor, M. H.: Phys. Rev. **141**, 873 (1966).
10. *Köpp, G., Söding, P.:* Phys. Letters **23**, 494 (1966).
11. Midpop III (unpublished), similar phases as *Arndt, R. A., MacGregor, M. H.:* Phys. Rev. **141**, 873 (1966).
12. *Köpp, G.:* Z. Physik **191**, 273 (1966). See also:
  Köpp, G., Kramer, G.: Phys. Letters **19**, 593 (1965) for a comparison with higher symmetry schemes and *Köpp, G.:* Rev. Mod. Phys. **39**, 640 (1967) for the extension to np scattering.
13. *Arndt, R. A., Bryan, R. A., MacGregor, M. H.:* Phys. Letters **21**, 314 (1966).
14. *Cini, M., Fubini, S.:* Ann. Phys. (N. Y.) **3**, 352 (1960).
15. *Amati, D., Leader, E., Vitale, B.:* Phys. Rev. **130**, 750 (1963).
16. *Durso, I. W., Signell, P.:* Phys. Rev. **135**, B 1057 (1964),
  Durso, I. W.: Phys. Rev. **149**, 1234 (1966). See
  Sugawara, H., von Hippel, F.: Phys. Rev. **172**, 1764 (1968); **185**, 2046 (1969) for similar conclusions in a nonrelativistic potential approach.
17. *Schierholz, G.:* Nucl. Phys. B **7**, 483 (1968).
18. *Chew, G. F., Goldberger, M. L., Low, F., Nambu, Y.:* Phys. Rev. **106**, 1337 (1957)
19. *Scotti, A., Wong, D. Y:* Phys. Rev. **138**, B 145 (1965);
  Ball, I. S., Scotti, A., Wong, D. Y.: Phys. Rev. **142**, 1000 (1966);
  Similar work has been done by *Dosch, H. G., Müller, V. F.:* Nuovo Cimento **39**, 886 (1965).
20. *Wong, D. Y.:* Rev. Mod. Phys. **39**, 622 (1967).
21. *Logunov, A. A., Tavkhelidze, A. N.:* Nuovo Cimento **29**, 380 (1963).
22. *Blankenbecler, R., Sugar, R.:* Phys. Rev. **142**, 1051 (1966).
23. *Bakamjian, B., Thomas, L. H.:* Phys. Rev. **92**, 1300 (1953).
  Foldy, L. L.: Phys. Rev. **122**, 275 (1961).
  Jordan, T. F., MacFarlane, A. I., Sudarshan, E. C. G.: Phys. Rev. **133**, B 487 (1964).
  Fong, R., Sucher, J.: J. Math. Phys. **5**, 456 (1964);
  Coester, F.: Helv. Phys. Acta **38**, 7 (1965).
24. *Schierholz, G.:* Nucl. Phys. B **7**, 432 (1968).

25. — Nucl. Phys. B **7**, 483 (1968).
26. For possible ways to define potentials from field theory see for example:
Charap, J. M., Fubini, S. P.: Nuovo Cimento **14**, 540 (1959), **15**, 73 (1960).
27. *Schierholz, G.:* (to be published).
28. *Bryan, R. A., Dismukes, R. C., Ramsay, W.:* Nucl. Phys. **45**, 355 (1963);
Bryan, R. A., Scott, B. L.: Phys. Rev. **135**, B 434 (1964), **164**, 1215 (1967), **177**, 1435 (1969).
Green, A. E. S., Sawada, T.: Nucl. Phys. B **2**, 267 (1967);
Ueda, T., Green, A. E. S.: Phys. Rev. **174**, 1304 (1968);
Gersten, A., Green, A. E. S.: Phys. Rev. **176**, 1199 (1968);
Sawada, T.: Dainis, A., Green, A. E. S.: Phys. Rev. **177**, 1541 (1969).
29. See for example the discussion of such methods in:
Morgan, D., Pisut, I., Roos, M.: Low Energy Pion-Pion Scattering (in this volume).
30. *Dolen, R., Horn, D., Schmid, C.:* Phys. Rev. **166**, 1768 (1968).

Prof. Dr. *G. Kramer*
II. Institut für Theoretische Physik der Universität Hamburg
D-2000 Hamburg-Bahrenfeld
Luruper Chaussee 149

Coupling Constants from PCAC

H. PILKUHN

Contents

1. The $\Lambda\Sigma\pi$ Coupling Constants 168
2. The $\Sigma\Sigma\pi$ Coupling Constant. $SU(3)$-Symmetry Prefers pv-Coupling 170
3. The Kaon Coupling Constants 171
References 173

1. The $\Lambda\Sigma\pi$ Coupling Constants

The strong πNN and $\pi\Lambda\Sigma$ coupling constants can be obtained from the matrix elements of $n \rightarrow pe\bar{\nu}$ and $\Sigma^{\pm} \rightarrow \Lambda e^{\pm}\nu$ decays by means of the Goldberger-Treimann relation,

$$-g_{\pi} f_{AB\pi}(t) = \tfrac{1}{2}(4\pi)^{-\frac{1}{2}} m_{\pi} g_{A,AB}(t), \tag{1.1}$$

where $g_{\pi} = (1.057 \pm 0.002)\ 10^{-6}/\text{GeV}$ is the pion decay constant (frequently written as $2^{-1/2} f_{\pi} g_{\nu}$), $g_{A,AB}$ is the axial decay matrix element, and t is the momentum transfer. In neutron decay we have $t = 0$ and $g_{A,np}(0) \equiv g_A = (1.23 \pm 0.01)\ g_{\nu}$.

The function $f_{AB\pi}(t)$ is the pion-baryon vertex function, $f_{AB\pi} \equiv f_{AB\pi}(m_{\pi}^2)$ being the pseudovector coupling constant. One normally writes the vertex function as the coupling constant times a form factor,

$$f_{AB\pi}(t) = f_{AB\pi} \cdot F_{AB\pi}(t), \qquad F_{AB\pi}(m_{\pi}^2) \equiv 1. \tag{1.2}$$

The presence of the form factor in (1.1) may be questioned. Numerically, if one inserts $f^2 \equiv f_{NN\pi}^2 = (0.0815 \pm 0.0016)$ [1], one gets

$$F_{NN\pi}(0) = 0.917 \pm 0.016. \tag{1.3}$$

Thus, if $F_{NN\pi}(t)$ is omitted, $f_{NN\pi}$ comes out $\sim(8 \pm 2)\%$ too small, which is still quite satisfactory. It thus appears that the theoretical uncertainty introduced by the Goldberger-Treiman relation is less than 10%.

With the Goldberger-Treiman relation thus checked in the $NN\pi$ case, we now turn to the determination of $f_{\Lambda\Sigma\pi}$.

The axial decay matrix element of $\Sigma^{\pm} \rightarrow \Lambda e^{\pm}\nu$ decays is easily obtained under the assumption $g_{\nu,\Sigma\Lambda} = 0$ (CVC hypothesis). Experimentally, one has $g_v/g_A = -0.29 \pm 0.20$ in this case [2], but our confidence in CVC is of course better than that. With the $\Sigma^- \to \Lambda e \bar{v}$ decay width known to about 10% (the corresponding Σ^+ width is less well known), $g_{A,\Sigma\Lambda}$ is thus known to about 5%. Normally experimentalists do not give $g_{A,\Lambda\Sigma}$ directly but make instead a best fit to all leptonic decays, assuming the validity of the Cabibbo theory. They then quote the fraction of axial d-type coupling and the axial Cabibbo angle [2],

$$\alpha = g_A^D/(g_A^D + g_A^F) = g_A^D/g_A = 0.613 \pm 0.023, \qquad \theta_A = 0.250 \pm 0.018 \quad (1.4)$$

from which one obtains $g_{A,\Sigma\Lambda}$ as $2\alpha g_A/\sqrt{3}$. If $g_{A,\Sigma\Lambda}$ is determined from $\Sigma \to \Lambda ev$ decays alone, the most recent analysis (private communication by M. Roos) yields

$$\alpha \equiv \sqrt{\tfrac{3}{4}} g_{A,\Sigma\Lambda}/g_A = 0.62 \pm 0.03 . \tag{1.5}$$

This value must be used if one wants to determine $f_{\Lambda\Sigma\pi}$ from PCAC alone, independent of any $SU(3)$-assumptions.

The average value of t in $\Sigma^\pm \to \Lambda e^\pm v$ decays is much closer to 0 than to m_π^2. If we assume the same form factor as in the $\pi N N$ case i. e.

$$f_{\Lambda\Sigma\pi}(t) = f_{\Lambda\Sigma\pi} \cdot F_{NN\pi}(t), \tag{1.6}$$

then we obtain the simple relation

$$f_{\Lambda\Sigma\pi}^2 = \tfrac{4}{3} f^2 \alpha^2 = 0.0418 \pm 0.0041 , \tag{1.7}$$

where the form factor has dropped out. The equivalent value of the pseudoscalar coupling constant is

$$G_{\Lambda\Sigma\pi}^2/4\pi = \frac{1}{m_\pi^2} (m_\Sigma + m_\Lambda)^2 f_{\Lambda\Sigma\pi}^2 = 11.4 \pm 1.2 . \tag{1.8}$$

If we add a theoretical uncertainty of 8% for the Goldberger-Treiman relation, the error of $f_{\Lambda\Sigma\pi}^2$ is raised to about 20%, which is still relatively small. Let me however remind you that the inclusion of the t-dependence of the pion-baryon vertex function in the Goldberger-Treiman relation is not completely ad hoc. One may estimate $F_{NN\pi}(t)$ from a number of peripheral reactions such as $\pi^- p \to \varrho^0 n$ [3], $\gamma p \to \pi^+ n$ [4] or $np \to pn$ [5] and always finds $F_{\pi NN}(0) < 1$. If one parametrizes $F(t)$ in a form suggested by the p-wave penetration factor of potential theory [6],

$$F_{\Lambda B\pi}^2(t) = \frac{1 + R^2 [(m_A - m_B)^2 - m_\pi^2]}{1 + R^2 [(m_A - m_B)^2 - t]} \tag{1.9}$$

then (1.3) requires $R^2 = 7/\text{GeV}^2$ [7] $(R = 0.53 \text{ fermi})$ which is quite reasonable. It is much smaller than the value $R^2 = 30/\text{GeV}^2$

$(R = 1.06$ fermi) found in Ref. [3] and slightly larger than the values of Refs. [4] and [5]. It thus appears that the actual uncertainty of the Goldberger-Treiman relation is considerably smaller than 8%.

2. The $\Sigma\Sigma\pi$ Coupling Constants.
$SU(3)$-Symmetry Prefers pv-Coupling

The $\Sigma\Sigma\pi$ coupling constant cannot be computed from PCAC, since the decay $\Sigma^- \to \Sigma^0 e v$ has not yet been observed. However, once $f_{\Sigma\Sigma\pi}$ is known from strong interactions one can use PCAC to say something about the $SU(3)$-properties of weak interactions.

Probably the best way to determine $f_{\Sigma\Sigma\pi}^2$ is to write down a super-convergence relation for the amplitude $B(s, t)$ of the double charge exchange reaction $\pi^- \Sigma^+ \to \pi^+ \Sigma^-$ [8]

$$G_{\Sigma\Sigma\pi}^2 - G_{\Lambda\Sigma\pi}^2 = \int_{(m_\Lambda + m_\pi)^2}^{\infty} \operatorname{Im} B(s', t) \, d s', \qquad (2.1)$$

and to saturate it with the contributions from the known hyperon resonances. Updating the work of *Everett* [9], *J. Engels* and myself obtained after the Lund Conference

$$\frac{G_{\Sigma\Sigma\pi}^2}{4\pi} = \frac{G_{\Lambda\Sigma\pi}^2}{4\pi} + 1 \pm 0.8 = 12.4 \pm 1.7 . \qquad (2.2)$$

The error is probably quite unrealistic in this case, but the possibility that $G_{\Sigma\Sigma\pi}^2$ is much smaller than $G_{\Lambda\Sigma\pi}^2$ appears to be ruled out. This allows an important test of $SU(3)$-invariance. Since the hyperons are heavier than the nucleons, one first has to specify whether one assumes $SU(3)$-invariance for the pseudoscalar coupling constant $G_{AB\pi}$ or for the pseudovector ones $f_{AB\pi}$. In the first case, one has

$$G_{\Lambda\Sigma\pi} = \frac{2\alpha_{ps}}{\sqrt{3}} G , \qquad G_{\Sigma\Sigma\pi} = 2(1 - \alpha_{ps}) G . \qquad (2.3)$$

From the first equation, Eq. (1.8) and the value $G^2/4\pi = 14.7$ one finds α_{ps}, from the second one then finds $G_{\Sigma\Sigma\pi}$:

$$\alpha_{ps} = 0.762 \pm 0.038 \quad G_{\Sigma\Sigma\pi}^2/4\pi = 3.4 \pm 1.1 , \qquad (2.4)$$

i. e. the test fails. In the case of pseudovector coupling, one has

$$f_{\Lambda\Sigma\pi} = \frac{2\alpha_{pv}}{\sqrt{3}} f , \qquad f_{\Sigma\Sigma\pi} = 2(1 - \alpha_{pv}) f , \qquad (2.5)$$

and (1.7) gives α_{pv}:

$$\alpha_{pv} = \alpha = 0.62 \pm 0.03, \qquad f_{\Sigma\Sigma\pi}^2 = 0.0475 \pm 0.0075. \qquad (2.6)$$

The value of $G_{\Sigma\Sigma\pi}^2$ is now given by

$$G_{\Sigma\Sigma\pi}^2/4\pi = \frac{4}{m_\pi^2} m_\Sigma^2 f_{\Sigma\Sigma\pi}^2 = 13.8 \pm 2.2 \qquad (2.7)$$

and now the test works beautifully. (The increase by a factor of 4 is partly due to the increase in $(1 - \alpha)^2$ and partly to the extra factor $m_\Sigma^2/m_p^2 = 1.61$). We conclude that the pion-baryon coupling constants satisfy $SU(3)$-invariance, provided the pseudovector coupling is chosen. To reach this conclusion it is of course essential that the baryon mass ratios in the Goldberger-Treiman relation (i. e. no extra masses in the form (1.1)) are uniquely determined by PCAC, up to effects of the order of the pion mass. This is true for all versions of PCAC.

Let us now return to the weak axial coupling constants $g_{A,AB}$. Since the strong coupling constants $f_{\pi AB}$ satisfy $SU(3)$-invariance, PCAC tells us that the $g_{A,AB}$ satisfy $SU(3)$-invariance with the same value of α, i. e. $\alpha = \alpha_{pv}$. Although this is exactly what everybody expects, it is not trivial. If for example the pseudoscalar coupling constants G had satisfied $SU(3)$ invariance, then the combinations $g_{A,AB}/(m_A + m_B)$ would satisfy $SU(3)$-invariance, with $\alpha = \alpha_{ps}$. This modification of the Cabibbo theory would make no difference for the two decays considered so far, although it would give a somewhat poorer fit for the hypercharge changing Σ and Λ decays. To sum up, the Goldberger-Treiman relations for pion coupling constants give us valuable information both about the strong and about the weak interaction.

3. The Kaon Coupling Constants

With α determined by (1.4), the $SU(3)$ values of the $N\Lambda K$ and $N\Sigma K$ coupling constants would be

$$f_{N\Lambda K}^2 = \tfrac{1}{3}(3 - 2\alpha)^2 f^2 = 0.089, \qquad G_{N\Lambda K}^2/4\pi = 18.5, \qquad (3.1)$$
$$f_{N\Sigma K}^2 = (1 - 2\alpha)^2 f^2 = 0.004, \qquad G_{N\Sigma K}^2/4\pi = 1.0. \qquad (3.2)$$

Here we have inserted

$$G_{ABi}^2/4\pi = \frac{1}{m_\pi^2} (m_A + m_B)^2 f_{ABi}^2 \qquad (3.3)$$

according to the definition $\dfrac{1}{m_\pi} \sqrt{4\pi} f_{ABi} \bar{u}_B \gamma_5 \gamma_\mu (P_A - P_B)^\mu u_A$ of the pv coupling. Unlike m_A and m_B which come from the use of the Dirac equation, the factor $1/m_\pi$ has been inserted only to make the f's dimensionless (to be more precise, $m_\pi \equiv m_{\pi^+}$ is used). There is no reason to define the pv ABi coupling with $1/m_i$ and expect $SU(3)$ for the resulting f_{ABi}. If one does that, all G^2 for kaons would be reduced by a factor $m_\pi^2/m_k^2 = 1/12.5$, which is certainly not the case.

From dispersion relation analyses of $K^\pm N$ scattering, one finds relatively accurately the combination $(G^2_{NAK} + 0.84\, G^2_{N\Sigma K})/4\pi$, which appears to be smaller than 10 (see the review by *Martin*). It thus appears that $SU(3)$-invariance fails at least for G^2_{NAK} by a factor of 2. On the other hand, the qualitative success of the Cabibbo theory for hypercharge changing decays indicates that $SU(3)$-symmetry must be present in one way or another also for the couplings of hypercharged mesons.

At this stage, one could introduce $SU(3)$ breaking, for example by replacing the ad hoc mass m_π in (3.3) by axial vector meson masses. However, one can also require that the vertex functions satisfy $SU(3)$-invariance at *fixed t*. More precisely, the values (3.1) and (3.2) would apply to $f_{NYK}(t)$ in the interval $0 < t < (m_\Sigma - m_N)^2$ covered by hyperon decays, but not to the coupling constants $f_{NYK}(m_K^2)$. This is equivalent to postulating kaon PCAC with a t-independent decay constant g_K analogous to g_π in (1.1) but with a form factor $F_{NYK}(t)$ which decreases by a factor $2^{-1/2}$ from $t = 0$ to $t = m_K^2$. For $t \approx 0$ and negative t, $F_{NYK}(t)$ should have the same t-dependence as $F_{NN\pi}(t)$. As a simple generalization of Eq. (1.9), one could take

$$F^2_{ABi} = \frac{1 + R^2[(m_A - m_B)^2 - m_i^2] + \varepsilon R^4[(m_A - m_B)^2 - m_i^2]^2}{1 + R^2[(m_A - m_B)^2 - t] + \varepsilon R^4[(m_A - m_B)^2 - t]^2} \quad (3.4)$$

with $R^2 \approx 7/\text{GeV}^2$ as before, and $\varepsilon = 1.3$ to make $F^2_{NAK}(0) \approx 2$. One could also omit $(m_A - m_B)^2$ in the factor of ε, in which case $\varepsilon = 1$ would do the job. The form (3.4) corresponds to increasing powers of the decay momentum p:

$$1 + R^2 p^2 + \varepsilon R^4 p^4. \quad (3.5)$$

Thus the new term is important only when large variations of p^2 occur. For the η-baryon couplings, (3.4) would give

$$F^2_\eta(0) = 1 - R^2 m_\eta^2 + \varepsilon R^4 m_\eta^4 = 3.33 \quad (3.6)$$

for $\varepsilon = 1$, i. e. the coupling constants $f^2_{AA\eta}(m_\eta^2)$ should be only 0.3 of their $SU(3)$-values at $t = m_\pi^2$. For the $Y_1^*(1382) \bar{K} N$ coupling constant on the other hand, the new term is negligible, since p^2 is very small there.

To conclude, it is very easy to remove the apparent $SU(3)$-violation of kaon couplings. For the pion couplings to N, Λ and $\Sigma, SU(3)$-invariance is satisfied.

References

1. *Samaranayake, V. K., Woolcock, W. S.:* paper no. 116 submitted to the Lund Conference.
2. *Eisele, F., et al.:* Z. Physik **225**, 383 (1969).
3. *Wolf, G.:* Phys. Rev. **182**, 1538 (1969).
4. *Bender, I., Dosch, H. G., Rothe, H. J.:* Nuovo Cimento **62**, 1026 (1969).
5. *Geicke, J., Mütter, K. H.:* Phys. Rev. **184**, 1551 (1969).
   Mütter, K. H., Tränkle, E.: Phys. Rev. **184**, 1555 (1961).
6. *Dürr, H. P., Pilkuhn, H.:* Nuovo Cimento **40**, 899 (1965).
7. *Pilkuhn, H., Swoboda, A.:* Lett. Nuovo Cimento **1**, 854 (1969).
8. *Babu, P., Gilman, F. J., Suzuki, M.:* Phys. Letters **24** B, 65 (1967).
9. *Everett, A. E.:* Phys. Rev. **177**, 2561 (1969).

Prof. Dr. *H. Pilkuhn*
Institut für Theoretische Kernphysik der Universität
D-7500 Karlsruhe 1, Kaiserstraße 12

Regge Residues

C. MICHAEL

Contents

1. Introduction 174
2. Regge Theory 174
3. Regge Phenomenology 180
   A. Meson Exchange 180
   B. Baryon Exchange 186
4. Summary 190
Acknowledgements 190
References 190

1. Introduction

The Regge pole model forms a basis for the discussion of forward and backward high-energy scattering data, particularly for two body and quasi-two body reactions. We discuss exchange degeneracy and absorptive corrections (or Regge cuts), both of which give extra clarity to the Regge pole theory. The implications of $SU(3)$ and vector-meson dominance for Regge analyses are also presented. From such an approach, we seek to distil the quantities which are well defined and well determined in Regge analyses – the "constants of nature". We shall concentrate upon the ratios of the residue functions of the same trajectory to different amplitudes, that is, either to different external particles or to different spin amplitudes. In particular at $t = 0$ and $u = 0$, these quantities are well determined and are rather insensitive to modifications in the model used in data analysis. We concentrate on pseudoscalar meson $-1/2^+$ baryon scattering and present numerical values for such pole residues and ratios of pole residues. The theoretical models are confronted with such empirical values and we find excellent qualitative agreement. We analyze the F/D mixing parameters for the different couplings considered and discuss resulting predictions and tests.

2. Regge Theory

The idea of exploiting the angular momentum plane analyticity to introduce a moving pole with trajectory $\alpha(t)$ and residue $\beta(t)$ is well discussed elsewhere [1, 2]. The basic properties of such poles are:

(i) phase-energy dependence correlation through the signature factor which arises on introducing crossing;

(ii) factorization;

(iii) connection of $\alpha(t)$ and $\beta(t)$ with particle masses and couplings at $t = m^2$.

Note that helicity flip and non-flip amplitudes have the same energy dependence and phase. Some elegant tests of these ideas are:

(i) the phase of the forward $\pi^- p \to \pi^0 n$ amplitude from the optical theorem agrees with the ϱ Regge phase from the s dependence;

(ii) the ratio of flip to non-flip amplitude of the ϱ Regge pole is of comparable size for πN as for KN scattering;

(iii) a) when no known particle has the quantum numbers exchanged (exotic exchange) there should be a strong suppression of the cross-section as for $K^- p \to pK^-$ or $\pi^- p \to K^+ \Sigma^-$;

b) the $\alpha(t)$ measured from the s dependence of the differential cross-section as a function of t (shrinking) should extrapolate to the particle pole value as is found for $\pi^- p \to \pi^0 n$.

These points of evidence indicate that there are processes dominated by the exchange of a single or small number of Regge poles. However, many indications exist for additional singularities beyond those expected naively — such as:

(i) non-zero polarization in $\pi^- p \to \pi^0 n$;

(ii) ω cross-over non-factorization problem [3];

(iii) π conspiracy approach;

(iv) missing dips at $t = -0.6\,\mathrm{GeV}^2$ in $\pi N \to \omega N, \gamma p \to \eta p$ and at $u = -0.2\,\mathrm{GeV}^2$ in $\gamma N \to N\pi$.

In order to stem the tide of additional singularities, the Regge pole parametrization must be restricted to allow data fitting with a sensible set of parameters. The usual approach is to admit that the additional singularities are Regge cuts and to pretend to be able to calculate them, more or less, by the absorption-eikonal-Glauber-multiple scattering prescription. If one takes these essentially equivalent prescriptions seriously, the cuts can be calculated with no extra parameters and one may even save parameters since the pole residues then take on a simpler aspect. This goal of simplifying the pole residues can be taken in two directions, either by using exchange degeneracy, or by taking structureless pole residues and varying the cut strength to induce the structure by cancellation of the pole and cut. The latter prescription is known by the name of *Michigan* [4] and has the disadvantage that it has rather little predictive power. The former prescription [5, 6] has much predictive power, however, since it combines exchange degeneracy predictions with modifications due to absorption corrections.

Exchange degeneracy was introduced by *Arnold* [7] from potential theory arguments but it has only appeared in full bloom when related to the concept of duality [8]. Combined with the *Harari-Freund* ansatz [9], this means for the imaginary part of the amplitude; resonances dual to Regge pole exchange and diffractive (Pomeron P) exchange dual to non-resonant background. Consider a process with an s channel of exotic quantum numbers (i. e., not a $q\bar{q}$ or qqq state), then it should have a Regge exchange amplitude with no imaginary part since no resonances are present. The real part can be thought of as coming from cross channel (u) states and is non-zero of course. As is very well known, such a constraint can only be satisfied for all s and t by t channel Regge poles if there are degenerate and opposite signatured trajectories exchanged:

$$\beta^+\left(\frac{-1-e^{-i\pi\alpha_+}}{\sin\pi\alpha_+}\right)\left(\frac{s}{s_0}\right)^{\alpha_+} - \beta^-\left(\frac{+1-e^{-i\pi\alpha_-}}{\sin\pi\alpha_-}\right)\left(\frac{s}{s_0}\right)^{\alpha_-} = \frac{-2\beta}{\sin\pi\alpha}\left(\frac{s}{s_0}\right)^{\alpha} \quad (1)$$

$$= \text{Real.}$$

The conclusion that $\beta^+ = \beta^-$ and $\alpha^+ = \alpha^-$ for all t is very restrictive. It gives as a consequence a simple way to resolve the thorny problem of the type of zero in $\beta^+(t)$ or in $\beta^-(t)$ as α passes through $0, -1$, etc. Since at $\alpha = 0$ a finite amplitude implies $\beta^+ = 0$, then $\beta^- = 0$ also. This is called nonsense choosing and ensures that the contribution of odd signature poles of any spin structure is zero at $\alpha = 0$. In principle β could have a higher order zero but the essential point is that $\beta^+/\sin\pi\alpha^+$ must remain non-infinite. The consequences of exchange degeneracy for Regge poles provide a very nice simplification and in general give qualitative agreement with data. Some features definitely need modification, however; in particular the prediction that all odd signatured trajectories decouple at $\alpha = 0$. In the $\pi^- p \rightarrow \pi^0 n$ differential cross-section, one has only a dip not a zero here ($t \approx -0.6$); the $\pi^+ p$ and $\pi^- p$ differential cross-sections are not equal at $t = -0.6$ although there is a cross-over (equality) at $t \sim -0.2$ and similarly for $K^+ p$ relative to $K^- p$ and pp relative to $\bar{p}p$ the anticipated cross-over is moved from $t = -0.6$ to $t \sim -0.15$. These feature may be explained by absorptive corrections and indeed such corrections do contribute in the correct sense although they are frequently insufficient in magnitude. In particular the cross-over zero is typically moved from -0.6 only to -0.4 unless the cut [6] corrections are enhanced. Phenomenologically, the (exchange-degenerate poles plus absorptive cuts) model is the most attractive approach available although the cut corrections are indicative rather than quantitative as presently understood.

In order to satisfy the reality condition in exotic channels for different reactions, the exchange-degenerate trajectories are related to each other. From $\pi^+\pi^+$ scattering, ϱ and f_0 exchanges are degenerate; then $K^+\pi^+$

scattering fixes further the ϱ and f_0 couplings to $\bar{K}K$ as well as $\pi\pi$; from KK scattering the degeneracies ϱ with A_2, f_0 with ω and f_0' with ϕ result. Thus $f_0, \omega, A_2, \varrho$ lie on a common trajectory and have equal couplings $(M\pi^+\pi^-)$ to $\pi\pi$ and equal couplings (MK^+K^-) to $K\bar{K}$ with $(M\pi^+\pi^-) = 2(MK^+K^-)$. One finds ϕ and f' together on a lower trajectory and $K_V^*(1^-)$ and $K_T^*(2^+)$ together on a third trajectory. This set of three separated trajectories for the vector V and tensor T mesons presents an attractive picture [10] for broken $SU(3)$ which includes also quark

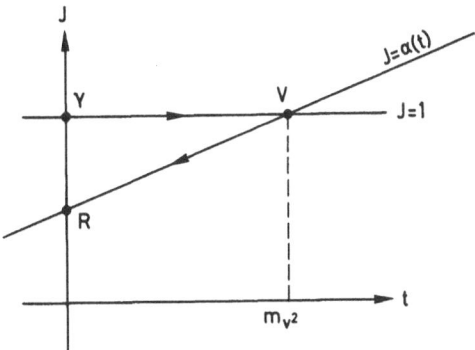

Fig. 1. *Chew-Frautschi* plot showing, for $B\bar{B}$ vector coupling, the extrapolation from real photon $\gamma(J = 1, t = 0)$ to vector meson $V(J = 1, t = m_V^2)$ and thence along the Regge trajectory to the scattering region $R(J = \alpha(t), t = 0)$.

model results — for instance the $\omega - \phi$ and $f' - f_0$ mixing is as in the quark model. In an explicit Veneziano model these relationships are maintained and one can further obtain $\alpha(m_\pi^2) = \frac{1}{2}$ for the $f_0 - \omega - \varrho - A_2$ trajectory [11]. Analysis of other meson-meson scattering processes is most easily exploited in an $SU(3)$ symmetric approach and leads [12] to nonets of axial mesons degenerate with nonets of pseudoscalar P mesons. Rejecting $VV \to VV$ constraints and keeping $PV \to PV$ yields [13] a more acceptable octet structure for the P mesons. The $V - T$ meson couplings in $MB \to MB$ and $BB \to BB$ have been studied by *Rosner* [14] using factorization and the usual exotic suppression. If the $f_0'\bar{N}N$ and $\phi\bar{N}N$ couplings are required to be zero, this leads [15] to $(f_0\bar{p}p) = (\omega\bar{p}p) = 3F - D$ and $(A_2\bar{p}p) = (\varrho\bar{p}p) = F + D$. The F/D ratio is not fixed and may be different for the two different spin couplings of these mesons to baryons. From a Veneziano-style approach one can go further and a value of $F/D = 1/3$ was predicted [16] for spin coupling corresponding to the A invariant amplitude in $PB \to PB$.

Vector dominance also leads to indications for the $V - T$ meson coupling to $\bar{B}B$. As shown in Fig. 1, extrapolations from $\gamma \to V \to R$ give information relevant to Regge pole exchange. For meson-baryon elastic scattering the relationship between the invariant amplitudes and quantities at γ and V is

$$A = -\gamma h$$

$$vB = \gamma(g + h)$$

$$(1 - t/4m^2)\, A' = (1 - t/4m^2)\, A + vB$$

$$= \gamma(g + th/4m^2)$$

where at γ; $g \to F_1$ the charge form factor and $h \to F_2$ the anomalous magnetic moment form factor. Thus $vB/A' = (g + h)/g = 4.7$ for $\varrho N\bar{N}$ and 0.9 for $\omega N\bar{N}$, where ϱ and ω denote isovector and isoscalar respectively. Furthermore $SU(6)$ implies $D/F = 0$ for F_1 and $F/D = 2/3$ for F_2 so that $D = 0$ for A'; $F/D = 2/3$ for vB and $F/D = (2(4.7) - 5)/3(4.7) = 0.31$ for A. At V, the t channel helicity flip and non-flip amplitudes are A' and vB, and g and h are related to the vector-spin and tensor-spin $B\bar{B}$ couplings of the vector-meson while γ contains the $\bar{M}M$ coupling. Note that $m_V^2/4m^2 \sim 0.17$. For the extrapolation from V to R, the A and B amplitudes enter naturally in a Veneziano approach. Since for ϱ, A' results from a large cancellation between A and B, it may be expected to be modified relatively more than for the ω. Thus vB/A' for ϱ and F/D for A' may be shifted from the predictions based on vector-meson dominance.

For the tensor mesons also, we shall quote coupling ratios to the t channel helicity amplitudes A' and vB and these may be related to suitably defined coupling constants.

For baryon spectra and couplings the exchange degeneracy scheme lends some clarity to the dense baryon jungle. For $PB \to PB$ scattering the conditions of eliminating imaginary parts in t and u exotic channels leads to the following two simple and independent exchange degenerate sequences [17]

$$8(F/D = 1)\ 1/2^+, 5/2^+, \ldots \Leftrightarrow 1 \oplus 8(F/D = 1)\ 3/2^-, 7/2^-, \ldots$$

$$10 \oplus 8(F/D = -1/3)\ 3/2^+, 7/2^+, \ldots \Leftrightarrow 8(F/D = -1/3)\ 5/2^-, 9/2^-, \ldots$$

These spectra are exhibited in Fig. 2.

We employ F/D rather than $\alpha = D/(F + D)$ to avoid confusion with the Regge intercept α; note $f = F/(F + D)$ is also a useful ratio. The F/D values are in broad agreement with experimental resonance decay analyses and by adding a $10(3/2^-, \ldots)$ one can perturb somewhat these ratios. Unwanted states of $10(1/2^+, 5/2^+, \ldots)$ coupled only to decuplet baryon channels and $8(3/2^+, 7/2^+, \ldots)$ coupled predominantly to such channels are present. It has been suggested to remove these states completely by deleting the t channel constraints [13]. No satisfactory broken symmetry-mixing approach for the Y states exists. The absence of a $1/2^-$ state requires the Igi zero at $\alpha = +1/2$ on the exchange degenerate

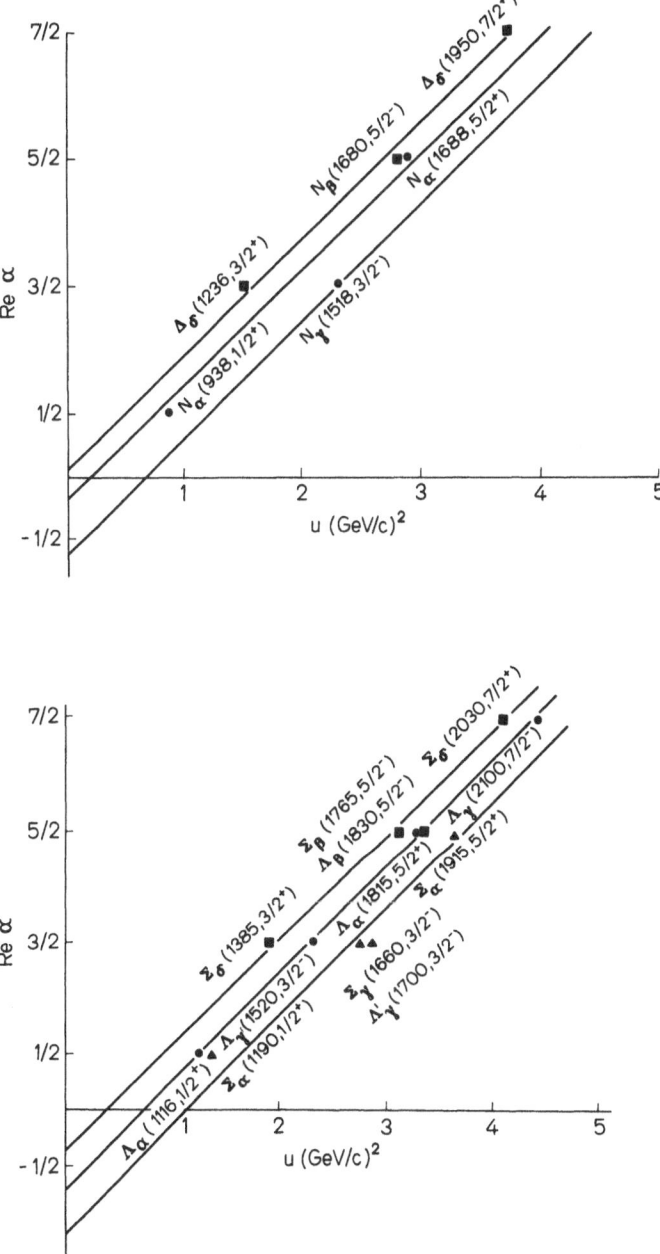

Fig. 2. Exchange degeneracy patterns for baryon spectra. The $\delta - \beta$ and $\alpha - \gamma$ sequences are supposed to be two independent degenerate trajectories.

12*

$\Delta(3/2^+, \ldots)$ trajectory. The relative couplings among the trajectories are predicted also [17] and will be useful in analyzing backward scattering data subsequently.

For baryon-baryon scattering the situation is bad [14] since the absence of dibaryon resonances implies exotic mesons coupled to $\bar{B}B$. *Rosner* has suggested [14] that such ($qq\bar{q}\bar{q}$) mesons might exist and do not couple to meson channels; otherwise cuts might be important. An easy way to visualize some of the difficulties is to consider $\bar{p}p \to \bar{\Xi}\Xi$ which has two exotic channels and thus the $\varrho, A_2, \omega, f_0$ mesons in the third channel must decouple from the $\bar{\Xi}\Xi$ vertex which is unreasonable.

One should mention here the duality diagrams [18] which are an elegant way of representing the $SU(3)$ crossing matrix arguments which underlie the above discussions.

3. Regge Phenomenology

A. Meson Exchange

We shall concentrate upon $PB \to PB$ scattering since this process has only two spin amplitudes and has good data available on cross-sections and polarization. The meson trajectories exchanged in the forward direction are the vector mesons $\varrho\omega K_V^* \phi$ and tensor mesons $A_2 f_0 K_T^* f'$. For other processes such as $PB \to VB$, etc., one has unnatural parity exchanges (π, B, A_1, etc.) in addition and, because these are on lower lying trajectories, they are more obscure to analyze. Problems also arise from the decoupling of the higher lying natural-parity vector and tensor trajectories in the forward direction for $PB \to VB$.

Returning to $PB \to PB$, we first discuss the contributions of the non-strange mesons to the non-flip amplitude A' at $t = 0$. The uniform parametrization we shall use is, for the Regge contribution,

$$\text{Im } A' = \beta \left(\frac{s - u}{2} \right)^\alpha = \beta(2\bar{m}v)^\alpha \tag{2}$$

where $\bar{m}$ is the mean baryon mass. The real part of A' is then found from the appropriate signature factor. The total cross-section $\sigma_{\text{tot}} = \text{Im } A'/k$ where k is the lab. momentum and $k \approx v = (s - u)/4\bar{m} \approx s/2\bar{m}$ at high energies.

A classic analysis was performed by *Barger et al.* [19] using total cross-sections, forward differential cross-sections and Coulomb inter-ference real/imaginary ratios. They fitted the trajectory intercepts α and residues β for P, f_0, ω, A_2 and ϱ to this data and the results are shown in Table 1. For ease of comparison the residues β have been shown as a product with one factor arising from the SU_3-exchange degenerate-universality limit. Many authors [20–24] have considered the four charge exchange processes $\pi^- p \to \pi^0 n$, $\pi^- p \to \eta^0 n$, $K^- p \to \bar{K}^0 n$ and $K^+ n \to K^0 p$ which have ϱ and A_2 exchanges. At $t=0$ one obtains the ratios of non-flip couplings and typical values [24] are

$$\frac{(\pi^- \pi^0)\,\varrho}{(\kappa^- \kappa^0)\,\varrho} = 0.84, \qquad \frac{(\pi^- \eta^0)\,A_2}{(\kappa^- \kappa^0)\,A_2} = 1.02.$$

These values agree with those in Table 1.

In order to compare residues of different Regge poles (i.e., ϱ with ω), it is important that the poles should have the same intercept α or else an energy v must be specified at which to compare βs^α. The β values in the Tables are implicitly evaluated at $s=1$ GeV, whereas v of 6 to 10 GeV would be more relevant in comparing the data since it is geometrically midway between a lower limit of 3 GeV where resonance structure appears and an upper limit of 30 GeV where possible cut effects modify poles as shown by the Serpukhov total cross-section data [25]. In order to resolve this question, we present a fit from *Berger* and *Fox* [26] which imposed equal intercepts of 0.5 for the mesons and of course 1.0 for the P. This is shown in Table 2. They present other fits also which allows some impression of the errors involved.

From the numbers in Tables 1 and 2, one sees that the theoretical ideas are an excellent first approximation since the quantities in brackets are all comparable. The most noticeable divergences are explained by taking $D_V \neq 0$ and $D_T \neq 0$. Then from Table 2, we relate

$$\frac{(K^- K^+)\,f_0(\bar{p}p)}{(K^- K^+)\,A_2(\bar{p}p)} = 4.8 = \frac{3F_T - D_T}{F_T + D_T} \tag{3}$$

$$\frac{(K^- K^+)\,\omega(\bar{p}p)}{(K^- K^+)\,\varrho(\bar{p}p)} = 3.5 = \frac{3F_V - D_V}{F_V + D_V} \tag{4}$$

and

$$\frac{(\pi^- \pi^+)\,f_0(\bar{p}p)}{(\pi^- \pi^+)\,\varrho(\bar{p}p)} = 4.7 = \frac{3F - D}{F + D} \tag{5}$$

These result in $D/F \approx -0.2$ for both vector and tensor mesons for consistency. A value of $D/F \sim -0.5$ had been obtained earlier by *Barger* and *Olsson* [27] but with unequal meson intercepts and comparing residues at $s=1$. For consistency with factorization, one also requires

$$2(\bar{p}p)\,f_0(\bar{p}p)/(\pi^- \pi^+)\,f_0(\bar{p}p) \approx (\bar{p}p)\,\omega(\bar{p}p)/(K^- K^+)\,\omega(\bar{p}p) \approx \frac{3F - D}{F + D} \text{ which}$$

gives $D/F \sim -0.3$ for the f_0 and $\lesssim 0$ for the ω. Generally the conclusion

Table 1. *Regge fit from Barger et al.* [17] *to* πN, KN, $\bar{K}N$, NN, *and* $\pi N \to \eta N$ *data at* $t = 0$. β *is defined as in Eq.* (2) *and is appropriate for exchange in the reaction* $M_1 B_1 \to M_2 B_2$ *if the t-channel coupling* $(M_1 \bar{M}_2)(\bar{B}_1 B_2)$ *is indicated in the table. The factors of* β *in brackets should all be equal in an exchange degenerate* $- SU(3) -$ *universality limit.*

Trajectory intercept α	Residue β in GeV^{-1}	
$\alpha_P = 1$	$(\pi^- \pi^+)(\bar{p}p)$	$29.0 \pm \ 0.4$
	$(K^- K^+)(\bar{p}p)$	$23.8 \pm \ 0.5$
	$(\bar{p}p)(\bar{p}p)$	$49.2 \pm \ 1.0$
$\alpha_{f_0} = 0.51 \pm 0.03$	$(\pi^- \pi^+)(\bar{p}p)$	$(27.2 \pm \ 1)$
	$(K^- K^+)(\bar{p}p)$	$(23.8 \pm \ 4)/2$
	$(\bar{p}p)(\bar{p}p)$	$(40.5 \pm \ 2)\,1.5$
$\alpha_\omega = 0.38 \pm 0.04$	$(K^- K^+)(\bar{p}p)$	$(34 \ \ \pm \ 4)/2$
	$(\bar{p}p)(\bar{p}p)$	$(38 \ \ \pm \ 5)\,1.5$
$\alpha_{A_2} = 0.34 \pm 0.03$	$(\pi^0 \eta)(\bar{p}p)$	$(21.4 \pm \ 0.4)/3\sqrt{3}$
	$(K^- K^+)(\bar{p}p)$	$(23 \ \ \pm \ 3)/6$
	$(\bar{p}p)(\bar{p}p)$	$\left(0 \begin{array}{c} +23 \\ -0 \end{array}\right)/6$
$\alpha_\varrho = 0.54 \pm 0.01$	$(\pi^- \pi^+)(\bar{p}p)$	$(14.7 \pm \ 0.8)/3$
	$(K^- K^+)(\bar{p}p)$	$(16.0 \pm \ 1.3)/6$
	$(\bar{p}p)(\bar{p}p)$	$\left(0 \begin{array}{c} +12 \\ -0 \end{array}\right)/6$

Table 2. *Regge fit from Berger and Fox* [26] *to* πN, KN, $\bar{K}N$, *and* NN *data at* $t = 0$. *Explanation as for Table 1. Note that* $\sigma \sim s^{\alpha - 1}$ *rather than* $(2\bar{m}v)^\alpha / k$ *was used, which makes a difference at lower energies.*

Trajectory intercept α	Residue β in GeV^{-1}	
$\alpha_{f_0} = 0.5$	$(\pi^- \pi^+)(\bar{p}p)$	(27.3)
	$(K^- K^+)(\bar{p}p)$	$(23.6)/2$
	$(\bar{p}p)(\bar{p}p)$	$(34.5)\,1.5$
$\alpha_\omega = 0.5$	$(K^- K^+)(\bar{p}p)$	$(24.2)/2$
	$(\bar{p}p)(\bar{p}p)$	$(24.8)\,1.5$
$\alpha_{A_2} = 0.5$	$(K^- K^+)(\bar{p}p)$	$(14.7)/6$
$\alpha_\varrho = 0.5$	$(\pi^- \pi^+)(\bar{p}p)$	$(17.3)/3$
	$(K^- K^+)(\bar{p}p)$	$(21.0)/6$

is that D/F is small and negative around -0.2 for V and T exchanges, although it may be nearer to zero for V exchange than for T exchange.

The intercepts α are not well known since conflicting evidence is usually available. For the best studied case of the ϱ, the classical fits [28] to $\pi^- p \to \pi^0 n$ energy dependence give $\alpha(0) = 0.57$ with small error. Requiring ϱ and A_2 to have the same trajectory intercept, *Jackson* [20] found $\alpha = 0.49$ from $\pi^- p \to \pi^0 n$, $\pi^- p \to \eta^0 n$ and $K^- p \to \bar{K}^0 n$. Adding absorptive cuts makes also a difference to the intercept and changes it [29] to $\alpha \sim 0.48$ to fit $\pi^- p \to \pi^0 n$. These latter values are close to *Lovelace*'s [11] theoretical value of 0.483.

The extension to $t \sim -1$ or $-2\ \text{GeV}^2$ of Regge fits has been made by many workers. However, the parametrization of the t dependence of the residue function introduces biases into the fit. Fits at each t value independently are needed to be objective. A more serious drawback is the need to introduce additional contributions in the form of secondary poles or cuts to explain features of the data. We feel that very few results in this field are stable against future developments and we concentrate on one particular ratio only, the ratio vB/A' of spin flip to non-flip at $t = 0$. Without experiments on R and A parameters (analysis of recoil polarization after scattering from a polarized target), vB/A' must be determined from a Regge analysis by fitting $d\sigma/dt$ and polarization P for $t > 0$.

For the famous quartet of reactions $\pi^- p \to \pi^0 n$, $\pi^- p \to \eta^0 n$, $K^- p \to \bar{K}^0 n$ and $K^+ n \to K^0 p$, the fits [21–24] are consistent with $SU(3)$ at the meson vertex, factorization and exchange degeneracy for $t > -1\ \text{GeV}^2$ and the same is true [23] for the reactions with Δ instead of N in the final state. For the A_2 this throws a rather oblique light on the "split $-A_2$", $-$ Regge does not need any special A_2 trajectory. Exchange degeneracy predicts a zero at $\alpha(t) = 0$ in $\pi^- p \to \pi^0 n$ which must be filled by cuts in any quantitative fit. Also K^+ and K^- charge-exchange are not equal at 2.3 and 3.0 GeV although there is evidence that the exchange-degeneracy prediction of equality is valid at 5.5 GeV [30]. Cuts may explain such a discrepancy also [30].

The values of vB/A' from such classical Regge fits are given in Table 3 [24, 28, 29, 31–34]. We also quote results from πN and KN finite energy sum rules (FESR) as well as for combined Regge pole fits to data and FESR. The values for ϱ and A_2 are large and are well determined [28, 32] from the forward dip in the charge-exchange differential cross-sections. For ω and f_0, the ratio vB/A' cannot be obtained from a pure Regge fit [35] without either FESR to relate to phase shifts or R and A measurements. The latter exist for πN [36] and check with the FESR Regge prediction [31]. For KN the FESR are less precise [34] and the Regge fit of *Dass et al.* [24] which employed FESR did not use the polarization data which have since become available. A fit to $K^\pm p$ cross-section and polarization data by *Daum et al.* [35] obtains vB/A' equal at $t \to 0$ for $(\varrho + \omega)$ and for $(f_0 + A_2)$ $-$ these combinations are dominated by the ϱ and A_2 spin flips, however. Note that for the P singularity, the ratio $vB/A' = +1$ corresponds to the s channel helicity flip amplitude being zero asymptotically, and such a theoretical picture has been advocated

Table 3. *Sample determinations of the spin flip to non-flip ratio* vB/A' *at the nucleon vertex for meson exchange at* $t = 0$.

Trajectory	vB/A' at $t = 0$	Comment
P	~1	πN FESR [33]
	0.6	$P + f_0 + \pi N$ FESR [31]
	0.8	$P + f_0 + P'' + \pi N$ FESR [31]
ϱ	11	Classical ϱ fit [28]
	11.6	$\varrho + \varrho' + \pi N$ FESR [31]
	7−8	$\varrho + $ cuts $+ \pi N$ FESR [29]
A_2	±9	Regge fit to $\pi^- p \to \eta^0 n$ [32]
	7−9	Regge fit $+ KN$ FESR [24]
f_0	~1	πN FESR [33]
	1.1	$P + f_0 + \pi N$ FESR [31]
	0.6	$P + f_0 + P'' + \pi N$ FESR [31]
ω	1−3	KN FESR [34]
	0.4	Regge fit $+ KN$ FESR [24]

Table 4. *Regge fit from Reeder and Sarma* [21] *to* $\bar{K}N \to \pi\Lambda, \pi\Sigma; \pi N \to K\Lambda, K\Sigma$ *data. The* $t = 0$ *nonflip residue and intercept are presented as for Table 1.*

Trajectory intercept α	Residue β in GeV^{-1}	
$\alpha_{K_V^*} = 0.35$	$(\pi^- \bar{K}^0)(\bar{p}\Lambda)$	$(-16)/\sqrt{6}$
	$(\pi^+ K^-)(\bar{p}\Sigma^+)$	$(-17)/3$
$\alpha_{K_T^*} = 0.24$	$(\pi^- \bar{K}^0)(\bar{p}\Lambda)$	$(10)/\sqrt{6}$
	$(\pi^+ K^-)(\bar{p}\Sigma^+)$	$(18)/3$

in general [37], supported in part by a similar observation of *s* channel helicity non-flip dominance in the photoproduction of ϱ mesons. For ω and ϱ the values may be compared with the vector dominance ratios and for the ϱ a ratio of 8–11 is found at R compared to 4.7 at γ. Both this vB/A' ratio and the F/D ratio for A' are consistent with a common explanation, namely the A'_ϱ contribution, which is anyway small, decreased further in passing from γ to R so that vB/A' increased from 5 to 8 and the coupling ratio $\varrho NN/\omega NN$ to A' decreased from 1/3 to 0.7/3.

For the $K_V^*(1^-)$ and $K_T^*(2^+)$ trajectories we can obtain information from data on $\bar{K}N \to \pi\Lambda, \pi\Sigma$ and $\pi N \to K\Lambda, K\Sigma$ reactions. *Reeder and Sarma* [21] performed such a fit and obtained the results shown in Table 4 for the non-flip amplitude A'. The sign of the vector amplitude is opposite to exchange-degeneracy expectations but this could be reversed without altering the fit unduly. The ratio of $(\bar{p}\Lambda)$ to $(\bar{p}\Sigma^+)$ is $(3F+D)/(F-D)\sqrt{6}$ so that $D/F \approx -0.1$ for V and -0.5 for T if universality $(D=0)$ is broken. Since the trajectory intercepts are different for the K^*'s and the non-strange mesons, any comparison needs a model for the t dependence of the residue functions — a Veneziano model has $\beta_0 s^{\alpha(t)} \sim \beta_0 (2mv)^{\alpha(t)}$ so $s=1$ GeV^2 is relevant for comparing β_0. With such a value the residues of Tables 2 and 4 are indeed comparable. The ratios of vB/A' from *Reeder* and *Sarma* [21] are unreliable since they had very poor polarization data available and they used an interference model to explain some features. The lack of any pronounced dip at $t=0$ means that $|vB/A'|$ is smaller than for the ϱ and A_2. *Barloutaud et al.* [38] also analyzed similar reactions at 3 GeV, but they assumed spin flip dominance of all charge and hypercharge exchange reactions which is unreasonable. We know that spin flip and non-flip amplitudes have different $SU(3)$ mixing ratios F/D a priori.

From Table 3 we can estimate the $SU(3)$ F/D ratio for the amplitudes A and B and then predict the ratios for K^* exchange. To fix ideas, we take $vB/A' = +1$ for f_0 and ω which corresponds to decoupling asymptotically of the A amplitude. Then $F/D = 1/3$ for the A amplitude coupling to $\bar{B}B$. For the B amplitude we evaluate from Tables 2, 3

$$\frac{(\bar{p}p)\,B_\varrho}{(\bar{p}p)\,B_\omega} = \frac{(\bar{p}p)_v\,B_\varrho}{(\bar{p}p)\,A'_\varrho} \cdot \frac{(\bar{p}p)\,A'_\varrho}{(\bar{p}p)\,A'_\omega} \cdot \frac{(\bar{p}p)\,A'_\omega}{(\bar{p}p)\,vB_\omega} = \frac{F+D}{3F-D} \tag{6}$$

$$= 10 \times \left(\frac{21.0}{3 \times 24.2}\right) \times 1 = 2.9 \,.$$

Thus $F/D \sim 1/2$ for B compared with the vector-meson expectation of $F/D \sim 2/3$ from $SU(6)$. Given general values of

$$R_\varrho = \frac{(\bar{p}p)\,vB_\varrho}{(\bar{p}p)\,A'_\varrho}$$

and of

$$R_\omega = \frac{(\bar{p}p)\,vB_\omega}{(\bar{p}p)\,A'_\omega}$$

one may estimate the $SU(3)$ mixing for B as

$$\frac{F}{D} = \frac{R_\varrho + 4R_\omega}{3R_\varrho - 4R_\omega} \tag{7}$$

where we have assumed $D/F = -0.2$ for the A' amplitude. Furthermore the predictions of $SU(3)$ for $R_A = (\bar{p}\Lambda)\,vB_{K^*}/(\bar{p}\Lambda)\,A'_{K^*}$ and

$$R_\Sigma = (\bar{p}\Sigma)\,vB_{K^*}/(\bar{p}\Sigma)\,A'_{K^*}$$

may be expressed

$$R_A = (4R_\omega + 3R_\varrho)/1 \sim 5$$
$$R_\Sigma = (4R_\omega - R_\varrho)/3 \sim -2$$

where the numerical values are obtained with $R_\omega = 1$, $R_\varrho = 10$ as previously. These ratios R_A and R_Σ are smaller than R_ϱ as needed to avoid a forward dip in $\pi N \to K Y$ and $\bar{K} N \to \pi Y$. Analyses employing polarization data and, more specifically, R and A parameter data can provide a test of such predictions for K^* exchange processes. Indeed, data indicate that the polarization has opposite sign between $\pi p \to K \Sigma$ and $\pi p \to K \Lambda$ and between $\bar{K} N \to \pi \Sigma$ and $\bar{K} N \to \pi \Lambda$ as required since R_Σ and R_A have opposite sign.

B. Baryon Exchange

As for the mesons we shall concentrate on $PB \to BP$ processes. Let us first review the present knowledge about πN backward scattering and then pass to discuss the ratios of N_α and Δ_δ Regge pole couplings in different processes, and possibilities to investigate other interesting trajectory coupling ratios.

Data exist for $\pi^- p \to p \pi^-$, $\pi^+ p \to p \pi^+$ and $\pi^- p \to n \pi^0$ differential cross-sections at high energies. The qualitative features are in accord with a strong N_α exchange having a zero at $u = -0.15 \text{ GeV}^2$ and a weaker and structureless Δ_δ exchange. Many fits have been made with these two Regge poles and a good description of the high energy data results [2, 26]. The one difficulty which remains is the charge-exchange data — at 6 GeV three experiments are in disagreement about the structure of the dip but they agree substantially for the backward point. Then from the isospin decomposition

$$\pi^- p \to p \pi^- = \Delta$$
$$\pi^+ p \to p \pi^+ = \frac{2}{3} N + \frac{1}{3} \Delta \qquad (8)$$
$$\pi^- p \to n \pi^0 = \frac{\sqrt{2}}{3} N - \frac{\sqrt{2}}{3} \Delta$$

the sign of the $N - \Delta$ interference term can be determined. The data [39] imply a relative destructive interference for $\pi^- p \to n \pi^0$ which corresponds to $\beta_N/\beta_A > 0$ where we define at $u = 0$

$$\operatorname{Im} \tilde{A} = \operatorname{Im}(A_s + m B_s) = \beta \left(\frac{s-t}{2} \right)^{\alpha - \frac{1}{2}} \qquad (9)$$

as the non-flip amplitude (recall that, near $u = 0$, the cross-section has asymptotically the form $d\sigma/du \sim |\tilde{A}|^2 - u|B|^2$). The real part of $\tilde{A}$ can

Table 5. *Regge fits to non-flip amplitude $\tilde{A}$ parametrized as $\mathrm{Im}\,\tilde{A} = \beta\,[(s-t)/2]^{\alpha-\frac{1}{2}}$ at $u=0$. The real parts are obtainable from the appropriate signature factors. The couplings shown are u-channel couplings $(\bar{M}_2 B_1)(B_2 \bar{M}_1)$ corresponding to backward exchange in $M_1 B_1 \rightarrow B_2 M_2$ scattering.*

Trajectory intercept	Residue β in GeV^{-1}		$SU(3)$ prediction	Reference
$\alpha_N = -0.38$	$(\pi N)_{\frac{1}{2}}(N\pi)_{\frac{1}{2}}$	-52.4		2
$\alpha_\Delta = 0.20$	$(\pi^+ p)(p\pi^+)$	$\mp\ 4.1$		2
α_N	$\left\lvert\dfrac{(K^+\Lambda)(n\pi^+)}{(\pi^0 p)(p\pi^-)}\right\rvert$	1.01 ± 0.2	$\dfrac{3F+D}{\sqrt{3}(F+D)}$	46
α_N	$\left\lvert\dfrac{(K^+\Sigma^0)(n\pi^+)}{(\pi^0 p)(n\pi^+)}\right\rvert$	$1.2??$	$\dfrac{F-D}{F+D}$	48
α_N	$\left\lvert\dfrac{(\eta^0 p)(n\pi^+)}{(\pi^0 p)(n\pi^+)}\right\rvert$	0.4 ± 0.1	$\dfrac{3F-D}{\sqrt{3}(F+D)}$	49
α_Δ	$\left\lvert\dfrac{(K^+\Sigma^+)(p\pi^+)}{(\pi^+ p)(p\pi^+)}\right\rvert$	~ 1	1	48,2
$\alpha_{\Delta\alpha} = \alpha_{\Delta\gamma} = -0.72$	$(K^- p)(pK^-)$	$\mp 41, \pm 41$		50

be found using the signature factors $(1 \pm e^{-i\pi(\alpha-\frac{1}{2})})$ for N, Δ respectively. However, the preliminary polarization data [40] at 2.5 to 3.75 GeV give a positive polarization for $\pi^+ p \rightarrow p\pi^+$ near $u=0$ which corresponds to $\beta_N/\beta_\Delta < 0$ using the signs of $MB/\tilde{A}$ to be discussed later. Theoretically a simple zero in the Δ residue at $\alpha = \frac{1}{2}$ is expected [17] from exchange degeneracy since it corresponds to the suppression of the particle pole at $\alpha = \frac{1}{2}$ on the degenerate N_β trajectory. Then from the residues at the particle poles one expects $\beta_N/\beta_\Delta < 0$ which agrees with the polarization but not with the charge-exchange data. Such a zero in the Δ residue is also necessary to have a reasonable chance of extrapolating from the magnitude [41] of the Δ residue (the width) to the magnitude of the backward $\pi^- p$ differential cross-section.

At lower energies the lack of a zero in the polarization at $\alpha_N = -\frac{1}{2}$ and the progressive filling in of the dip in $\pi^+ p \rightarrow p\pi^+$ indicate the presence of additional singularities. The $\alpha_N = -\frac{5}{2}$ and $\alpha_\Delta = -\frac{3}{2}$ dips are also absent empirically. The N_γ and N_β Regge poles [26], absorption cuts [42] and kinematic cuts [43] have each been discussed in this context. These contributions make a significant difference to the interpretation of u dependent features of the cross-section but should leave the backward (or effectively $u=0$) point least modified. In Table 5 we give results for the magnitudes of the residues found in πN Regge fits. The interpretation of the Δ_δ residue sign was given above.

The ratio of spin flip to non-flip $MB/\tilde{A}$ is fixed in most Regge fits by the requirement of suppressing the lowest *MacDowell* symmetric parity-doublet state on the trajectory. This leads to $MB/\tilde{A} = +1$ for N_α and $-(M/M_\Delta)$ for Δ respectively. A u dependence of this ratio may also be incorporated. *Barger* and *Cline* [2] fitted this ratio from data to find 1.6 for N_α and -1.1 for Δ. Note that one may suppress several such states simultaneously in the residue but that the limit of such a series needs careful treatment [44]. Other approaches to parity doublet suppression are: (i) to distort the trajectory from linear so as to pass through observed resonance states or (ii) to introduce a kinematic square root branch point at $l = \alpha(0)$ in the l plane [43]. The latter approach gives $MB/\tilde{A}$ a function of energy with asymptotically $\tilde{A}/MB = 0$ at $u = 0$. Polarization data are particularly necessary to resolve this problem; FESR are rather unreliable although a study by *Barger et al.* [45] found $MB/\tilde{A} \gtrsim +1$ for the N_α at $u = 0$. Polarization data exist for $\pi^- p \to \Lambda^0 K^0$ and may be interpreted as due to interference between an exchange degenerate $\Sigma_\alpha - \Sigma_\gamma$ trajectory and an exchange degenerate $\Sigma_\delta - \Sigma_\beta$ trajectory — the sign is in accord with the ratios $MB/\tilde{A}$ from parity doublet suppression together with the relative residue sign which follows from a zero at $\alpha_\delta = \frac{1}{2}$ as predicted by exchange degeneracy.

From the above discussion the uncertainties should be rather clear. However, ratios of couplings of a dominant Regge pole to different processes may still be significant. In order to estimate what poles will be dominant the $SU(3)$ duality scheme discussed previously will be used. Since $SU(3)$ is broken, and trajectories are not empirically degenerate ($N_\alpha - N_\gamma$ for example) the predictions from such a scheme must be used with care. The expectations for exchanges in four processes of interest are as follows [17]

	N_α	N_γ	N_β	Δ_δ
$\pi N \to N\pi$	big	small	small	(small)
$K^- n \to \Lambda \pi^-$	big	small	small	–
$\bar{K}N \to \Sigma\pi$	small = small		small = (small)	
$\pi^- p \to n\eta^0$	small	small	bigger	–

The $I_u = \frac{1}{2}; \frac{3}{2}$ combinations may be isolated for $\pi N \to N\pi$ as $\sigma(\pi^+ p \to p\pi^+) + \sigma(\pi^- p \to n\pi^0) - \frac{1}{3}\sigma(\pi^- p \to p\pi^-)$ and $\sigma(\pi^- p \to p\pi^-)$ and for $\bar{K}N \to \Sigma\pi$ as $\sigma(K^- n \to \Sigma^- \pi^0) + \sigma(K^- p \to \Sigma^- \pi^+) - \frac{1}{3}\sigma(K^- p \to \Sigma^+ \pi^-)$ and $\sigma(K^- p \to \Sigma^+ \pi^-)$ respectively; other reactions $\bar{K}N \to \Lambda\eta$ and $\bar{K}N \to \Sigma\eta$ also proceed by $S = 0$ baryon exchange but data are not available. The most favourable case for comparison is the N_α couplings to $K^- n \to \Lambda\pi^-$ and $\pi N \to N\pi$ and the result is given in Table 5. An error estimate in this work [46] comes from (i) the statistical error, (ii) normalization error of the 3.9 GeV $K^- n \to \Lambda\pi^-$ data, (iii) possible deviation of the data at this energy from the energy average due to resonance effects. A simultaneous fit was made to $\pi N \to N\pi$ and $K^- n \to \Lambda\pi^-$ with N_α and

Δ_δ poles and corrections to N_α exchange at these energies will be significant. However, N_γ pole contributions, absorptive cut corrections and kinematic cut corrections should each be proportional to the N_α contribution itself (for the N_γ since it has F/D similar to N_α), and so in the ratio of Regge residues these effects cancel. Present data are not sufficiently accurate to permit a u dependence of the ratio to be fitted and so it was taken constant. Indeed the $\Lambda K/\pi N$ coupling ratio of the N_α trajectory is known at $\alpha = 1/2$ to be 0.5 to 1.0 from $g_{\Lambda KN}/g_{NN\pi}$ using KN dispersion relations and at $\alpha = 5/2^+$ from the $SU(3)$ analysis [47] of the $5/2^+$ octet to be 1.1. Unfortunately the $5/2^+$ N_α resonance is very near ΛK threshold so that any direct determination of the $\Lambda K/\pi N$ branching ratio depends strongly on the F wave resonance shape used in the analysis. These values are consistent with the ratio 1.0 ± 0.2 for $-1 < u < 0$ from the Regge analysis and so a constant u dependence is acceptable. With such a constant u dependence the Regge result corresponds to $g_{\Lambda KN}^2/4\pi = 14.5 \pm 5$.

The analysis [48] of $\bar{K}N \to \Sigma\pi$ leads to a Δ_δ coupling to $K^+\Sigma^+/\pi^+p$ compatible with the equality predicted by $SU(3)$ and for the N_α trajectory a ratio $K^+\Sigma^0/\pi^0 p \sim 1$. This latter value is in considerable disagreement with $SU(3)$, but the highly exploratory fit by *Barger* [48] did not include the N_γ and N_β contributions and used poor data at 3 GeV. From very crude backward η production data [49], the ratio $\eta^0 p/\pi^0 p$ was obtained, again neglecting sizeable N_γ and N_β effects. Further data on these processes, particularly at higher energies, will allow a thorough analysis of these exchange mechanisms and a reliable test of $SU(3)$ and of the duality scheme presented earlier.

The data on $K^+p \to pK^+$, $K^0p \to nK^+$, $\pi^+p \to \Sigma^+K^+$, $K^-p \to \Xi^-K^+$ $\pi^-p \to \Lambda^0 K^0$ etc., may similarly be related by the Λ and Σ trajectories introduced previously. The $\Lambda_\alpha - \Lambda_\gamma$ trajectory fits [50] well $K^+p \to pK^+$ data and the residue is quoted in Table 5. $\pi^-p \to \Lambda^0 K^0$ is the only other reaction with good high-energy data and it may be explained with $\Sigma_\alpha - \Sigma_\gamma$ and $\Sigma_\delta - \Sigma_\beta$ trajectories. The $\Sigma_\alpha - \Sigma_\gamma$ trajectory decouples in the $F/D = +1$ duality limit but empirical evidence suggests that $F/D < 1$ for Σ_α and $F/D > 1$ for Σ_γ and these states are suppressed rather than decoupled from $\bar{K}N$.

Many other processes have been measured in the backward direction and give hints about residue couplings while not allowing a definitive analysis of exchanged Regge poles. For instance, $K^-p \to \Lambda\omega$ relative to $K^-p \to \Lambda\phi$ is consistent [51] with $N_\alpha N\phi \ll N_\alpha N\omega$ in agreement with the quark model; $K^-n \to \Sigma^-\omega$ relative to $K^-p \to \Lambda\omega$ is consistent [51] with $N_\alpha K\Sigma < N_\alpha K\Lambda$; $\pi N \to N\varrho, N\omega, N\phi$ should give a wealth of information about vector meson couplings to nucleons also; $\varrho - \omega$ interference could even be used to investigate the relative phase of these production processes.

4. Summary

We have not mentioned Regge fits to ϱ, ω and K^* production, photo-production, $np \to pn$ and $\bar{p}p \to \bar{n}n$, etc., since there are substantial un-certainties in such analyses. The unnatural parity exchanges are not well understood and even for π exchange there are difficulties in including absorption in such a way as to have quantitative predictions.

For vector and tensor meson exchanges near $t = 0$ the picture is much clearer. Ratios of couplings to different processes should be reliable, and spin flip − non-flip ratios are of interest even though they will be more sensitive to cut effects. The results are basically in agreement with exchange degeneracy and $SU(3)$ and correspond to $F/D \sim -5$ for the A' coupling and $F/D \approx \frac{1}{2}$ for vB. Tests in hypercharge exchange will be possible.

For the N_α and Δ_δ baryon exchanges the $SU(3)$ ratios seem consistent with $F/D \approx 1$ for the N_α but much improved data at higher energies will be necessary before full confidence can be gained. Very many pre-dictions can be made for such processes.

Acknowledgements

I wish to thank Dr. *R. J. N. Phillips* and Professor *V. Barger* for introducing me to subtleties of Regge polology, and Professor *H. Pilkuhn* for the invitation to attend the *Ruhestein* meeting where this review was presented and discussed.

References

1. *Collins, P. B. D., Squires, E. J.:* Regge Poles in Particle Physics. Springer Tracts in Modern Physics, Vol. 45. Berlin Heidelberg-New York: Springer 1968.
2. *Barger, V., Cline, D.:* Phenomenological Theories of High-Energy Scattering. New York: W. A. Benjamin Inc., 1969.
3. — *Durand III, L.:* Phys. Rev. Letters **19**, 1295 (1967).
4. *Henyey, F., Kane, G. L., Pumplin, J., Roos, M.:* Phys. Rev. Letters **21**, 946 (1968); — Phys. Rev. **182**, 1579 (1969).
5. *Arnold, R. C., Blackmon, M.:* Phys. Rev. **176**, 2082 (1968). *Blackmon, M., Goldstein, G.:* Phys. Rev. **179**, 1480 (1969).
6. *Lovelace, C.:* Nucl. Phys. B **12**, 253 (1969).
7. *Arnold, R. C.:* Phys. Rev. Letters **14**, 657 (1965).
8. *Dolen, R., Horn, D., Schmid, C.:* Phys. Rev. **166**, 1768 (1968). *Schmid, C.:* Phys. Rev. Letters **20**, 689 (1968). *Veneziano, G.:* Nuovo Cimento **57** A, 190 (1968).
9. *Freund, P. G. O.:* Phys. Rev. Letters **20**, 235 (1968). *Harari, H.:* Phys. Rev. Letters **20**, 1395 (1968).
10. *Chiu, C. B., Finkelstein, J.:* Phys. Letters **27** B, 510 (1968).
11. *Lovelace, C.:* Phys. Letters **28** B, 264 (1968).
12. *Mandula, J., et al.:* Phys. Rev. Letters **22**, 1147 (1969).
13. — *Weyers, J., Zweig, G.:* Phys. Rev. Letters **23**, 266 (1969).

14. *Rosner, J.:* Phys. Rev. Letters **21**, 950 (1968).
15. — *Rebbi, C., Slansky, R.:* Phys. Rev. **188**, 2367 (1969).
16. *Auvil, P., Halzen, F.: Michael, C.:* CERN Preprint TH. 1164 (1970).
17. *Barger, V., Michael, C.:* Phys. Rev. **186**, 1592 (1969).
18. *Harari, H.:* Phys. Rev. Letters **22**, 562 (1969).
    Rosner, J.: Phys. Rev. Letters **22**, 689 (1969).
19. *Barger, V., Olsson, M., Reeder, D.:* Nucl. Phys. B **5**, 411 (1968).
20. *Jackson, J. C.:* Phys. Rev. **174**, 2098 (1968).
21. *Reeder, D., Sarma, K.:* Phys. Rev. **172**, 1566 (1968).
22. *Derem, A., Smadja, G.:* Nucl. Phys. B **3**, 628 (1967).
23. *Mathews, R. D.:* Nucl. Phys. B **11**, 339 (1969).
24. *Dass, G. V., Michael, C., Phillips, R. J. N.:* Nucl. Phys. B **9**, 549 (1969).
25. *Allaby, J. V., et al.:* Phys. Letters **30** B, 500 (1969).
26. *Berger, E., Fox, G.:* Phys. Rev. **188**, 2120 (1969).
27. *Barger, V., Olsson, M.:* Phys. Rev. **146**, 1080 (1966).
28. *Höhler, G., Baacke, J., Eisenbeiss, G.:* Phys. Letters **22**, 203 (1966).
29. *Michael, C.:* Nucl. Phys. B **8**, 431 (1968).
30. *Clive, D., Matos, J., Reeder, D.:* Phys. Rev. Letters **23**, 1318 (1969).
    Michael, C.: Nucl. Phys. B **13**, 644 (1969).
31. *Barger, V., Phillips, R. J. N.:* Phys. Rev. **187**, 2210 (1969).
32. *Phillips, R. J. N., Rarita, W.:* Phys. Rev. Letters **15**, 807 (1965).
33. *Barger, V., Phillips, R. J. N.:* Phys. Letters **26** B, 730 (1968).
34. *Dass, G. V., Michael, C.:* Phys. Rev. **175**, 1774 (1968).
35. *Daum, C., Michael, C., Schmid, C.:* Phys. Letters **31** B, 222 (1970).
36. *Amblard, B., et al.:* Lund Conference (1969).
37. *Gilman, F., Pumplin, J., Schwimmer, A., Stodolsky, L.:* Phys. Letters **31** B, 387 (1970).
38. *Barloutaud, R., et al.:* Nucl. Phys. B **10**, 683 (1969).
39. *Boright, J. P., et al.:* Phys. Rev. Letters **24**, 964 (1970).
40. *Miller, R., Yokasawa, A.:* ANL Preprint (1969).
41. *Igi, K., Matsuda, S., Oyanagi, Y., Sato, H.:* Phys. Rev. Letters **21**, 580 (1968).
42. *Amman, R. F.:* EFINS Preprint 69–100 (1969).
    Kelly, R. L., Kane, G. L., Heneyey, F.: Phys. Rev. Letters **24**, 1511 (1970).
43. *Carlitz, R., Kislinger, M.:* Phys. Rev. Letters **24**, 186 (1970).
    Halzen, F., Kumar, A., Martin, A. D., Michael, C.: Phys. Letters, **32** B, 111 (1970).
44. *Schmid, C.:* Private communication.
45. *Barger, V., Michael, C., Phillips, R. J. N.:* Phys. Rev. **185**, 1852 (1969).
46. *Martin, A. D., Michael, C.:* Phys. Letters **32** B, 297 (1970).
47. *Plane, D. E., et al.:* CERN D. Ph. II/Phys. 70–9 (1970).
48. *Barger, V.:* Backward Peaks and Regge Phenomenology. Proceedings of Coral Gables Conference (1969).
49. *Chase, R. C., et al.:* Phys. Letters **30** B, 659 (1969).
50. *Barger, V.:* Phys. Rev. **179**, 1371 (1969).
51. *Derrick, M.:* Backward Peaks. Proceedings of CERN Topical Conference, CERN 68–7, Vol. 1 (1968).

Dr. *C. Michael*
CERN
CH-1211 Genf 23

Vector Mesons in Electromagnetic Interactions

M. GOURDIN

Contents

I.   Vector Meson Dominance Model 193
II.   Time-like Photon 194
III.   Ambiguities on Assumptions H_2 and H_3 197
IV.   Discussion of the Photon Vector Meson Analogy 198
V.   Photoproduction of Charged π Mesons 200
VI.   Photoproduction of Vector Mesons 202
VII.   Electromagnetic Sum Rules 203
VIII. The $\omega - \varphi$ Mixing 207
IX.   Strong Decay of Vector Mesons 209
References 211

The purpose of this talk is a review of the role played by vector mesons in electromagnetic interactions. Because of the lack of time such a review cannot be complete and we apologize for that and also for the incompleteness of the list of references.

Interesting problems have not been discussed as for instance

a) For time-like photons: the $\omega - \varrho$ electromagnetic mixing; the isobaric model for K-meson isoscalar form factors; the high-energy behaviour of the electron positron annihilation cross section.

b) For space-like photons; the nucleon electromagnetic form factors; the electroproduction of pseudoscalar and vector mesons; the inelastic lepton nucleon scattering.

c) For light-like photons: the photoproduction of π^0 meson, isobaric resonances, strange particles.

We consider only the following topics

I   Vector meson dominance model assumptions
II   Electron-positron annihilation into mesons
III   Discussion of the VMD model assumptions
IV   The validity of the vector meson photon analogy
V   The difficulties in charged-π meson photoproduction
VI   Photoproduction of vector mesons
VII   The electromagnetic sum rules from storage ring experiments
VIII   The $\omega - \varphi$ mixing
IX   The strong decay of vector mesons into pseudoscalar mesons.

The general aim is more a discussion of the theoretical ideas implied by the notion of vector meson dominance than a detailed quantitative comparison between theory and experiment expecially for parts V and VI.

Our general impression is that the vector meson dominance model is certainly correct for time-like photons in the region where it reduces simply to an isobaric model. Perhaps the extrapolation for zero vector meson masses makes sense at high energy in some specific cases like for instance the photoproduction of vector mesons. Other applications seem to be very doubtful. In the space-like region the nucleon electromagnetic form factor cannot be explained by a simple vector meson dominance model. In the high energy time-like region future storage ring experiments will give an answer to the problems of new vector mesons and also will measure the various cross sections. We will then have an idea if all the speculations made for instance in section III about sum rules can survive or not.

I. Vector Meson Dominance Model

1. The vector meson dominance (VMD) model is formulated in the semi phenomenological approach of the vector meson-photon analogy [1] using a direct vector meson-photon coupling [2, 3, 4].

The matrix element of the electromagnetic current

$$M_\mu = \langle f | J_\mu^{\text{em}}(0) | N \rangle$$

is assumed to be well represented by the sum of its vector meson contributions. Graphically

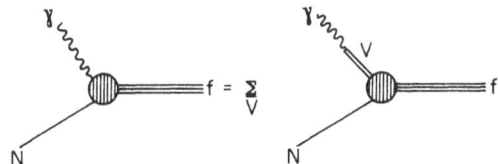

and for the amplitude M_μ the expression

$$M_\mu = \sum_V \frac{e}{f_V} \frac{m_V^2}{W_V(q^2) + q^2} \langle f | J_\mu^V(0) | N \rangle$$

where

q_μ is the photon energy momentum four vector

f_V is the inverse $\gamma - V$ coupling constant

$W_V(q^2) + q^2$ the inverse vector meson propagator.

2. Now Comes the First Assumption

> ASSUMPTION H_1 Coupling constant $\gamma - V$.

The vector meson-photon coupling constants f_V are independent of the photon mass q^2. They can be obtained from storage ring experiments (S. R.) where they are measured for vector mesons on mass shell.

We must be careful with such a problem of extrapolation (remember for instance the case of P. C. A. C. where an extrapolation from 0 to m_π^2 can produce changes of 10% to 20% in the coupling constants). Therefore a weaker form of assumption H_1 can be: The vector meson-photon coupling constants are slowly varying functions of q^2 which must be the same for all the phenomena induced by photons of the same q^2.

3. Physical Interpretation

The matrix element of the vector meson current $\langle f | J_\mu^V(0) | N \rangle$ describes an off mass shell amplitude for a vector meson reaction $V + N \Rightarrow f$ and the extrapolation corresponds to a zero mass vector meson of helicity ± 1.

We then have to describe such an extrapolation in order to use the information given by the observable reaction with a physical vector meson of mass m_V and spin one. We introduce two more assumptions:

> ASSUMPTION H_2 Mass extrapolation.

The transition amplitude $A(V + N \Rightarrow f)$ is a slowly varying function of q^2.

> ASSUMPTION H_3 Frame of reference.

The vector mesons are transversally polarized in a given frame of reference.

As will be clear in a moment, the assumptions H_2 and H_3 are empty if the conditions of application are not completely defined.

II. Time-like Photon [5]

1. Electron-positron annihilation experiment provides the best source of information about properties of a time-like virtual photon. The one-photon exchange approximation is assumed

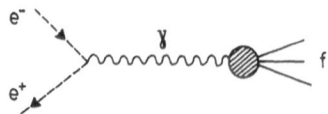

and the reaction $e^+ + e^- \Rightarrow f$ involves the current matrix element $\langle f | J_\mu^{em}(0) | 0 \rangle$.

2. In the time-like region around the vector meson masses where experiments have been performed, the VMD model reduces simply to an isobaric model. We assume the $e^+ e^-$ annihilation amplitude to be well described by the sum of the vector meson contributions:

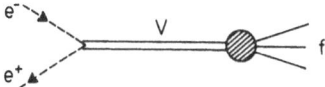

At the mass $q^2 = -m_V^2$ the total cross section $e^+ e^- \Rightarrow f$ has the simple structure

$$\sigma_{tot}(e^+ e^- \Rightarrow V \Rightarrow f) = \frac{12\pi}{m_V^2} \frac{\Gamma(V \Rightarrow e^+ e^-)\Gamma(V \Rightarrow f)}{\Gamma_V^2} . \qquad (1)$$

The assumptions needed to obtain this formula are

a) Only one vector meson contribution is important in the neighborhood of $q^2 \simeq -m_V^2$.

b) For $q^2 = -m_V^2$ the function W_V is given by

$$W_V(-m_V^2) = -i m_V \Gamma_V$$

which is a model-independent statement about the vector meson propagator.

Formula (1) has been used to deduce from experiment the value of the radiative widths $\Gamma(V \Rightarrow e^+ e^-)$. The Orsay results are [6]

$$\Gamma(\varrho \Rightarrow e^+ e^-) = (7.4 \pm 0.7)\,\text{keV},$$
$$\Gamma(\omega \Rightarrow e^+ e^-) = (1.0 \pm 0.18)\,\text{keV},$$
$$\Gamma(\varphi \Rightarrow e^+ e^-) = (1.41 \pm 0.12)\,\text{keV}.$$

In the one-photon exchange model the decay width $\Gamma(V \Rightarrow e^+ e^-)$ is related to the coupling constant f_V by

$$\Gamma(V \Rightarrow e^+ e^-) = \frac{\alpha^2}{3} m_V \frac{4\pi}{f_V^2} .$$

We then obtain

$$\frac{f_\varrho^2}{4\pi} = 1.85 \pm 0.17, \quad \frac{f_\omega^2}{4\pi} = 13.9 \pm 2.5, \quad \frac{f_\varphi^2}{4\pi} = 12.85 \pm 1.10.$$

13*

3. The π meson electromagnetic form factor F_π is measured in the experiment $e^+ + e^- \Rightarrow \pi^+ + \pi^-$. The VMD model for F_π is simply

$$F_\pi(q^2) = \frac{1}{f_\varrho} \frac{m_\varrho^2}{W_\varrho(q^2) + q^2} f_{\varrho\pi\pi} \tag{2}$$

where the decay coupling constant $f_{\varrho\pi\pi}$ is related to the $\varrho \Rightarrow 2\pi$ width by

$$\Gamma(\varrho \Rightarrow 2\pi) = \frac{2}{3} \frac{f_{\varrho\pi\pi}^2}{4\pi} \frac{k^3(m_\varrho^2)}{m_\varrho^2}$$

with $k(-q^2) = \frac{1}{2}(-q^2 - 4m_\pi^2)^{\frac{1}{2}}$.

The ϱ-meson parameters have been measured in the Orsay experiment and found to be

$$m_\varrho = (773.5 \pm 5.3)\,\text{MeV} \qquad \Gamma_\varrho = (110.7 \pm 5.3)\,\text{MeV}$$

so that

$$\frac{f_{\varrho\pi\pi}^2}{4\pi} = 2.11 \pm 0.13\,.$$

In the dispersion relation language, we assume an unsubtracted dispersion relation for F_π to be saturated by the ϱ meson contributions. In particular the form factor F_π must satisfy the normalization condition $F_\pi(0) = 1$. From the model (2) we deduce

$$1 = \frac{1}{f_\varrho} \frac{m_\varrho^2}{W_\varrho(0)} f_{\varrho\pi\pi}. \tag{3}$$

The π meson electric charge is entirely given by the ϱ meson term.

In the narrow width approximation [2] $W_\varrho(0) = m_\varrho^2$ and we recover the famous universality relation $f_\varrho = f_{\varrho\pi\pi}$ compatible with experiment.

If now we use more elaborate models for $W_\varrho(q^2)$ one can take into account non zero width corrections. For instance in the *Gounaris-Sakurai* approach of the π-meson electromagnetic form factor we have [7, 8]

$$W_\varrho(0) = m_\varrho^2 \left[1 + a_\varrho \frac{\Gamma_\varrho}{m_\varrho} \right]$$

and the normalization condition becomes

$$\frac{f_{\varrho\pi\pi}}{f_\varrho} = \frac{W_\varrho(0)}{m_\varrho^2} = 1 + a_\varrho \frac{\Gamma_\varrho}{m_\varrho}\,.$$

Squaring the equality and comparing with experiment we obtain

$$\left(\frac{f_{\varrho\pi\pi}}{f_{\varrho}}\right)^2 = 1.14 \pm 0.11$$

$$\left(1 + a_{\varrho}\frac{\Gamma_{\varrho}}{m_{\varrho}}\right)^2 = 1.14.$$

The agreement between theory and experiment is impressive if not too much significant.

III. Ambiguities on Assumptions H_2 and H_3

1. Let us now consider the problem of the frame of reference. We begin with a linearly polarized real photon and we introduce three, two by two, orthogonal axes Ox, Oy, Oz

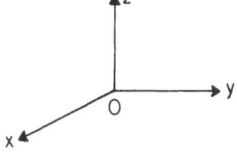

$Ox\,Oy$ production plane

Ox photon momentum

Oz normal to the production plane

The linear polarizations are defined as follows
    1. parallel polarization along Oy in the production plane $\gamma_{\parallel}$
    2. normal polarization along Oz orthogonal to the production plane $\gamma_{\perp}$

Now to go to a vector meson in its rest mass system we have to perform a Lorentz transformation not uniquely defined and such a non-uniqueness is closely related to the gauge invariance of physical photons.

The simplest procedure seems to perform a pure Lorentz transformation along the momentum q and to take the frame $Ox\,Oy\,Oz$ for the vector meson at rest. In particular longitudinal vector mesons will correspond to a polarization along Ox and transverse vector mesons to a polarization in the $Oy\,Oz$ plane. The result is the so-called helicity frame (H).

After that we have the freedom of an arbitrary space rotation of the $Ox\,Oy\,Oz$ system which induces a change in the definition of the matrix elements of the density operator

$$d^R_{\lambda\lambda'} = \sum_{\mu\mu'} R^{\mu}_{\lambda} d_{\mu\mu'} R^{-1\,\mu'}_{\lambda'}$$

where R describes the rotation.

In fact the other frames of reference popular in the literature are the *Gottfried-Jackson* (GJ) frame and the *Donohue-Hogaasen* frame (DH) both associated to specific rotations around Oz so that, in particular the density matrix element d_{zz} has the same expression in all the cases.

2. The current conservation relation is written as

$$q_0 \langle |J_0| \rangle = |\boldsymbol{q}| \langle \hat{q} \cdot \boldsymbol{J}| \rangle$$

where $\hat{q}$ is a unit vector along $\boldsymbol{q}$.

The assumption H_2 is then incompatible with the current conservation relation because the ratio $q_0/|\boldsymbol{q}|$ is not a slowly varying function of q^2.

We are in fact faced with two ambiguities

a) Choice of the current components where to apply the VMD relation with the assumption H_2

b) Choice of the scalar variables to keep fixed in the q^2 extrapolation.

3. In the helicity frame of reference we can relate the transverse polarizations with the transverse space components of the electromagnetic current. We use the assumption H_2 for these two components. The situation is more complex for the longitudinal polarization because one can use either the longitudinal space component $\hat{q} \cdot \boldsymbol{J}$ or the time component J_0 or something more complicated. Of course the VMD predictions are different in each case.

4. The matrix element of the electromagnetic current for an off mass shell photon $(q^2 \neq 0)$

$$M_\mu = \langle f | J_\mu^{em}(0) | N \rangle$$

can be decomposed on a set of Lorentz covariants

$$M_\mu = \sum_j I_\mu^j A_j(q^2; \ldots).$$

The scalar functions A_j depend on q^2 only for vertex functions (form factors) and there is no ambiguity in the choice of the scalar variables. But in all other cases, beside q^2 we can define a set of scalar variables independent of q^2, two for the photoproduction of mesons or nucleons, one for the form factors in inelastic lepton scattering. We have a priori a complete freedom to decide which variables will be kept fixed in the q^2 extrapolation. Again the VMD model predictions depend on the choice.

IV. Discussion of the Photon Vector Meson Analogy [9–13]

1. A photon state can be considered as a superposition of a pure photonic component and of hadronic components of mass m_V.

If q_0, the photon energy is assumed to be large with respect to m_V, the $\gamma - V$ conversion corresponds to an energy fluctuation

$$\Delta E = \sqrt{q_0^2 + m_V^2} - q_0 \simeq \frac{m_V^2}{2 q_0}.$$

The fluctuation has a life time given by the uncertainty principle

$$\Delta t \simeq \frac{2 q_0}{m_V^2}$$

and can traverse a distance δ

$$\delta \simeq \frac{2 q_0}{m_V^2}.$$

Such a distance can become very large at high energy, for instance with $q_0 = 20$ GeV and $m_V = m_\varrho$ we obtain $\delta \simeq 13.2 f$.

For fixed m_V the probability to find the photon in the hadronic state m_V is an increasing function of energy. Therefore, only if the electromagnetic current spectrum is dominated by low mass hadronic states, the photon can behave like a hadron at high energy.

2. As an example let us consider a nucleon target of radius R. The condition for observing diffractive effects is simply

$$\delta \gg 2R.$$

For instance if $m_V = m_\varrho$ and $R = 0.8$ fermi we obtain $q_0 \gg 2.5$ GeV.

Experimentally the photoproduction of ϱ^0 mesons on nucleons shows diffractive features for incident photon energies larger than 3 GeV in agreement with the previous estimate.

3. Let us go now to the photoabsorption by nuclei and we denote by λ the mean free path in matter of the hadronic component m_V.

a) If $\delta \ll \lambda$ we expect only a small absorption of the hadronic component and the contribution to photoabsorption will go like the volume R^3 of the nuclei, e. g. like A, the number of nucleon in the nuclei.

b) If $\delta \gg \lambda$ we expect a strong absorption of the hadronic component and the contribution to photoabsorption will go now like the surface R^2 of the nuclei, e. g. like $A^{2/3}$.

Experimentally the total photoabsorption cross sections $\sigma^{\gamma A}$ show a small departure of the linear A dependence [14].

If this result is confirmed it can be considered as an evidence about the existence of large ranges of interaction of high energy photons with hadrons increasing linearly with the energy.

Similar conclusions have also been obtained from high energy inelastic lepton-nucleon scattering data [11].

4. All these considerations can be extended to space-like virtual photons. Now the extrapolation is larger and the mass m_V^2 is simply replaced by $q^2 + m_V^2$ in all the previous formulae. The diffractive condition is simply written as

$$q_0 \gg R(q^2 + m_V^2).$$

We then obtain the condition of validity of the VMD model in inelastic lepton-nucleon scattering on the form [13]

$$q_0 \gg 3 + 5q^2$$

in unit 1 GeV and with $m_V = m_\varrho$, $R = 1$ f.

V. Photoproduction of Charged π Mesons

1. The VMD model combined with time-reversal invariance relates the photoproduction of $\pi^\pm$ mesons on nucleons with polarized photons to the neutral vector meson production in π-nucleon collisions

$$\gamma + p \Rightarrow \pi^+ + n,$$
$$\Leftrightarrow \pi^- + p \Rightarrow V^0 + n,$$
$$\gamma + n \Rightarrow \pi^- + p.$$

The comparison with experiment of the VMD predictions [16] is not very successful and shows clearly a disagreement between experiment and the strong version of the VMD model in the intermediate energy domain $E_\gamma \simeq 3$ GeV.

2. Before abandoning the model we must look with some care at the assumption H_2 (mass extrapolation) and H_3 (frame of reference) in order to make the conditions of applicability of the VMD model, if they exist, precise.

The question of the mass extrapolation has received recently a considerable interest [16, 17, 18, 19, 20] and we briefly review the situation.

As in the previous section the photoproduction amplitude is decomposed into a set of six Lorentz covariants assuming parity conservation

$$M_\mu = \sum_{j=1}^{j=6} I_\mu^j A_j(q^2; s, t)$$

where s and t are the usual Mandelstam variables.

Fraas and *Schildknecht* [16] study the variation with q^2 of the invariants I_μ^j disregarding the possible variation with q^2 of the invariant amplitudes A_j.

For such a reason the work of *Le Bellac* and *Plaut* [17] is a priori more convincing. These authors use for the scalar amplitudes A_j dynamical models of the Regge type in order to have some idea about the q^2 dependence of the amplitudes A_j at high energy.

Potter and *Sullivan* [19] write fixed s and t dispersion relations in q^2 for the invariant amplitudes $A_j(q^2; s, t)$ and they assume the dominance of the vector meson poles contributions.

In all these approaches the question of the choice of the best set of invariant amplitudes where to apply the VMD model has not been completely answered. The ambiguity about the choice of variables to keep fixed in the q^2 extrapolation is also discussed in Ref. [19] but not resolved. In practice, s and t are taken at the same value for both processes.

Nevertheless all the authors agree with the following conclusion:

At high energy to *leading order in s* the amplitudes for charged π-meson photoproduction are proportional to the amplitudes for π production of vector mesons with transverse polarization in the *helicity frame*.

An analogous result cannot be obtained in the DH frame.

3. The same problems associated to assumptions H_2 and H_3 have also been studied using more or less realistic models for the amplitude M_μ.

The electric Born model of *Cho* and *Sakurai* [21, 22] reproduces correctly photoproduction data in the near forward direction $|t| \lesssim m_\pi^2$. The VMD relations can be obtained at high energy provided the vector meson polarizations are defined in the helicity frame; moreover, the amplitudes for production of vector mesons with longitudinal polarization exhibit a q^2 dependence even at high energy. Again the DH frame is excluded.

Schmidt [23] writes fixed t and q^2 dispersion relations in s for the invariant amplitudes $A_j(q^2; s, t)$ and incorporates the complete Born approximation and the N_{33}^* contributions in a crossing symmetric way. An explicit mass dependence for the amplitudes is found and appears to be essentially due to

a) the mass dependence of the kinematic factors entering the dispersion integrals (in particular because of crossing symmetry)

b) the variation of the coupling constants of the *Born* terms (because of the difficulties to apply the VMD model to the nucleon electromagnetic form factors).

In particular a correct mass dependence is computed for the asymmetry in photoproduction with polarized photons.

4. The conclusions concerning the assumptions H_1, H_2 and H_3 can be the following

H_3 (frame of reference) all the theoretical arguments converge for using the VMD relations in the helicity frame. The final decision will be given by experiment.

H_2 (mass extrapolation) at the actual experimental energies $E_\gamma \simeq 3.4$ GeV a q^2 dependence is certainly present in the transverse amplitude and we can only compute it in a model-dependent way. The VMD relations are expected to become exact only in the high energy limit s⟹∞ |t| fixed.

H_1 (coupling constant) it is difficult to separate now a violation of the VMD relations due to H_1 or H_2. If the statements for H_2 are correct, f_V can be given by its SR value.

Of course very useful information concerning the problem of the mass extrapolation can certainly be given by the electroproduction of $\pi^\pm$ mesons for which q^2 is space-like [24].

VI. Photoproduction of Vector Mesons

1. The vector meson dominance model, combined with the optical theorem, relates the forward photoproduction of vector mesons on hydrogen to the total cross sections for vector meson nucleon collisions with a transverse vector meson.

Unfortunately we do not have at our disposal vector meson beams to measure directly these total cross section. Nevertheless if one believes in the VMD model with the assumption H_1 about the f_V coupling constant, the photoproduction data can be used to estimate the total $V - N$ cross sections. The results agree very well with the quark model predictions [25].

2. Photoproduction experiments of vector mesons have been performed using complex nuclei targets from deuterium to uranium. Of course, nuclear physics enters in the analysis of the experimental data via elaborate optical models and the difficulty is increased by a lot of theoretical uncertainties.

Nevertheless the various results from experiments performed at DESY, SLAC and CORNELL seem to converge towards an agreement with the VMD model for f_V and with the quark model predictions for σ^{VN}.

As an illustration the last DESY experiment [26] on ϱ^0 photoproduction at $E_\gamma = 7.5$ GeV on a set of 14 nuclei targets gives

$$\frac{f_\varrho^2}{4\pi}(q^2 = 0) = 2.28 \pm 0.40 \qquad \sigma_T^{\varrho^0 p} = (26.7 \pm 2.0) \text{ mb}$$

the VMD and quark model predictions are⋆ [27]

$$\frac{f_{\varrho\pi\pi}^2}{4\pi} = 2.11 \pm 0.15 \qquad \sigma_{QM}^{\varrho^0 p} = (26.55 \pm 0.2) \text{ mb}.$$

⋆ We use the condition $\dfrac{1}{f_\varrho}\dfrac{m_\varrho^2}{W_\varrho(0)} = \dfrac{1}{f_{\varrho\pi\pi}}$ previously obtained in Eq. 3.

3. The total photoabsorption on nucleons has been experimentally measured and the data agrees well with the VMD predictions for values of the coupling constants f_V compatible with the assumption H_1.

VII. Electromagnetic Sum Rules

We now apply the VMD model in the time-like region to the spectral function sum rules one can deduce from current algebra or asymptotic $SU(3)$ symmetry [28, 29, 30, 31, 5].

1. First *Weinberg* sum rule

Starting from the *Gell Mann* commutation relations for the components of the $SU(3)$ vector currents and assuming the *Schwinger* terms to be finite, *Weinberg* derived a spectral function sum rule which is convenient to write in the form $(q^2 + s = 0)$

$$\int_0^\infty s\sigma_{\text{Tot}}(e^+ e^- \Rightarrow I = 1)ds = 3 \int_0^\infty s\sigma_{\text{Tot}}(e^+ e^- \Rightarrow I = 0)ds \qquad (4)$$

where $\sigma_{\text{Tot}}(e^+ e^- \Rightarrow I)$ is the total cross section for electron positron annihilation into a final hadronic state of total isospin I. Of course the electromagnetic current is assumed to belong to an $SU(3)$ octuplet.

We now use the VMD model and saturate the integrals with the ϱ^0 meson $(I = 1)$ ω and φ mesons $(I = 0)$ contributions. In the narrow width approximation the result is the *Das-Mathur-Okubo* sum rule [29–32]

$$m_\varrho \Gamma(\varphi \Rightarrow e^+ e^-) = 3[m_\omega \Gamma(\omega \Rightarrow e^+ e^-) + m_\varphi \Gamma(\varphi \Rightarrow e^+ e^-)] \qquad (5)$$

which, in terms of coupling constants, reduces simply to

$$\frac{m_\varrho^2}{f_\varrho^2} = 3\left[\frac{m_\omega^2}{f_\omega^2} + \frac{m_\varphi^2}{f_\varphi^2}\right]. \qquad (6)$$

We then have a current mixing model for the $SU(3)$ breaking [4–29, 32]. A good way to compare the sum rules with experiment is to introduce the parameter δ

$$\delta = \frac{[I = 1] - [I = 0]}{[I = 1] + [I = 0]}$$

where $[I]$ is the isotopic spin I contribution to the sum rule. From the Orsay measurements we obtain

$$\delta_W^0 = (-7.3 \pm 6.1) 10^{-2}. \qquad (7)$$

The agreement with experiment is good.

The *Das-Mathur-Okubo* sum rule is a zero width approximation of the sum rule (4) and it is interesting to compute finite width corrections going back to the original form in the VMD model. With the auxiliary parameters

$$x_V = \frac{1}{12\pi^2} \frac{\int_0^\infty s\sigma_{\mathrm{Tot}}(e^+ e^- \Rightarrow V)\, ds}{m_V \Gamma(V \Rightarrow e^+ e^-)}$$

the sum rule (5) becomes

$$x_\varrho m_\varrho \Gamma(\varrho \Rightarrow e^+ e^-) = 3[x_\omega m_\omega \Gamma(\omega \Rightarrow e^+ e^-) + x_\varphi m_\varphi \Gamma(\varphi \Rightarrow e^+ e^-)] \quad (8)$$

and in terms of coupling constants f_V we now have

$$x_\varrho \frac{m_\varrho^2}{f_\varrho^2} = 3\left[x_\omega \frac{m_\omega^2}{f_\omega^2} + x_\varphi \frac{m^\varphi}{f_\varphi^2}\right]. \quad (9)$$

For the ϱ meson, the *Gounaris-Sakurai* model can be used to compute x_ϱ and the result is

$$x_\varrho \simeq 0.93.$$

For the ω meson the zero width approximation is good and $x_\omega \simeq 1$ but for the φ meson x_φ can differ appreciably from unity because if the width is small, the threshold $K\bar{K}$ is very close to the φ meson mass and the K meson momenta are small. A calculation by *Renard* [34] gives

$$x_\varphi \simeq 1.15.$$

The value of δ_W obtained including these finite widths corrections

$$\delta_W = (-15.5 \pm 6)\, 10^{-2} \quad (10)$$

is less good than (7) but it is not possible to conclude to a disagreement with experiment.

2. Charge renormalization sum rule

If one uses now the commutation relations of the *Sugawara* model [30, 31, 33] and if one assumes the hadronic part of the charge renormalization to be finite, one can derive a spectral function sum rule

$$\int_0^\infty \sigma_{\mathrm{Tot}}(e^+ e^- \Rightarrow I = 1)\, ds = 3\int_0^\infty \sigma_{\mathrm{Tot}}(e^+ e^- \Rightarrow I = 0)\, ds. \quad (11)$$

The physical interpretation of this sum rule is the equality of the iso-scalar and isovector contributions to the charge renormalization. As previously, we use the VMD model in the narrow width approximation [32]

$$\frac{\Gamma(\varrho \Rightarrow e^+ e^-)}{m_\varrho} = 3 \left[\frac{\Gamma(\omega \Rightarrow e^+ e^-)}{m_\omega} + \frac{\Gamma(\varphi \Rightarrow e^+ e^-)}{m_\varphi} \right] \qquad (12)$$

which, in terms of the coupling constants reduces to

$$\frac{1}{f_\varrho^2} = 3 \left[\frac{1}{f_\omega^2} + \frac{1}{f_\varphi^2} \right]. \qquad (13)$$

We then have a mass mixing model for the $SU(3)$ breaking [4]. The comparison with experiment of equation (12) gives

$$\delta_G^0 = (+9.2 \pm 6.4) \, 10^{-2}. \qquad (14)$$

The agreement with experiment is good.

Again, we compute the finite width corrections. The parameter y_V is now defined by

$$y_V = \frac{1}{12\pi^2} \frac{\int_0^\infty \sigma_{\text{Tot}}(e^+ e^- \Rightarrow V) \, ds}{m_V^{-1} \, \Gamma(V \Rightarrow e^+ e^-)}.$$

Numerical estimates for y_V are

$$y_\varrho \simeq 0.95 \qquad y_\omega \simeq 1 \qquad y_\varphi \simeq 1.075.$$

The sum rule (12) and the relation (13) become

$$y_\varrho \frac{\Gamma(\varrho \Rightarrow e^+ e^-)}{m_\varrho} = 3 \left[y_\omega \frac{\Gamma(\omega \Rightarrow e^+ e^-)}{m_\omega} + y_\varphi \frac{\Gamma(\varphi \Rightarrow e^+ e^-)}{m_\varphi} \right] \qquad (15)$$

$$\frac{y_\varrho}{f_\varrho^2} = 3 \left[\frac{y_\omega}{f_\omega^2} + \frac{y_\varphi}{f_\varphi^2} \right]. \qquad (16)$$

Comparing that with experiment we obtain

$$\delta_G \simeq (+4.7 \pm 6.3) \, 10^{-2}. \qquad (17)$$

3. *Sugawara* sum rule

In the framework of the *Sugawara* model the first *Weinberg* sum rule cannot be exact and it must be corrected with a specific factor corresponding to the particular breaking of the $SU(3)$ symmetry. In the VMD version of the sum rule this factor is a mass square ratio $(m_3/m_8)^2$

one computes using a *Gell Mann-Okubo* formula

$$m_3^2 = m_\varrho^2 \qquad m_8^2 = \tfrac{1}{3}(4 m_{K^*}^2 - m_\varrho^2)$$

and instead of the *Das-Mathur-Okubo* sum rule we obtain the *Sugawara* sum rule [30]

$$m_\varrho \, \Gamma(\varrho \Rightarrow e^+ e^-) = 3 [m_\omega \, \Gamma(\omega \Rightarrow e^+ e^-) + m_\varphi \, \Gamma(\varphi \Rightarrow e^+ e^-)] \frac{3 m_\varrho^2}{4 m_{K^*}^2 - m_\varrho^2}. \quad (18)$$

The agreement with experiment, including (δ_s) or not (δ_s^0) finite width corrections

$$\delta_s^0 = (11.1 \pm 6.7) \, 10^{-2}, \qquad\qquad\qquad\qquad (19)$$

$$\delta_s = (2.9 \pm 6.6) \, 10^{-2} \qquad\qquad\qquad\qquad (20)$$

is satisfactory.

4. Hadronic contributions to the charge renormalization

The hadronic contributions to the charge renormalization are given by [35]

$$\frac{\delta e_0^2}{e^2} \text{(hadrons)} = \frac{1}{\pi e^2} \int \sigma_{\text{Tot}}(e^+ e^- \Rightarrow \text{hadrons}) \, ds.$$

In the narrow width approximation of the VMD model we have

$$\frac{\delta e_0^2}{e^2} \text{(hadrons)} = \frac{3}{\alpha} \sum_V \frac{\Gamma(V \Rightarrow e^+ e^-)}{m_V} = \alpha \sum_V \frac{4\pi}{f_V^2}.$$

Using the Orsay experimental data, we obtain, taking finite width corrections

$$\frac{\delta e_0^2}{e^2} \text{(hadrons)} \simeq (4.88 \pm 0.35) \, 10^{-3}.$$

5. Hadronic contributions to the muon anomalous magnetic moment

The hadronic contributions to the muon anomalous magnetic moment $a_\mu = \tfrac{1}{2}(g_\mu - 2)$ due to vacuum polarization corrections has the structure [36, 37, 38]

$$a_\mu(\text{hadrons}) = \frac{1}{\pi e^2} \int \sigma_{\text{Tot}}(e^+ e^- \rightarrow \text{hadrons}) \, K_\mu^{(2)}(s) \, ds$$

where $K_\mu^{(2)}(s)$ is the second order vertex function.

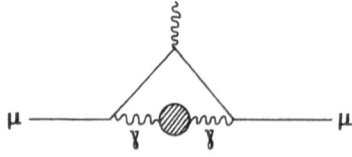

We now apply the VMD model. The ϱ^0 meson contribution has been computed with a numerical integration of the integral and for the ω and φ meson contributions we use a narrow width approximation. The result is [39]

$$a_\mu(\text{hadrons}) = (6.5 \pm 0.5)\, 10^{-8}.$$

The theoretical prediction including 2nd, 4th, 6th order calculation is now given by [40]

$$a_\mu = (116587 \pm 2)\, 10^{-8}.$$

The last experimental value is [41]

$$a_\mu = (116616 \pm 31)\, 10^{-8}.$$

Therefore the agreement between theory and experiment is very good

$$a_\mu(\text{th}) - a_\mu(\text{exp}) = (-29 \pm 34)\, 10^{-8}.$$

VIII. The ω—φ Mixing

1. Let us denote by ω_1 and φ_8 the isoscalar vector mesons belonging respectively to the singulet and octuplet representations of $SU(3)$. Because of the $SU(3)$ breaking ω_1 and φ_8 are not the physical mesons and we must introduce a configuration mixing to describe the ω and φ mesons

2. In the $SU(3)$ theory the electromagnetic current is assumed to transform like a U spin scalar of an adjoint representation. Therefore

$$\langle 0|J_\mu^{\text{em}}(0)|\omega_1\rangle = 0. \tag{21}$$

The baryonic current is an $SU(3)$ invariant and belongs to a singulet representation

$$\langle 0|J_\mu^{\text{B}}(0)|\varphi_8\rangle = 0. \tag{22}$$

The $SU(3)$ relation in the octuplet is

$$\langle 0|J_\mu^{\text{em}}(0)|\varphi_8\rangle = \frac{1}{\sqrt{3}} \langle 0|J_\mu^{\text{em}}(0)|\varrho\rangle. \tag{23}$$

The three relations (21) (22) (23) are the basis of the application of the $\omega - \varphi$ mixing theory to electromagnetic interactions.

3. Let us define as θ the mixing angle as deduced from the old-fashionable formalism with a *Gell Mann-Okubo* type relation

a) for the squared vector meson masses [42]

$$4m_{K^*}^2 - m_\varrho^2 = 3[m_\varphi^2 \cos^2\theta + m_\omega^2 \sin^2\theta],$$

$$\theta = 40.2°.$$

b) for the inverse squared masses [43]

$$4m_{K^*}^{-2} - m_{\varrho}^{-2} = 3[m_{\varphi}^{-2}\cos^2\theta + m_{\omega}^{-2}\sin^2\theta],$$
$$\theta = 29.6°.$$

The direct coupling constants of the vector mesons with the electromagnetic and baryonic currents are defined, for stable mesons by

$$\langle 0|J_\mu^{em}(0)|V\rangle = \frac{m_V^2}{f_V}e_\mu,$$

$$\langle 0|J_\mu^B(0)|V\rangle = \frac{m_V^2}{g_V}e_\mu$$

where e_μ is the vector meson polarization vector.

4. We are now in position to write the $SU(3)$ relations (21), (22) and (23) for the coupling constants f_V and g_V. Unfortunately because of the large mass differences between the vector mesons and because of their instability it remains some ambiguities in the precise meaning of these relations.

We apply the relations (21) (22) and (23) to the quantities $z_V m_V^\alpha/f_V$ where z_V is a finite width correction factor and α an arbitrary integer. We then obtain obtain three relations

$$\sin\theta\left[z_\varphi\frac{m_\varphi^\alpha}{f_\varphi}\right] - \cos\theta\left[z_\omega\frac{m_\omega^\alpha}{f_\omega}\right] = 0, \tag{24}$$

$$\cos\theta\left[z_\varphi\frac{m_\varphi^\alpha}{g_\varphi}\right] + \sin\theta\left[z_\omega\frac{m_\omega^\alpha}{g_\omega}\right] = 0, \tag{25}$$

$$\cos\theta\left[z_\varphi\frac{m_\varphi^\alpha}{f_\varphi}\right] + \sin\theta\left[z_\omega\frac{m_\omega^\alpha}{f_\omega}\right] = \frac{1}{\sqrt 3}\left[z_\varrho\frac{m_\varrho^\alpha}{f_\varrho}\right]. \tag{26}$$

In particular, equations (24) and (26) can be replaced by

$$z_\varphi\frac{m_\varphi^\alpha}{f_\varphi} = \frac{1}{\sqrt 3}\cos\theta\, z_\varrho\frac{m_\varrho^\alpha}{f_\varrho}, \tag{27}$$

$$z_\omega\frac{m_\omega^\alpha}{f_\omega} = \frac{1}{\sqrt 3}\sin\theta\, z_\varrho\frac{m_\varrho^\alpha}{f_\varrho}. \tag{28}$$

It is convenient to introduce two auxiliary parameters θ_Y and θ_B

$$\frac{f_\varphi}{f_\omega} = \tan\theta_Y = \frac{z_\varphi}{z_\omega}\left(\frac{m_\varphi}{m_\omega}\right)^\alpha\tan\theta,$$

$$-\frac{g_\omega}{g_\varphi} = \tan\theta_B = \frac{z_\omega}{z_\varphi}\left(\frac{m_\omega}{m_\varphi}\right)^\alpha\tan\theta$$

with the relation

$$\tan\theta_Y \tan\theta_B = \tan^2\theta .$$

Eliminating θ between equations (27) and (28), we obtain the sum rule

$$z_\varrho^2 \frac{m_\varrho^{2\alpha}}{f_\varrho^2} = 3\left[z_\varphi^2 \frac{m_\varphi^{2\alpha}}{f_\varphi^2} + z_\omega^2 \frac{m_\omega^{2\alpha}}{f_\omega^2}\right].$$

The VMD model for the first *Weinberg* sum rule corresponds to Eq. (9)

$$z_V^2 = x_V , \qquad \alpha = 1 .$$

and we have a current mixing model.
The VMD model for the charge renormalization sum rule corresponds to Eq. (16)

$$z_V^2 = y_V , \qquad \alpha = 0 .$$

We have a mass mixing model and in the zero width approximation $\theta_Y = \theta = \theta_B$.

5. Comparison with experiment [6]

We compute the partial decay widths $\Gamma(\omega \Rightarrow e^+ e^-)$ and $\Gamma(\varphi \Rightarrow e^+ e^-)$ with, as an input, the experimental value of $\Gamma(\varrho \Rightarrow e^+ e^-) = (7.4 \pm 0.7)\,\text{keV}$ and we look at the angle θ_Y.

		$\Gamma(\omega \Rightarrow e^+ e^-)_{\text{keV}}$	$\Gamma(\varphi \Rightarrow e^+ e^-)_{\text{keV}}$	$\tan^2\theta_Y$
Current mixing	$\theta = 40.2°$	0.94 ± 0.09	0.88 ± 0.08	1.39
$z_V^2 = x_V \ \alpha = 1$	$\theta = 29.6°$	0.52 ± 0.05	1.16 ± 0.11	0.58
Mass mixing	$\theta = 40.2°$	0.99 ± 0.10	1.68 ± 0.16	0.77
$z_V^2 = y_V \ \alpha = 0$	$\theta = 29.6°$	0.55 ± 0.05	2.21 ± 0.21	0.32
Experiment		(1.0 ± 0.18)	1.41 ± 0.12	0.92 ± 0.17

The models with a mixing angle $\theta = 29.6°$ seem to be excluded on the basis of the ω meson data.
The mass mixing model is slightly favourable.

IX. Strong Decay of Vector Mesons

1. In the isobaric vector meson model the strong decay coupling constant $f_{\varphi K \bar{K}}$ is measured from the value of the total cross section $\sigma_{\text{Tot}}(e^+ e^- \Rightarrow K \bar{K})$ at the meson mass. For the $K^+ K^-$ mode we have

$$\left(\frac{f_{\varphi K^+ K^-}}{f_\varphi}\right)^2 = \frac{3}{\pi\alpha^2} \Gamma_\varphi^2 \frac{\sigma_{\text{Tot}}(e^+ e^- \Rightarrow K^+ K^-)_{s=m_\varphi^2}}{[1 - (4m_{K^+}^2/m_\varphi^2)]^{3/2}}$$

and using the Orsay data, we obtain

$$\left|\frac{f_{\varphi K^+ K^-}}{f_\varphi}\right| = 0.349 \pm 0.025 .$$

The $\gamma - \varphi$ coupling constant is deduced from the set of experiments

$$e^+ e^- \Rightarrow K^+ K , \qquad e^+ e^- \Rightarrow K^0 \bar{K}^0 , \qquad e^+ e^- \Rightarrow \pi^+ \pi^- \pi^0$$

assuming the other φ meson decay modes to be small and the result is

$$\frac{f_\varphi^2}{4\pi} = 12.85 \pm 1.10 .$$

It follows

$$\frac{f_{\varphi K^+ K^-}^2}{4\pi} = 1.55 \pm 0.18 .$$

Because of the large experimental errors, we can at this stage neglect the Coulomb corrections [44] and use the same numerical value for $f_{\varphi K \bar{K}}^2/4\pi$ in both charge states.

Combining that now with

$$\frac{f_{\varrho\pi\pi}^2}{4\pi} = 2.11 \pm 0.13$$

given by the Orsay experiment, we obtain

$$\frac{f_{\varphi K \bar{K}}^2}{f_{\varrho\pi\pi}^2} = 0.728 \pm 0.095 . \tag{29}$$

2. In the simplest version of exact $SU(3)$ symmetry, we have the relation

$$\frac{f_{\varphi K K}^2}{f_{\varrho\pi\pi}^2} = \frac{3}{4} \cos^2 \theta$$

with $\theta = 40.2°$ the numerical value of this ratio is 0.44 in disagreement with the experimental result (29).

It seems necessary to introduce a first order perturbation with respect to the medium strong interactions. Unfortunately, we have the choice between three independent ways for the breaking of $SU(3)$ symmetry and it is not possible to make predictions except if a universal vector meson coupling à la *Sakurai* is assumed. In that particular case [45]

$$\frac{f_{K^* K\pi}^2}{f_{\varrho\pi\pi}^2} = \frac{3}{8}\left(\frac{1 - \alpha/2}{1 + \alpha}\right)^2 , \qquad \frac{f_{\varphi K \bar{K}}^2}{f_{\varrho\pi\pi}^2} = \frac{3}{4} \cos^2 \theta \left(\frac{1 - \alpha}{1 + \alpha}\right)^2 .$$

The value of the mixing parameter α is computed from the experimental K^* width [33]

$$\alpha = -0.138 \pm 0.017$$

and the prediction of the broken $SU(3)$ symmetry is now

$$\frac{f^2_{\varphi K \bar{K}}}{f^2_{\varrho \pi \pi}} = 0.760 \pm 0.057$$

in good agreement with experiment.

References

1. *Roos, M., Stodolsky, L.:* Phys. Rev. **149**, 1172 (1966).
2. *Gell-Mann, M., Zachariasen, F.:* Phys. Rev. **124**, 953 (1961).
   — *Sharp, D., Wagner, W. A.:* Phys. Rev. Letters **8**, 261 (1962).
3. *Nambu, Y., Sakurai, J. J.:* Phys. Rev. Letters **8**, 79 (1962).
4. *Kroll, N., Lee, T. D., Zumino, B.:* Phys. Rev. **157**, 1376 (1967).
5. *Gourdin, M.:* Boulder Lecture Notes 1969.
6. *Augustin, J., et al.:* Phys. Letters **28**B, 508, 513, 517 (1969).
7. *Frazer, W. R., Fulco, J.:* Phys. Rev. Letters **2**, 365 (1959); **117**, 1603 (1960).
8. *Gounaris, G., Sakurai, J. J.:* Phys. Rev. Letters **21**, 244 (1968).
9. *Gribov, V. N.:* SLAC translation 102 (1969).
10. *Brodsky, S. J., Pumplin, J.:* Phys. Rev. **182**, 1595 (1969).
11. *Ioffe, B. L.:* Phys. Letters **30**B, 123 (1969).
12. *Gottfried, K.:* preprint CNLS 87 (1969).
13. *Nieh, H. T.:* preprint Stony Brook (1970).
14. *Caldwell, D. O., et al.:* Phys. Rev. Letters **23**, 1256 (1969).
15. See for instance *Schildknecht, D.:* DESY 69/10 and 69/41.
16. *Fraas, H., Schildknecht, D.:* Nucl. Phys. B **6**, 395 (1968).
17. *Le Bellac, M., Plaut, G.:* Nuovo Cimento **64** A, 95 (1969).
18. *Meiere, F. T.:* Phys. Letters **30** B, 44 (1969).
19. *Potter, W. T., Sullivan, J. D.:* preprint Illinois (1969).
20. *Brown, S. G.:* preprint MIT (1969).
21. *Cho, C. F., Sakurai, F. J.:* EFI 69/73.
22. *Diederich, S.:* quoted by D. Schildknecht in DESY 69/41.
23. *Schmidt, W.:* Cornell preprint CLNS 72 (1969).
24. *Iso, C., Yoshii, H.:* Ann. Phys. (N. Y.) **51**, 490 (1969).
    — *Schildknecht, D.:* preprint DESY 70/11.
25. *Joos, H.:* Phys. Letters **24** B, 103 (1967).
26. *Alvensleben, H., et al.:* Phys. Rev. Letters **24**, 786 (1970).
27. *Fox, G. C., Quigg, C.:* UCRL 20001.
28. *Weinberg, S.:* Phys. Rev. Letters **18**, 507 (1967).
29. *Das, T., Mathur, V. S., Okubo, S.:* Phys. Rev. Letters **19**, 470 (1967).
30. *Sugawara, H.:* Phys. Rev. Letters **21**, 772 (1968).
31. *Cremmer, E.:* Nucl. Phys. B **15**, 131 (1970).
32. *Oakes, R. J., Sakurai, J. J.:* Phys. Rev. Letters **19**, 1266 (1967).
33. *Gourdin, M.:* Invited paper Intern. Conf. on Quark Models, Detroit 1969.
34. *Renard, F. M.:* Nucl. Phys. B **15**, 267 (1970).

35. *Källén, G.:* Helv. Phys. Acta **25**, 417 (1957).
36. *Bouchiat, C., Michel, L.:* J. Phys. Radium **22**, 121 (1961).
37. *Brodsky, S. J., de Rafael, E.:* Phys. Rev. **168**, 1620 (1968).
38. *Lautrup, B. E., de Rafael, E.:* Phys. Rev. **174**, 1835 (1968).
39. *Gourdin, M., de Rafael, E.:* Nucl. Phys. B **10**, 667 (1969).
40. *Aldins, J., et al.:* Phys. Rev. Letters **23**, 441 (1969).
41. *Barley, J., et al.:* Phys. Letters **28** B, 287 (1968).
42. *Sakurai, J. J.:* Phys. Rev. Letters **9**, 472 (1962).
43. *Coleman, S., Schnitzer, H. J.:* Phys. Rev. **134** B, 863 (1964).
44. *Cremmer, E., Gourdin, M.:* Nucl. Phys. B **12**, 383 (1969).
45. *Gourdin, M.:* Unitary Symmetries, Amsterdam-London: North-Holland Publ. Co. 1967.

Prof. Dr. M. Gourdin
Laboratoire de Physique Théorique
Bâtiment 211, Faculté des Sciences
F-91 Orsay, France

Coupling Parameters of Pseudoscalar Meson Photoproduction on Nucleons

W. Pfeil and D. Schwela

Contents

I. Introduction 213
II. Photoproduction of Single Pions in the Region of the First Pion Nucleon
Resonance. Dispersive Calculations 214
III. Isobar Model Calculations 220
IV. Threshold Photoproduction and Pion Nucleon Scattering Lengths 223
V. Predictions for the $\gamma N \Delta$ (1230) Transition Moment 224
VI. Decay Half Widths of the Higher Nucleon Resonances in π Photoproduction
below 1.5 GeV 228
VII. Photoproduction of η- and K-mesons 233
VIII. Conclusions 235
References 236

I. Introduction

This review is devoted to a discussion of the processes

$$\gamma p \to \begin{array}{c} p\pi^0 \\ n\pi^+ \end{array}, \tag{1a}$$

$$\gamma n \to p\pi^- , \tag{1b}$$

$$\gamma p \to p\eta , \tag{2}$$

$$\gamma p \to \begin{array}{c} \Lambda K^+ \\ \Sigma^0 K^+ \end{array} \tag{3}$$

which can be represented by the graph of Fig. 1.

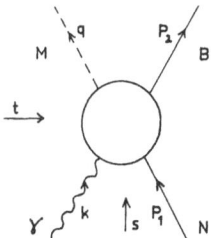

Fig. 1. Photoproduction of a pseudoscalar meson M of fourmomentum $q = (q_0, \boldsymbol{q})$ and a baryon B of fourmomentum $p_2 = (E_2, \boldsymbol{p}_2)$ on a nucleon N of fourmomentum $p_1 = (E_1, \boldsymbol{p}_1)$, $k = (k_0, \boldsymbol{k})$ is the fourmomentum of the incident photon

The investigation of these processes has turned out an important means for the determination of the decay widths for the transitions of the excited s-channel nucleon resonances into nucleon and photon. Moreover, the coupling parameters of the one particle exchange in the t-channel

$$\gamma \bar{M} \to B \bar{N} \tag{4}$$

and the u-channel

$$\gamma \bar{B} \to M \bar{N} \tag{5}$$

can be determined, at least in principle, from an investigation of these processes.

In practice, however, the extraction of these parameters from experimental data suffers from several serious deficiencies:

A. In any theoretical description the separation of background and resonance structure is strongly model dependent. This implies the known uncertainty of the coupling constants of the t-channel exchanges in the processes (1) and (2) and of the t- and u-channel exchanges in process (3).

B. The coupling parameters of the strong vertex

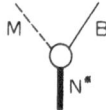

are not well known, especially for the higher nucleon resonances.

C. Experimental data for the processes (1b), (2) and (3) are few and not very accurate. Also for the processes (1a) at energies above 450 MeV photon lab. energy data are inaccurate and sometimes even contradictory.

For these reasons we confine ourselves to the determination of decay widths of the nucleon resonances exchanged in the s-channel of the processes (1a, 1b, 2, 3) in the region between threshold and 1.5 GeV photon lab. energy.

II. Photoproduction of Single Pions in the Region of the First Pion Nucleon Resonance. Dispersive Calculations

In the last years a large amount of experimental and theoretical work has been done in pion photoproduction for photon lab. energies below 500 MeV.

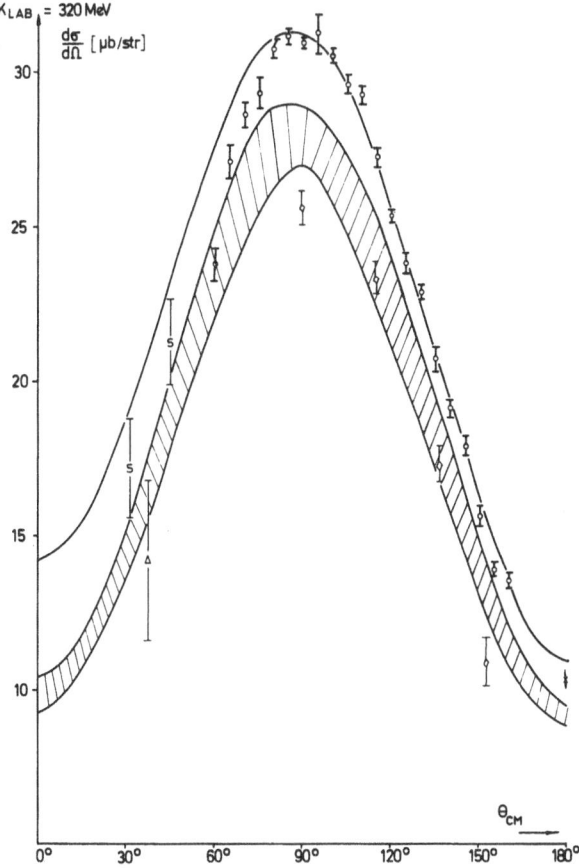

Fig. 2. Angular distribution for the π^0 differential cross section on protons at 320 MeV lab. photon energy. The solid line and the shaded area are theoretical curves taken from Refs. [12] and [13] respectively. For the data see Refs. [1] and [17a, b]

A summary of the experimental data existing in 1969 has been given by R. L. *Walker* [1] at the Liverpool Conference. Very accurate and abundant data exist now for the π^0 and π^+ differential cross sections for photon lab. energies between 200 and 440 MeV. Very recently, a new differential cross section measurement of π^0 photoproduction at forward angles has been completed at Bonn [2]. Typical examples of the π^0 and π^+ angular distributions near the $\varDelta$ (1230) resonance are shown in Figs. 2 and 3.

In Fig. 4 we display some asymmetry ratios for π^0 and π^+ photoproduction with polarized photons [3, 4].

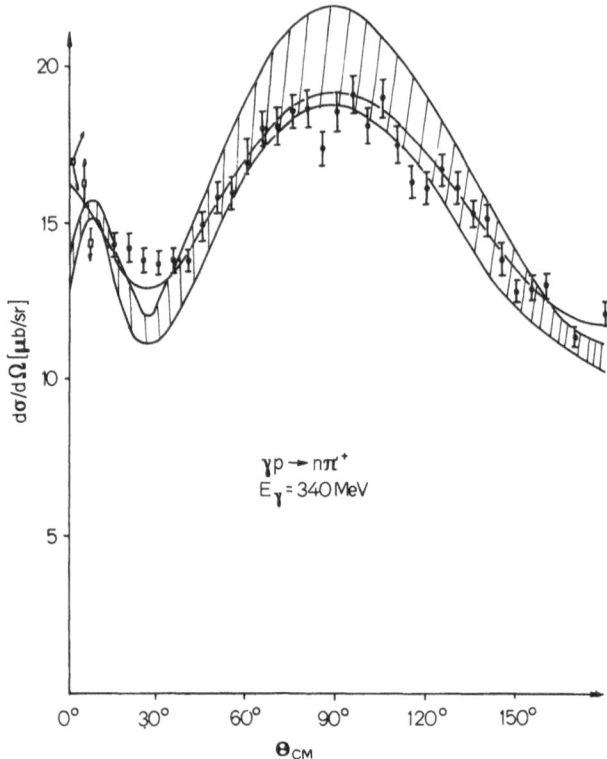

Fig. 3. Angular distribution for the π^+ differential cross section at 340 MeV lab. photon energy. The solid line and the shaded area are explained in Fig. 2. For the data see Refs. [1] and [17c, d]

Very recently, the results of four measurements for the differential cross section of $\gamma n \to p\pi^-$ in the resonance region have been published. Two of them, Refs. [5, 6] measure the inverse reaction $\pi^- p \to \gamma n$, while the other ones [7a, b] have investigated the process $\gamma d \to 2p\pi^-$. In the latter experiments the neutron data were obtained by use of the spectator model. While the results of Ref. [6] measured at energies above resonance agree with those of the two experiments [7a, b] within the error bars, large discrepancies between Ref. [5] and Refs. [7a, b] at small pion CM angles for energies between 260 and 420 MeV are present indicating a limitation of the validity of the spectator model. In Fig. 5 we show an example for the angular distribution of this reaction near the $\Delta(1230)$ resonance. At $\Theta_{CM}^\pi = 90°$ also some new asymmetry ratios for polarized photons are available [8, 39]; they are also shown in Fig. 4.

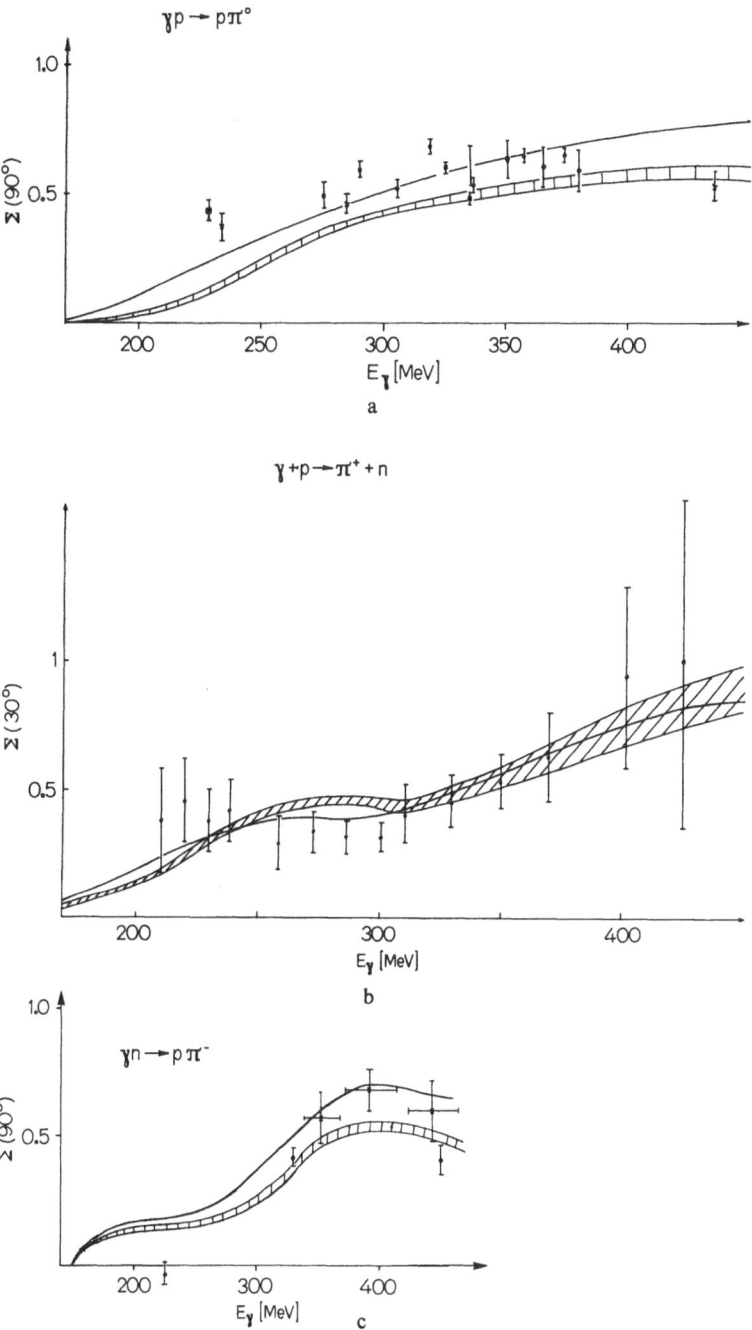

Fig. 4a–c. Excitation curves π^0 and $\pi^{\pm}$ asymmetry ratios for polarized photons. The solid line and the shaded area are explained in Fig. 2. For the data see Refs. [1, 3, 4, 8]

For the theoretical description of the pion photoproduction mainly two approaches have been developed to some degree of refinement, viz.

(a) the dispersive approach
(b) the isobar model, which will be treated in chapter III.

The starting point of the dispersive approach is fixed $-$ t-dispersion relations for the invariant amplitudes $A_i(s, t)$. From these equations one projects in the usual manner to a system of dispersion relations for the multipole amplitudes $M_{l\pm}$, $E_{l\pm}$ and obtains for a particular amplitude $M_i(=M_{l\pm}$ or $E_{l\pm})$

$$M_i(W) = B_i(W) + \frac{1}{\pi} \int_{M+m_\pi}^{\infty} dW' \frac{\operatorname{Im} M_i(W')}{W' - W - i\varepsilon}$$

$$+ \sum_j \int_{M+m_\pi}^{\infty} dW' K_{ij}(W, W') \operatorname{Im} M_j(W') \tag{6}$$

$B_i(W)$ denotes the Born approximation of M_i, $K_{ij}(W, W')$ are explicitely known, nonsingular kernels, W is the CMS energy.

This procedure is applied because of the Fermi-Watson-theorem, which is a consequence of the unitarity and time reversal invariance of the S-matrix. It states that the phase of a particular multipole with definite isospin is equal to the corresponding πN scattering phase, i. e.

$$M_i(W) = \pm |M_i(W)| e^{i\delta_i(W)} \tag{7a}$$

or equivalently

$$\operatorname{Im} M_i(W) = e^{-i\delta_i(W)} \sin \delta_i(W) M_i(W). \tag{7b}$$

This theorem is strictly valid only below the two pion threshold. As long as the phase shifts of πN scattering are elastic, this theorem is usually applied also above the 2π-threshold at 320 MeV photon lab. energy.

Two serious difficulties occur in Eqs. (6), (7b): One has to know the multipole phase on the whole physical cut and for all multipoles. Two mainly different approaches have been developed to overcome this problem.

I. Here one considers only the lowest multipole amplitudes ($l \leq 2$), and either introduces a cut off (*Engels et al.*, Ref. [9]) or makes assumptions about the high energy behaviour of the multipole phases (*Adler*, Ref. [10], *von Gehlen*, Ref. [11], *Schwela et al.*, Ref. [12]).

II. In the approach of Ref. [13] conformal mapping techniques are applied to reduce the system of singular integral equations to a system of linear algebraic equations. The system is then truncated and solved using only the experimentally known πN phases.

A further important difference between the approaches I. and II. is the occurence of free parameters in Refs. [9, 12], which reflect the influence of the high energy behaviour on the low energy part of the multipole amplitude. Formally, these parameters correspond to the existence of homogeneous solutions of Eq. (6) (the well known CDD-poles, Ref. [14]) as has been discussed in detail in the talk of *Rollnik* [15] at the Heidelberg Conference and in Ref. [12]. In contrast to this situation the solution of *Berends et al.* [13] is unique.

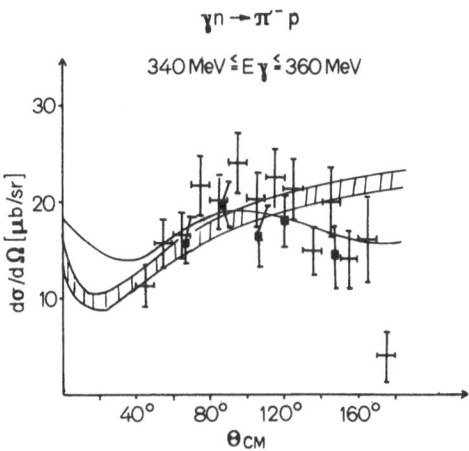

Fig. 5. Angular distribution of π^- differential cross section on neutrons at approximately 350 MeV photon lab. enery. The solid line and the shaded area are explained in Fig. 2. For the data see Refs. [6, 7]

In Figs. 2–5 we compare the experimental data of $\pi^0, \pi^{\pm}$-photoproduction with the results of the theoretical calculations of Refs. [12] (solid line) and [13] (shaded area). The agreement between theory and experiment is somewhat better for Ref. [12] than for Ref. [13]. It should also be mentioned that the agreement between experiment and the other theoretical approaches is fair.

All theoretical work in the past years has confirmed the conclusions of the basic work of *Chew et al.* [16] of a predominant $M_{1+}^{(3/2)}$ in π^0 photoproduction and a large M_{1+} and E_{0+} in $\pi^{\pm}$ photoproduction. The calculations of the last few years have been undertaken mainly in order to determine quantitatively the small multipole amplitudes $E_{0+}^{\pi^0}, E_{1+}$, M_{1-} and $M_{1+}^{(1/2)}$. However, in spite of the elaborate and sophisticated calculations we are still not able to determine these small quantities more reliably than CGLN did. So, a striking new theoretical idea is needed and/or more and more accurate data which allow a reliable multipole analysis. For the latter point the new Bonn and Orsay data [17] are promising.

A deficiency of all these theories is the dependence of the calculations on the πN phase shifts. The experiment of *Bugg et al.* [18] which has been discussed on this meeting, shows the uncertainty in the (large!) P_{33} phase shift. The difference between the new and the old πN data may alter the small phase shifts quite substantially, in consequence also the small photoproduction multipoles will change.

III. Isobar Model Calculations

The isobar model has been first used in pion photoproduction by *Gourdin* and *Salin* [19]. The model has been refined and extended by several people, the most recent being the work of one of the present authors [20]. In the isobar model approach one calculates — using quantum field theoretic methods — the contributions of the Feynman graphes of Fig. 6, where N^* denotes any nucleon resonance allowed.

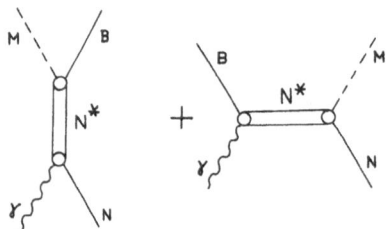

Fig. 6. Feyman graphs defining the isobar model

One starts by considering the nucleon resonances as stable particles with definite spin and real masses M^* and defines effective Hamiltonian interactions for the various vertices. However, the field theoretic description contains many sources of trouble as no satisfactory free field theory for spin $\geq 3/2$ particles is known. This means essentially that the propagator of the intermediate particles with spin $\geq 3/2$ cannot be determined in a unique manner.

Consequently the contributions of the Feynman graphs to the nonresonant partial waves cannot be fixed.

Even for the exchange of nucleon resonances of spin 1/2, where a unique propagator exists a difficulty occurs due to different equivalent forms of the derivative coupling of the isobar to nucleon and meson. Again the nonresonant multipoles cannot be fixed.

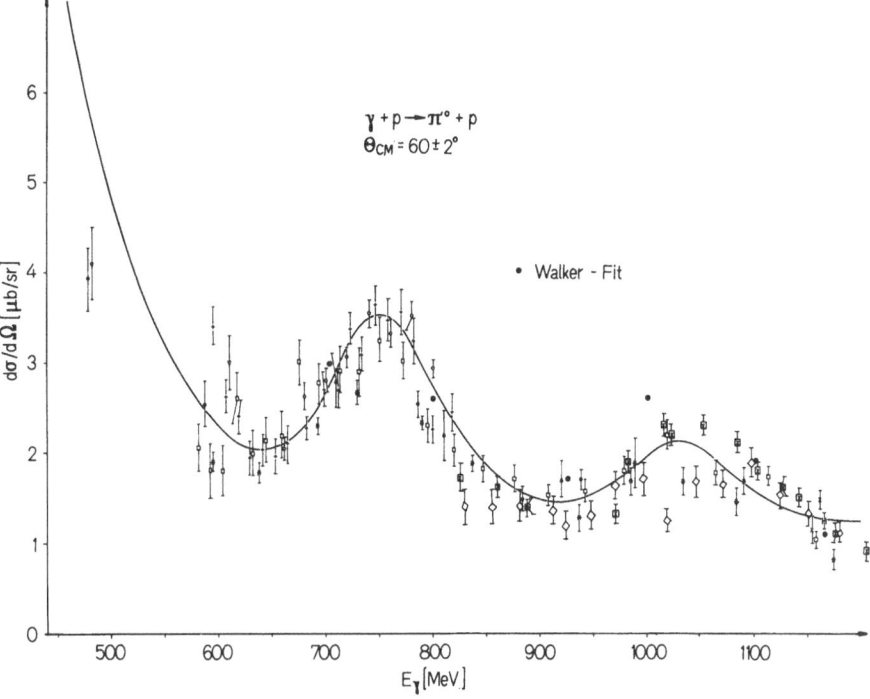

Fig. 7. Excitation curve at $\theta_{CM} = 60°$ for the π^0 differential cross section. The solid line is the fitted curve taken from Ref. [20a]. For the data see Refs. [35] and [1]

For a thorough discussion of all these problems we refer the reader to Ref. [20] and the review talk given by *Rollnik* at the Hercegovni Summer School 1967 [21].

In order to circumvent these troubles a pole approximation for the invariant amplitudes $A_i(s, t)$ is made

$$A_i(s, t) = \frac{a_i(s = M^{*2}, t)}{s - M^{*2} + iM^*\Gamma(s)} . \tag{8}$$

In this formula we have neglected crossed channel terms and already generalized the ansatz to unstable particles with finite width $\Gamma(s)$ in an obvious and straightforward way.

In addition the pole approximation guarantees an acceptable high energy behaviour of the invariant amplitudes.

Strictly speaking, the assumptions made can only be justified by the success of the model in representing the experimental data in an appropriate way. To illustrate that this is really the case we show in Figs. 7

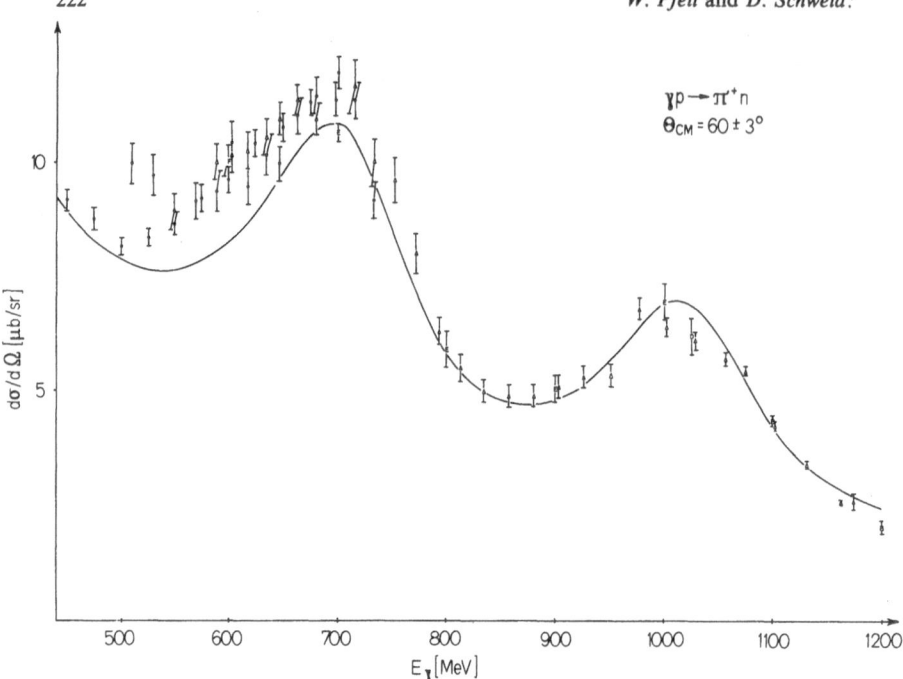

Fig. 8. Excitation curve at $\theta_{CM} = 60°$ for the π^+ differential cross section on protons taken from Ref. [20b]. For the data see Ref. [1]

and 8 excitation curves for the π^0 and π^+ differential cross sections, taken from Ref. [20]. The solid line is a fit with the seven isobars $P_{33}(1230)$, $P_{11}(1470)$, $D_{13}(1520)$, $S_{31}(1670)$, $F_{15}(1690)$, $D_{15}(1680)$, $S_{11}(1710)$, ex-changed in the s and u channel. The inclusion of u channel resonances leads to crossing symmetric amplitudes. In this connection it should be stressed that crossing symmetry correlates strongly the background and resonance contributions. This correlation plays a definite role in the determination of the resonance parameters from experimental data.

As can be seen from the figures, the isobar model in the formulation of Ref. [20] gives a satisfactory representation of the experimental differential cross section data in π^0 and π^+ photoproduction in the energy range from threshold up to 1.5 GeV. Even the asymmetry ratios for polarized photons in the whole energy range and the nucleon polarization above 500 MeV are described by the model, although these data have not been included in the determination of the parameters. The model shows some deviations from the experimental π^+ differential cross section data and from the proton polarization data in π^0 production in the region of the first resonance. These discrepancies are correlated with the fact that the Fermi-Watson theorem is badly violated for the nonresonant multipole amplitudes which — through the occurrence of interference terms — play a larger role in these experimental quantities.

As compared to the dispersion method of chapter II the isobar model for the processes (1 a, b) gives only a crude approximation to the dispersion integrals. But for photon energies above 500 MeV for processes (1 a, b) and in the region of several hundred MeV above threshold for the processes (2) and (3) the sophisticated isobar treatment is the only model until now available for the description of the photoproduction of pseudoscalar mesons.

IV. Threshold Photoproduction and Pion Nucleon Scattering Lengths

Between the electric dipole $E_{0+}^{\pi^-}$ at threshold and the difference of the s-wave πN scattering lengths we have the well known connection

$$P|E_{0+}^{\pi^-}|^2 = \frac{2}{9} v_0 \frac{M+1}{2M+1} (a_3 - a_1)^2 \tag{9}$$

where $P = 1.533 \pm 0.021$ [49] is the Panofsky ratio known from experiment, v_0 is the relative velocity of the π^0, M the nucleon mass and a_1, a_3 are the s-wave pion nucleon scattering lengths with isospin 1/2, 3/2 respectively. Given $E_{0+}^{\pi^-}$ and P, we can predict the difference of the s-wave scattering lengths through

$$|a_3 - a_1| = (7.99 \pm 0.06) |E_{0+}^{\pi^-}|_{\text{threshold}} . \tag{10}$$

In the following table we present the results for $E_{0+}^{\pi^-}|_{\text{thr.}}$ and $|a_3 - a_1|$, calculated from the papers [9, 13, 12].

Table 1.

Author	$E_{0+}^{\pi^-}|_{\text{thr.}} \cdot 10^2$	$|a_3 - a_1|$
Engels et al.	-3.39 ⎫	0.271 ± 0.015
BDW	-3.40 ⎬ ± 0.2	0.272 ± 0.015
Schwela	-3.10 ⎭	0.248 ± 0.016

The first two values are in agreement with the values of the s-wave scattering lengths calculated directly from πN dispersion relations [22]. The last one is somewhat smaller but within the error bars compatible with the result from the pole fits to backward dispersion relations [23].

V. Prediction of the $\gamma N \Delta$ (1230) Transition Moments

The well known $SU(6)$ prediction of the $M1$ photoexcitation $\gamma N \rightarrow \Delta(1230)$ has been written down by *Bég et al.* [24]

$$\mathcal{M} = (\Delta, m = \tfrac{1}{2} | M_z | p, m = \tfrac{1}{2}) = \tfrac{2}{3}\sqrt{2}\,\mu_p \tag{11}$$

with M_z the z-component of the magnetic moment operator and $\mu_p = 2{,}79\, e/2M$. Using $SU(6)_W$ *Harari* and *Lipkin* [25] have shown that the electric quadrupole transition $\mathscr{E}$ is forbidden. The same results have also been obtained in the quark model [26, 27].

Since the time these results were derived, people were always concerned with the problem of reasonable comparison of these predictions with experiment. If the photoproduction process at the energy of the $\Delta(1230)$ resonance were entirely due to resonance production with no background present, a measurement of the decay products N and π would immediately allow the extraction of the transition moments. In practice, however background effects are important, so a precise definition of the transition moments is necessary.

Let us review and discuss briefly the work on this subject that has been done in the past.

In the work of *Dalitz* and *Sutherland* [28] a very heuristic connection of the magnetic transition moment (which in contradistinction to our later definition we now denote by $\tilde{\mu}^*$) with the experimental quantities is given based mainly on the assumption that background effects in π^0 photoproduction are small and may be accounted for by subtracting from the total π^0 cross section a small term estimated in a handwaving way.

At resonance the total π^0 cross section is approximated by a Breit-Wigner form

$$\sigma(\gamma p \rightarrow p\pi^0) = \frac{8}{3}\frac{\pi}{k^{*2}}\frac{\Gamma_\gamma^p \Gamma_{\pi N}}{\Gamma_0^2} \simeq \frac{8}{3}\frac{\pi}{k^{*2}}\frac{\Gamma_\gamma^p}{\Gamma_0} \tag{12}$$

where k^* is the CMS photon momentum at resonance, $\Gamma_0 = \Gamma_{\pi N} + \Gamma_\gamma$ the total width of the $\Delta(1230)$, Γ_γ^p the partial width for the decay $\Delta \rightarrow p + \gamma$ and $\Gamma_\gamma = \tfrac{3}{2}\Gamma_\gamma^p$. In terms of the transition element (11), Γ_γ^p turns out to be

$$\Gamma_\gamma^p = \frac{M k^{*3}}{2\pi M^*}\,\tilde{\mu}^{*2} \tag{13}$$

where M^* is the resonance mass, α the fine structure constant. Using the most recent π^0 results [29], we obtain, repeating the calculations of Ref. [28]

$$\Gamma_\gamma^p = (0.69 \pm 0.02)\,\text{MeV} \qquad \tilde{\mu}^* = (1.32 \pm 0.02)\,\tfrac{2}{3}\sqrt{2}\,\mu_p$$

which differs slightly from the result of *Dalitz* and *Sutherland*.

If one writes, on the other hand, the total cross section in terms of the resonant multipole amplitudes $M_{1+}^{(3/2)}$ and $E_{1+}^{(3/2)}$ neglecting any background amplitudes, one obtains (q^* is the pion CMS momentum at resonance)

$$\sigma(\gamma p \to p \pi^0) = \frac{32\pi}{9} \frac{q^*}{k^*} \{|M_{1+}^{(3/2)}|^2 + 3|E_{1+}^{(3/2)}|^2\} \tag{14}$$

and consequently, from a comparison with Eq. (12)

$$\Gamma_\gamma^p = \tfrac{4}{3} q^* k^* \Gamma_0 \{|M_{1+}^{(3/2)}|^2 + 3|E_{1+}^{(3/2)}|^2\} \tag{15}$$

and

$$\tilde{\mu}^* = \sqrt{\frac{8\pi M^* q^* \Gamma_0}{3 M k^{*2}} \{|M_{1+}^{(3/2)}|^2 + 3|E_{1+}^{(3/2)}|^2\}} \,. \tag{16}$$

The ansatz always made in the literature for the connection between $\mathscr{M}$ and the magnetic dipole amplitude is obtained from this equation with $E_{1+}^{(3/2)}$ neglected.

$$\tilde{\mu}^* = \sqrt{\frac{8\pi M^* q^* \Gamma_0}{3 M}} \frac{|M_{1+}^{(3/2)}|}{k^*} \,. \tag{17}$$

Somewhat different analyses of experimental data have been done by *Ash et al.* [30] and *Grilli et al.* [31].

Ash et al. assume only the magnetic $J = 3/2^+$ dipole amplitude to contribute in the π^0 differential cross section, and they extract $\tilde{\mu}^*$ from the measured cross section using Eq. (17). Their result differs substantially from that of *Dalitz* and *Sutherland*

$$\tilde{\mu}^* = (1.15 \pm 0.02) \tfrac{2}{3} \sqrt{2} \, \mu_p \,.$$

But this result is not very reliable because in differential cross sections interference (and background) terms play a nonnegligible role.

In contrast to this determination of $\tilde{\mu}^*$, *Grilli et al.* analyze the π^+ differential cross section for polarized photons. The background term which accounts for approximately one third of the cross section, is calculated in the Born approximation. Solving for $\tilde{\mu}^*$ they obtain

$$\tilde{\mu}^* = (1.23 \pm 0.03) \tfrac{2}{3} \sqrt{2} \, \mu_p \,.$$

This result also depends strongly on the assumption for the background term.

While, in principle (from Eq. (16)) the electric quadrupole $E_{1+}^{(3/2)}$ could contribute to $\tilde{\mu}^*$, in the determination of *Grilli et al.* $E_{1+}^{(3/2)}$ does not occur at all, because it cancels out.

These examples show the need for a more careful discussion of the transition moments. Our determination of these moments will be based on the Lagrangian model. The idea is as follows:

Assume that a complete experimental information on the photo-production process in the resonance region has been obtained so that a multipole analysis is available. Then we consider only the multipoles with the quantum numbers of the resonance and represent these multipoles with a relativistic Breit-Wigner model in which the resonance width and the decay coupling parameters of the resonance into N and π are taken from πN scattering experiments. The matrix element for the decay of the isobar at rest into nucleon and photon then determines the transition moments. Let us look at this in some more detail:

We describe the $\gamma N \Delta$-vertex by the Hamiltonian

$$H_{\gamma N\Delta} = -eg_M \bar{\Psi}_\mu \gamma_\nu \Psi F^{\mu\nu} + ieg_E \bar{\Psi}_\mu \gamma_5 \overset{\leftrightarrow}{\partial}_\nu \Psi F^{\mu\nu} + \text{h. c.} \qquad (18)$$

Here $F_{\mu\nu} = \partial_\mu A_\nu - \partial_\nu A_\mu$ denotes the electromagnetic field operator, $e^2 = 4\pi/137$, $\overset{\leftrightarrow}{\partial}_\nu = \overset{\rightarrow}{\partial}_\nu - \overset{\leftarrow}{\partial}_\nu$, and Ψ_μ and Ψ denote the isobar and nucleon field, respectively[*]. The two terms in Eq. (18) correspond to the magnetic dipole and electric quadrupole amplitudes for a spin $J^P = 3/2^+$ particle. In order to see this in detail we calculate from Eq. (18) the Hamiltonian for the decay of the isobar at rest into nucleon and photon (for the moment we neglect any isospin factors). The result is, see Ref. [20a]

$$H_{\gamma N\Delta} = e\sqrt{\frac{E_1 + M}{M}} \left\{ g_M (\chi^\dagger \times \sigma) B(k)\chi + \frac{g_M + 2M g_E}{E_1 + M} E(k)\, \chi^\dagger(\sigma \cdot k)\, \chi \right\} \quad (19)$$

Here $B(k)$ and $E(k)$ are the Fourier components of the magnetic and electric field strengths, χ is the Pauli spinor-vector of the Δ-particle and χ is the Pauli spinor of the nucleon, σ denote the Pauli matrices, E_1 is the energy of the nucleon. From (19) we obtain for the magnetic transition moment the expression

$$\mathcal{M} = eg_M \sqrt{\frac{E_1 + M}{3M}} \cdot \sqrt{\frac{2}{3}}^{\;[**]} \qquad (20)$$

and for the electric quadrupole transition moment

$$\mathcal{E} = \frac{e(g_M + 2M g_E)}{3M(E_1 + M)} \sqrt{\frac{2}{3}}. \qquad (21)$$

---

[*] The connection between g_M and g_E and the familiar C_3 and C_4 of *Gourdin* and *Salin* [32] is given by $g_M = C_3$, $g_E = \frac{1}{2}C_4$.

[**] Note that the last factor $\sqrt{2/3}$ is an isospin factor due to the definition of $\mathcal{M}$ in Eq. (11). This factor has been omitted in most publications on this subject.

An expansion of $H_{\gamma N_\Delta}$ in Eq. (19) into partial waves yields the connection between g_M, g_E and $\mathcal{M}$, $\mathcal{E}$ (s. Ref. [20a]) and consequently, (from Eqs. (20)) between $\mathcal{M}$, $\mathcal{E}$, and the multipole amplitudes $M_{1+}^{(3/2)}$, $E_{1+}^{(3/2)}$. The result is

$$\mathcal{M} = \frac{M_{1+}^{(3/2)} + E_{1+}^{(3/2)}}{i} \frac{\Gamma_0}{k^*} \sqrt{\frac{4\pi q^* M^*}{M\Gamma_{\pi N}}} \sqrt{\frac{2}{3}} = \mu^* \tag{22}$$

and

$$\mathcal{E} = \frac{i E_{1+}^{(3/2)} \Gamma_0}{k^{*2}} \sqrt{\frac{16\pi q^* M^*}{M\Gamma_{\pi N}}} \sqrt{\frac{2}{3}}. \tag{23}$$

Quite naturally, also the electric quadrupole amplitude $E_{1+}^{(3/2)}$ occurs in our result for the magnetic transition moment, Eq. (22).

In the following table we compare the predictions for Γ_γ^p, $\tilde{\mu}^*$ and μ^* from different dispersive models (Refs. [9, 10, 11, 12, 13]) and from the results of the isobar model Ref. [20] and the multipole analyses of R. L. Walker [47] and the Bonn group [48].

Table 2

$\operatorname{Im} M_{1+}^{(3/2)}(M^*)$ $\times 10^2 \left[\frac{1}{m_\pi}\right]$	$\dfrac{E_{1+}^{(3/2)}}{M_{1+}^{(3/2)}}$	Γ_γ^p [MeV]	$\tilde{\mu}^* \left[\frac{2}{3}\sqrt{2}\mu_p\right]$	$\mu^* \left[\frac{2}{3}\sqrt{2}\mu_p\right]$	Reference
3.53	−0.045	0.69	1.30	1.25	[9]
3.41	−0.0155	0.59	1.25	1.24	[10]
3.43	−0.045	0.64	1.27	1.22	[11]
3.64	−0.01	0.69	1.35	1.34	[12]
3.38	−0.035	0.63	1.25	1.21	[13]
3.55	−0.059	0.59	1.31	1.25	[20]
3.37	−0.045	0.61	1.25	1.20	[47]
3.51	−0.01	0.66	1.30	1.29	[48]

From the Table 2 we infer the following facts:

1) The values for the imaginary part of the magnetic dipole $\operatorname{Im} M_{1+}^{(3/2)}$ evaluated or fitted by different authors, are nearly equal.

2) The values for the electric quadrupole $E_{1+}^{(3/2)}$ or rather the ratio $E_{1+}^{(3/2)}/M_{1+}^{(3/2)}$ differ substantially up to a factor of six.

3) The differences between $\tilde{\mu}^*$ and μ^* are small.

4) All different models yield values for μ^* which differ by $25-30$ percent from the $SU(6)$ prediction.

Due to the preponderance of the magnetic dipole amplitude the different definitions Eq. (17) and (22) do not lead to much different results. But the procedure outlined above plays a more important role in the electroproduction of pions where $E^{(3/2)}_{1+}$ becomes larger in relation to $M^{(3/2)}_{1+}$ for higher virtual photon masses.

VI. Decay Half Widths of the Higher Nucleon Resonances in π Photoproduction below 1.5 GeV

For energies E_γ between 0.5 and 2. GeV theoretical models for the photoproduction of π mesons have not been studied to the same extent as for lower energies. The reasons for this are twofold:

1) No convincing theoretical concept exists.

2) Up to the year 1969 data has been very scarce and even contradictory.

As far as the second point is concerned the situation has become better during the last year: New cross section data for π^- [33], π^+ [34] and π^0 [35] have been presented. Here we want to mention only those data, which have been published after the Liverpool Conference. We refer the reader to the review of *R. L. Walker* [1] presented at the Liverpool Conference for the older data. New cross section data at extreme backward direction $\Theta_{CM} = 180°$ for the π^+ production have been measured by *Fischer et al.* [36] at energies up to 1.5 GeV. New polarizations of the recoil nucleon for π^0 photoproduction have been measured by *Blüm et al.* [37] at Bonn. Moreover people have started with the investigation of photoproduction using polarized targets.

As mentioned in the introduction, the analysis of photoproduction of pions is at present the best way to determine the radiative widths of the nucleon resonances for the decay of the excited nucleon states to the ground state, proton or neutron. The radiative width Γ_γ can be expressed by the multipole amplitudes $E_{l\pm}, M_{l\pm}$ introduced in chapter II. For unnatural parity $P = (-1)^{l+1}$ of the resonance we get

$$\Gamma_\gamma = \frac{k^* q^* \Gamma_0^2}{\Gamma_{\pi N}} (l+1) \left[l |M_{l+}|^2 + (l+2) |E_{l+}|^2 \right] . \star \qquad (24)$$

For natural parity $P = (-1)^l$

$$\Gamma_\gamma = \frac{k^* q^* \Gamma_0^2}{\Gamma_{\pi N}} l \left[(l-1) |E_{l-}|^2 + (l+1) |M_{l-}|^2 \right] . \star \qquad (25)$$

In these equations, Γ_0 is the total width, $\Gamma_{\pi N}$ the partial πN decay width and k^*, q^* are the respective CMS momenta of the photon and the meson. Γ_γ has to be multiplied by an isospin-factor if one wants to evaluate the partial γN widths for a particular process.

    $\star$ To be precise we have denoted by $E_{l\pm}, M_{l\pm}$ the electric and magnetic multipole amplitudes of definite isospin at the resonance energy $W = M^*$ in the CMS system.

As the data is still insufficient for energies above 400 MeV, an independent determination of the masses and total decay widths from photoproduction data, supplementing the results obtained from πN scattering, is not very accurate.

Apart from the isobar model discussed in chapter III only purely phenomenological methods have been used. We may divide the latter methods into two classes.

I. The energy-independent phenomenological multipole analysis. In this method one makes no special separation between the resonant and background terms but fits the total multipole amplitudes. In our opinion duality yields a strong argument favouring this approach, as the precise separation between the two parts has become very difficult for intermediate energies.

This method, however, has the unpleasant feature that it is very tedious to find a unique solution and to continue this solution from energy to energy.

Such an analysis has been first tried by *Berends* and *Donnachie* [38] up to energies of one GeV. These authors still assumed some nonresonant multipole amplitudes to be given explicitly from dispersive calculations. But dispersion theory can yield only crude values above 0.5 GeV approximately (because of the divergence of the Legendre series).

II. The second approach may be phrased as the energy dependent resonance-fit procedure.

In this method one splits the multipole amplitudes or, equivalently, the partial helicity amplitudes M^{total} into two parts

$$M^{\text{total}} = M^{\text{resonance}} + M^{\text{background}} \qquad (26)$$

Assuming a particular energy dependence, the resonance part is parametrized with some constants which are fitted together with the non-resonant part from the experimental data.

The following ansätze for the resonant term have been used in the literature:

a) *Walker* [47] assumes

$M^{\text{res.}}$ = simple Breit-Wigner formula with energy dependent decay widths.

With this assumption *Walker* has analysed processes (1a) and (1b) up to photon lab. energy of 1.2 GeV. He included the six resonances $P_{33}(1230)$, $P_{11}(1470)$, $S_{11}(1560)$, $D_{13}(1530)$, $D_{15}(1680)$ and $F_{15}(1690)$.

b) *Schmidt* and *Schwiderski* [40] used the ansatz

$$M^{\text{res.}} = \text{const.} \times f$$

where f denotes the πN partial wave amplitude corresponding to $M^{\text{res.}}$. The dominant resonances $D_{13}(1530)$ and $F_{15}(1690)$ have been included in their analysis.

c) The ansatz

$$M^{\text{res.}} = \text{generalized Breit-Wigner resonance}$$

$$= (\text{Breit-Wigner}) \times \exp(i\Phi)$$

has been tried by *Moorhouse* and coworkers [41a]. Φ is a resonance coupling phase which is either taken from πN scattering or allowed to vary. This ansatz was applied to pion photoproduction on protons (process (1a)) in the photon energy region between 550 and 850 MeV, including the $P_{11}(1450)$, $S_{11}(1560)$ and $D_{13}(1520)$ resonance. Using a perturbed K-Matrix parametrization for the dominant s-wave contribution *Moorhouse et al.* fit the π and η photoproduction simultaneously. The analysis of the process (1b) with the same ansatz has recently been completed by *Proia* and *Sebastiani* [41b].

From Eqs. (24) and (25) one then obtains, introducing the multipoles determined by one of the above fitting procedures, the decay widths of the various nucleon resonances into γN. The results are summarized in Table 3. In this table we have included only those references from which an extraction of the half widths from the multipole amplitudes was feasible (as values of the multipoles were available only in these papers).

From Table 3 we see that the various values of the half widths of the $P_{33}(1230)$, $D_{13}(1530)$ and $F_{15}(1690)$ resonances are in rough agreement. But the widths of the other resonances, P_{11}, D_{15} and $S_{11}(1700)$ are completely undetermined as the vastly different values in the table show. However, concerning the $S_{11}(1560)$ and the $F_{37}(1920)$ the results give some evidence for the existence of these resonances in photoproduction of π-mesons.

We would like to stress the following points:

1. *different* models have bee used,

2. different authors have used different experimental data as input,

3. new analyses of the new data (s. Ref. [33, 34, 35, 36, 37]) may change the result got from the analysis of the older data; especially the isoscalar contribution of the $F_{15}(1672)$ has to be changed significantly as *Walker* [1] pointed out at the Liverpool Conference.

For the present situation it is very useful to complete these results by an investigation of the η and K-photoproduction, which will be done in the next section.

Table 3

Resonance exchanged	in the photo-production of	M^* MeV	Γ MeV	Fraction Γ_{MB}/Γ	$EM = E^\alpha/M^\alpha$ $ME = M^\alpha/E^\alpha$ [%]	$\Gamma^\alpha_{\gamma N}$ MeV	$\dfrac{E^{(0)}}{E^{(1/2)}}$ [%]	$\dfrac{M^{(0)}}{M^{(1/2)}}$ [%]	References
P_{33}	π^0, π^+, π^-	1236	120	0.994	EM $-$ 4.6	0.96			Walker [47]
	π^+	1230	107		$-$ 3.1	0.66			Kim [20]
	π^0	1230	108		$-$ 5.9	0.88			Pfeil [20]
D_{13}	π^+, π^0	1519	102	0.5	ME 48.	1.35	0.5	6.6	Walker [47]
	π^-	1519	102		33.	1.0			Walker [47]
	π^+	1518	150		23.	2.51			Kim [20]
	π^0	1530	138		42.	1.56			Pfeil [20]
	π^+, π^0	1499	126		34.	1.9	4.9	5.2	Moorhouse [41a]
	π^-	1499	126		34.	0.97			Proia [41b]
F_{15}	π^+, π^0	1672	104	0.6	ME 50.	0.93	3.0	3.0	Walker [47]
	π^-	1672	104		50.	0.65			Walker [47]
	π^+	1695	130		41.	0.87			Kim [20]
	π^0	1692	191		$-10.$	0.92			Pfeil [20]
S_{11}	π^+, π^0	1560	180	0.34		1.25	$-$ 3.4		Walker [47]
	π^-	1560	180			1.91			Walker [47]
	π^+	1530	159			0.69	-12.0		Moorhouse [41a]
	π^-	1531	159			3.16			Proia [41b]
	η	1570	130	0.66		0.15			Beisenherz [42]
	η	1570	113			0.14			Deans [42]
F_{37}	π^0	1924	170	0.45	EM $-60.$	0.22			Bloom [50]
	$k^+ \Sigma^0$	1917	143	0.06	$+ 24.$	0.27			Meyer zu Hörste [46]
P_{11}	$K^+ \Lambda$	1700	210	0.08		0.021			Orito [43]

Table 3 (continued)

Resonance exchanged	in the photo-production of	M^* MeV	Γ MeV	Fraction Γ_{MB}/Γ	$EM = E^\alpha/M^\alpha$ $ME = M^\alpha/E^\alpha$ [%]	$\Gamma_{\gamma N}^\alpha$ MeV	$\dfrac{E^{(0)}}{E^{(1/2)}}$ [%]	$\dfrac{M^{(0)}}{M^{(1/2)}}$ [%]	References
P_{11}	$\pi^+\pi^0$	1471	200	0.6		0.089		33.0	Walker [47]
	π^-	1471	200			0.0			Walker [47]
	π^+	1412	159			0.069			Kim [20]
	π^0	1489	242			2.0			Pfeil [20]
	π^+,π^0	1469	243			1.7			Moorhouse [41a]
	π^-	1469	243			4.57		4.4	Proia [41b]
D_{15}	π^+,π^0	1652	134	0.42	EM $-50.$	0.09	0.	0.	Walker [47]
	π^-	1652	134		$-50.$	0.09			Walker [47]
	π^+	1656	130		13.2	0.16			Kim [20]
	π^0	1714	249		-49.6	0.99			Pfeil [20]
S_{11}	π^+	1750	228	0.7		0.016			Kim [20]
	π^0	1711	229			4.27			Pfeil [20]
	$K^+\Lambda$	1713	130	0.06		0.013			Orito [43]

Radiative decay widths of nucleon resonances with mass M^* and total decay width Γ. Γ_{MB} means the decay width of the resonance for the decay into the final state, meson and baryon respectively.

EM and ME are the ratios of the electric and magnetic multipole amplitudes at the CMS energy $W = M^*$. α is an isospin index indicating the following connection with the isospin I of the resonances:

$I = 3/2$ $M^\alpha = M^{(3/2)}$ for photoproduction of π and $K^+\Sigma$

$I = 1/2$ $M^\alpha = 3M^{(0)} + M^{(1/2)}$ for photoproduction of $\pi^+, \pi^0, K^+\Sigma$

           $M^\alpha = 3M^{(0)} - M^{(1/2)}$ for photoproduction of π^-

           $M^\alpha = M^{(0)} + M^{(1/2)}$ for photoproduction of $\eta, K^+\Lambda$

From these amplitudes $\Gamma_{\gamma N}$ has been calculated.

VII. Photoproduction of η- and K-mesons

In this section we give a brief review of the photoproduction of η- and K-mesons.

1. η-production

Several isobar model analyses including a smaller or larger number of resonances have been carried out [42]. A critical study of these analyses shows, that only the contribution of the $S_{11}(1560)$ resonance has been observed with true significance; the decay width Γ_γ is stated in Table 3. For completeness we mention that for the coupling constant $g_{\eta NN}/\sqrt{4\pi}$ one obtains roughly

$$\frac{g_{\eta NN}}{\sqrt{4\pi}} = 1.15 \, .$$

2. K-photoproduction

Recently *Orito* [43] has made a phenomenological energy dependent multipole analysis for $K^+\Lambda$-photoproduction. Besides a Breit-Wigner-Ansatz for the resonances, *Orito* fixed the energy dependence of the background by the inclusion of barrier penetration factors. Only a significant $P_{11}(1750)$ resonance contribution has been observed. The indication for the contribution of a $S_{11}(1700)$ resonance occuring in the process $\pi N \to K\Lambda$ [44], is weak.

Very recently new differential cross section data for $K^+\Lambda$ and $K^+\Sigma^0$ at 1.3 and 1.45 GeV have been measured at Bonn [45]. Therefore, the analysis of the data has been repeated at Bonn [46]. The authors included the Born approximation, the K^* exchange and various resonances in simple Breit-Wigner form and fitted the $K^+\Lambda$ and $K^+\Sigma^0$ production processes simultaneously, see Figs. 9, 10, and 11. It turned out, that the value of the coupling constants $g_{N\Lambda K}/\sqrt{4\pi}$ and $g_{N\Sigma K}/\sqrt{4\pi}$ are relatively stable against variations of the resonance parts in the ansatz and even against changing the $\Lambda\Sigma$-transition moment in reasonable limits of 30% of the $SU(3)$ value. The absolute values of these constants are nearly equal to each other:

$$\frac{g_{N\Lambda K}}{\sqrt{4\pi}} = 1.34 \, ,$$

$$\frac{g_{N\Sigma K}}{\sqrt{4\pi}} = -1.89 \, .$$

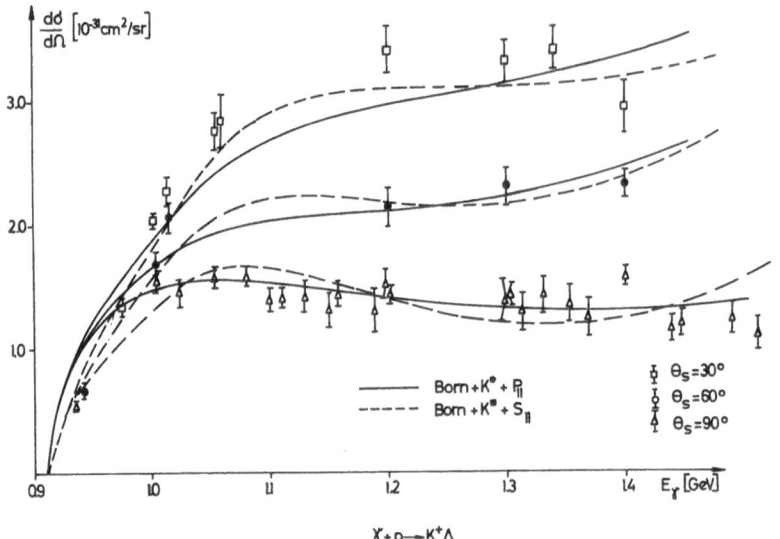

Fig. 9. Excitation curves at $\Theta_{CM} = 30°, 60°$ and $90°$ for the differential cross section in $K^+ \Lambda$ photoproduction. For the data see Ref. [45]. The solid and dashed lines are fit curves taken from Ref. [46]

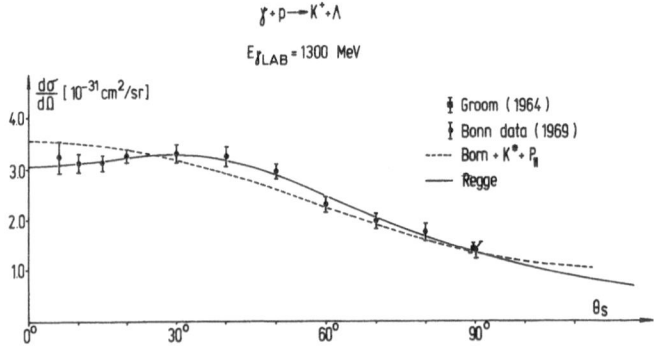

Fig. 10. Angular distribution for $K\Lambda$ differential cross section at 1.3 GeV lab. photon energy. Solid and dashed lines are fit curves taken from Ref. [46]

The analysis of the authors [46] shows that in $K^+ \Lambda$ production the P_{11} is not so well established as predicted by *Orito*. The contributions of the imaginary parts for the M_{1-} or E_{0+} multipoles could not be well separated. However there is strong evidence for the contribution of the $F_{37}(1950)$ in $K^+ \Sigma^0$ production.

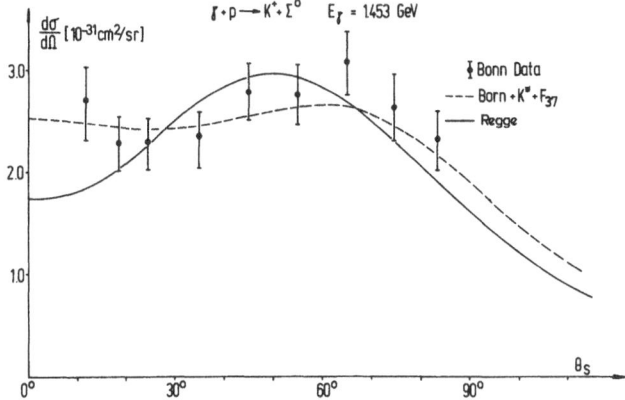

Fig. 11. Angular distribution at $E_y = 1.453$ GeV of the differential cross section in $K^+ \Sigma^0$ photoproduction. The solid and dashed lines are fit curves taken from Ref. [46]

VIII. Conclusions

We have given a general view of the coupling parameters of the photoproduction of pseudoscalar mesons on nucleons. We have shown that the magnetic transition moment for the $\Delta(1230)$ resonance is quite well established; the electric quadrupole transition is very small and different calculations yield values between 0 to -6% of the magnetic transition. Moreover, the decay widths of the transitions of the excited nucleon states $D_{13}(1530)$, $F_{15}(1690)$ and to a smaller reliability also of $F_{37}(1950)$ and $S_{11}(1560)$ can be determined from the presently available data of pion photoproduction.

In η-photoproduction where data are scarce, the $S_{11}(1560)$ contribution is well established; the corresponding decay width is more reliably determined than that following from pion photoproduction.

In $K^+ \Lambda$-photoproduction the contributions from the $P_{11}(1700)$ and $S_{11}(1700)$ resonances are not well determined.

The $F_{37}(1950)$ resonance plays a dominant role in $K^0 \Sigma$ photoproduction.

More and more accurate data and more refined theoretical models are needed for a better determination of the coupling parameters discussed above. Hopefully, then also further resonances which have been observed as yet in πN scattering will also be seen in photoproduction of pseudoscalar mesons. It is encouraging that the different analyses mentioned above converge to compatible results for the dominant resonances.

References

1. *Walker, R. L.:* Proceedings of the 4th International Symposium on Electron and Photon Interactions, Liverpool 1969, p. 23.
2. *Hilger, E., Roegler, H., Tonutti, M.:* Contribution to the Kiev Conference, 1970.
3. *Barbiellini, G., Bologna, G., Capon, G., de Zorzi, G., Fabbri, F. L., Murtas, G. P.,* and *Sette, G., Diambrini, G.,* and *de Wire, J.:* Phys. Rev. **184**, 1402 (1969).
4. *Grilli, M., Spillantini, P., Soso, F., Nigro, M., Schiavuta, E.,* and *Valente, V.:* Nuovo Cimento **54**, 877 (1968).
5. *Favier, J., Alder, J. C., Joseph, C., Vaucher, B., Schinzel, D., Zupančič, Č., Bressani, T., Chiavassa, E.:* CERN 1970 (subm. to Phys. Letters). *D. Schinzel,* private communication.
6. *Berardo, P., Haddock, R. P., Nefkens, B. M. K., Verhey, L. J., Zeller, M. E., Parsons, A. S. L., Truoel, P.:* Phys. Rev. Letters **24**, 419 (1970).
7. a) Aachen-Berlin-Bonn-Hamburg-Heidelberg-München Collaboration: Nucl. Phys. B **8**, 539 (1968).
   b) *Carbonara, F., Fivre, L., Gialanella, G., Lodi-Rizzini, E., Mantovani, G. C., Napolitano, M., Piazza, A., Piazzoli, A., Rinzivillo, R., Rossi, V., Susinno, G.:* Frascati Preprint, LNF-70/16 (1970).
8. *Nishikawa, T., Hiramatsu, S., Kimura, Y., Kobayashi, M., Kondo, K., Okumura, S., Suzuki, T., Takikawa, K., Tsuru, T., Yoshida, H.:* Phys. Rev. Letters **21**, 1288 (1968).
9. *Engels, J., Müllensiefen, A., Schmidt, W.:* Phys. Rev. **175**, 1951 (1968).
10. *Adler, S. L.:* Ann. Phys. (N. Y.) **50**, 189 (1968).
11. *von Gehlen, G.:* Nucl. Phys. B **20**, 102 (1970).
12. *Schwela, D., Rollnik, H., Weizel, R., Korth, W.:* Z. Physik **202**, 452 (1967).
    Schwela, D., Weizel, R.: Z. Physik **221**, 71 (1969).
    Schwela, D.: Z. Physik **221**, 158 (1969).
13. *Berends, F., Donnachie, A., Weaver, D.:* Nucl. Phys. B **4**, 1 (1969).
14. *Castillejo, L., Dalitz, R. H., Dyson, F. J.:* Phys. Rev. **101**, 453 (1956).
15. *Rollnik, H.:* Talk given at the Heidelberg International Conference on Elementary Particles, 1967, p. 400.
16. *Chew, G. F., Goldberger, M. L., Low, F. E., Nambu, Y.:* Phys. Rev. **106**, 1345 (1957).
17. a) *Fischer, G., Fischer, H., von Holtey, G., Kämpgen, H., Knop, G., Schultz, P., Wessels, H., Braunschweig, W., Genzel, H., Wedemeyer, R.:* Nucl. Phys. B **16**, 93 (1970).
    b) *Morand, R., Erickson, E. F., Pahin, J. P., Croissiaux, M. G.:* Phys. Rev. **180**, 1299 (1969).
    c) *Fischer, G., Fischer, H., Heuel, M., von Holtey, G., Knop, G., Stümpfig, J.:* Nucl. Phys. B **16**, 119 (1970).
    d) *Bizot, J., Perez, J. P., Jorba, Y., Treille, D.:* Phys. Letters **25** B, 489 (1969).
18. *Bugg, D. V., Bussey, P. J., Carter, A. A., Dance, D. R., Williams, J. R.:* Rutherford Laboratory Preprint RPP/H/70.
19. *Gourdin, M., Salin, P.:* Nuovo Cimento **27**, 193 (1963).
20. a) *Pfeil, W.:* To be published Z. Phys.
    b) *Kim Yong Taik:* Diplomarbeit, Bonn 1968.
21. *Rollnik, H.:* In: Methods in Subnuclear physics, Vol. III, p. 666 ff. (Ed. by M. Nikolic) 1968.
22. See e. g. *Ebel, G., Pilkuhn, H., Steiner, F.:* Nucl. Phys. B **17**, 1 (1970).
23. *Lovelace, C.:* Pion Nucleon Scattering, ed. by G. L. Shaw and D. J. Wong. New York: Wiley Interscience 1969.
24. *Beg, M. A., Lee B. W., Pais, A.:* Phys. Rev. Letters **13**, 173 (1964).
25. *Harari, H., Lipkin, H. J.:* Phys. Rev. **140**, B 1617 (1965).
26. *Becchi, G., Morpurgo, G.:* Phys. Letters **17**, 352 (1965).
27. *Dalitz, R. H.:* Quarkmodels for the "Elementary Particles". In: High Energy physics, Les Houches Lectures (Ed. by Dewitt and M. Jacob): Gordon & Breach 1965.

28. *Dalitz, R. H., Sutherland, D. G.:* Phys. Rev. **146**, 1180 (1966).
29. see ref. [17] a).
30. *Ash, W. W., Berkelman, K., Lichtenstein, C. A., Ramanauskas, A., Sieman, R. H.:* Phys. Letters **24** B, 165 (1967).
31. *Grilli, M., Nigro, M., Schiavuta, E.:* Nuovo Cimento **49**, 326 (1967).
32. *Gourdin, M., Dufour, J.:* Nuovo Cimento **27**, 1410, (1963). also see ref. [19].
33. *Scheffler, P. E., Walden, P. L.:* Phys. Rev. Letters **24**, 952 (1970).
    Ito, A., Loe, R., Loh, E., Ramanauskas, A., Ritchie, D., Schmidt, W.: Cornell preprint 1970.
34. *Alvarez, R. A., Cooperstein, G., Kalata, K., Lanza, R. C., Luckey, D.:* To be published Phys. Rev.
    See also Phys. Rev. Letters **21**, 1019 (1968).
35. *Holt, J. R.,* (University of Liverpool) private communication.
36. *Fischer, G., Fischer, H., Lenike, M., Wessels, H.:* Bonn, University PIB 1–100 (1970) preprint.
37. *Blüm et al.,* (University Bonn) private communication.
38. *Berends F. A., Donnachie, A.:* Contribution No 140 to the 4th International Symposium on Electron and Photon interactions at high energies, Liverpool, 1969.
39. *Kondo, K., Nishikawa, T., Suzuki, T., Takikawa, K., Yoshida, H., Kimura, Y., Kobayashi, M.:* J. Phys. Soc. of Japan **29**, 13 (1970).
40. *Schwiderski, G., Schmidt, W.:* Karlsruhe report 3/67–1 (1968) (unpublished)
    Schwiderski, G.: Thesis, Karlsruhe 1967.
41a. *Chau, Y. C., Dombey, N., Moorhouse, R. G.:* Phys. Rev. **163**, 1632 (1970).
    Moorhouse, R. G., Rankin, W. A.: Daresbury preprint 1970.
41b. *Proia, A., Sebastiani, F.:* Lettere al Nuovo Cimento III, 483 (1970).
42. *Srinivasan, S. K., Achuthan, P., Sarkar, R. N.:* Nuovo Cimento **59**, 171 (1969).
    Deans, S. R., Holladay, W. G.: Phys. Rev. **169**, 1886 (1968).
    Beisenherz, H. G.: Diplomarbeit, Bonn 1969.
43. *Orito, S.:* University of Tokyo preprint 1969.
    Thom, H.: Phys. Rev. **151**, 1322 (1966).
44. *Iliopoulos, J., Lovelace, C., Wagner, F.:* Contribution No. 256 to the Vienna Conference, 1968.
45. *Bleckmann, A., Herda, S., Opara, U., Schulz, W., Schwille, W. J., Urbahn, H.:* University of Bonn, PI 1–92 (1970), preprint.
46. *Meyer zu Hörste, V.:* Diplomarbeit, Bonn 1970.
47. *Walker, R. L.:* Phys. Rev. **182**, 1729 (1969).
48. *Noelle, P., Pfeil, W., Schwela, D.:* Contribution to the Kiev Conference.
49. *Cocconi, V. T., Fazzini, T., Fidecaro, G., Legros, M., Merrison, W.:* Nuovo Cimento **22**, 494 (1961).
50. *Bloom, E. D.:* Thesis, CALTECH (1967).
    Bloom, E. D., Heusch, C. A., Prescott, C. Y., Rochester, L. S.: Phys. Rev. Letters **19**, 671 (1967).

Dr. *W. Pfeil*
Dr. *D. Schwela*
Physikalisches Institut der Universität
D-5300 Bonn, Nußallee 6

Compilation of Coupling Constants and Low-Energy Parameters

(September 1970 edition)

G. Ebel, D. Julius, A. Müllensiefen, H. Pilkuhn, W. Schmidt,
F. Steiner, Kernforschungszentrum and University Karlsruhe

G. Kramer, G. Schierholz, DESY, Hamburg

B.R. Martin, University College London

J. Pišút, M. Roos, CERN, Geneva

G. Oades, Rutherford Laboratory

J.J. de Swart, University of Nijmegen

CONTENTS

1. Introduction and credits

2. General formulae and definitions
2.1 Cross sections and T-matrix elements, threshold behaviour
2.2 Born terms, coupling constants and dispersion relations
2.3 Isospin and SU(3) properties of coupling constants
2.4 Electromagnetic form factors
2.5 Unstable particles and resonances
2.6 The ϵ and f couplings

3. Strong interactions
3.1 πN scattering
3.2 KN, $\bar{K}$N, $\pi\Sigma$ and $\pi\Lambda$ channels
3.3 NN scattering
3.4 YN scattering
3.5 Meson-meson scattering

4. Electromagnetic interactions
4.1 Photon couplings
4.2 π and η photoproduction

5. Weak interactions

5.1 Muon decay

5.2 Beta decay

5.3 Pion decay

5.4 Goldberger-Treiman relations

5.5 Cabibbo angles from leptonic hyperon decays

5.6 Pionic hyperon decays

6. SU(3) comparison

6.1 pseudoscalar meson-baryon octet (PBB)

6.2 pseudoscalar meson-baryon decuplet (PBB*)

6.3 Vector mesons

1. INTRODUCTION AND CREDITS

This work is the continuation and extension of two previous compilations (Ebel 69,70). The intention is similar
to that of the Particle Data Group (PDG 70, 70a), namely to
facilitate access to the original literature. The responsa-
bilities are distributed as follows: Chapter 2 collabora-
tive, edited by Pilkuhn, sections 3.1 Steiner, 3.2 Martin,
3.3 Kramer and Schierholz, 3.4 de Swart, 3.5 Pisút, chapter
4 Ebel, Julius, Müllensiefen, Schmidt, chapter 5 Roos,
chapter 6 Pilkuhn.

Some of the references to older publications have been
dropped. For the first time we have recommended best values
for some of the coupling constants. The reference system
has been adapted to that of the Particle Data Group. References are found at the end of each chapter, in chapter 3
at the end of each subchapter. Our list of Journal abbreviations is almost identical to that of PDG:

ADVP Advances in Physics

ANP   Annals of Physics

ARNS Annual Review of Nuclear Science

CNPP Comments on Nuclear and Particle Physics

FP   Fortschritte der Physik

JETP English translation of Soviet Physics JETP

JETL	JETP Letters
NC	Nuovo Cimento
NCL	Nuovo Cimento Letters
NP	Nuclear Physics
PL	Physics Letters
PPSL	Proceedings of the Physical Society of London
PR	Physical Review
PRL	Physical Review Letters
PRSL	Proceedings of the Royal Society of London
PTP	Progress of Theoretical Physics
RMP	Reviews of Modern Physics
SJNP	Soviet Journal of Nuclear Physics
ZP	Zeitschrift für Physik

The following conference abbreviations are used:

Lund Conf.   Lund International Conference on Elementary
                Particles, 1969 (Proc. ed. by G. von Dardel.
                Berlingska Boktryckeriet, Lund, Sweden)

Daresbury Conf. 4.th International Symposium on Electron and
                Photon Interactions at High Energies, 1969
                (Proc. ed. by D.W. Graben. Daresbury Nuclear
                Physics Lab.)

Kiev Conf.   15.th International Conference on High-
                Energy Physics, Kiev 1970.

References:

Ebel	69	Lund Conf.	G. Ebel, H. Pilkuhn, F. Steiner
Ebel	70	NP B17,1	G. Ebel, H. Pilkuhn, F. Steiner
PDG	70	RMP 42,87	Particle Data Group
PDG	70a	PL 33B,1	Particle Data Group

2. GENERAL FORMULAE AND DEFINITIONS

2.1 Cross sections, matrix elements, threshold behaviour

(i) Kinematics and differential cross sections

For a general two-particle reaction ab→cd, the Mandelstam variables are defined as follows

$$s = (P_a+P_b)^2 = (P_c+P_d)^2 = (\text{total c.m. energy})^2,$$

$$t = (P_a-P_c)^2 = (P_b-P_d)^2 = (\text{4-momentum transfer})^2, \quad (1.1)$$

$$u = (P_a-P_d)^2 = (P_b-P_c)^2 = m_a^2+m_b^2+m_c^2+m_d^2-s-t.$$

When the target particle b is at rest in the lab, $s = m_a^2+m_b^2+2E_a^{lab}m_b$. The initial and final c.m. momenta are denoted by q and q':

$$4q^2s = (2p_a^{lab}m_b)^2 = \lambda(s,m_a^2,m_b^2), \quad (1.2)$$

$$\lambda(a,b,c) \equiv a^2+b^2+c^2-2(ab+ac+bc) = \left[a-(\sqrt{b}+\sqrt{c})^2\right]\left[a-(\sqrt{b}-\sqrt{c})^2\right], (1.3)$$

where $d\Omega = d\phi\ d\cos\theta$ denotes the c.m. differential solid angle. Then the differential cross section for unpolarized particles is expressed in terms of the T-matrix elements

$$\frac{d\sigma}{dt} = \frac{\pi}{qq'}\frac{d\sigma}{d\Omega} = \frac{\pi}{4q^2s}\frac{1}{(2S_a+1)(2S_b+1)}\sum_{\lambda_a,\lambda_b,\lambda_c,\lambda_d}\left|T_{\lambda_a\lambda_b\lambda_c\lambda_d}\frac{(s,t)}{4\pi}\right|^2,$$

$$(1.4)$$

where each particle's helicity λ_i varies between $+S_i$ and $-S_i$.

(ii) Expansion of helicity amplitudes in partial waves of total angular momentum J

$$T_{\lambda_a\cdots\lambda_d}(s,t) = 8\pi\sqrt{s}\ e^{i(\lambda-\lambda')\phi}\sum_J(2J+1)d_{\lambda\lambda'}^J(\cos\theta)T_{\lambda_a\cdots\lambda_d}^J(s),$$

$$\lambda \equiv \lambda_b-\lambda_a, \quad \lambda' \equiv \lambda_d-\lambda_c, \quad d_{\lambda\lambda'}^J(\cos\theta) \equiv (e^{-i\theta J_y})_{\lambda\lambda'}. \quad (1.5)$$

For fixed J, the $T^J_{\lambda_a \cdots \lambda_d}$ form the $(2S_a+1)(2S_b+1) \times$ $(2S_c+1)(2S_d+1)$ matrix $T^J(ab \rightarrow cd)$ of the reaction $ab \rightarrow cd$. To formulate unitarity, we introduce a large matrix T^J which comprises the partial-wave T-matrices of all communicating open channels. The S-matrix is defined by

$$S^J = 1 + 2i\sqrt{Q}\ T^J \sqrt{Q}, \qquad (1.6)$$

where Q = diagonal matrix of q's of all open channels. Then unitarity requires

$$S^{J\dagger}S^J = 1, \quad \text{Im } T^J = T^{J\dagger}QT^J, \qquad (1.7)$$

where time-reversal invariance $(T^J = T^J_{\text{transposed}})$ has been assumed. Parity conservation allows one to split T^J into two matrices, each of which fulfills (1.7) separately.

(iii) <u>One-meson-one-baryon channels</u>

When only one-meson-one-baryon channels are open (e.g. πN and ηN, or $\pi\Lambda$, $\pi\Sigma$ and $\bar{K}N$), the two-parity submatrices can be distinguished by their orbital angular momentum ℓ

$$T^J_{\frac{1}{2}, \pm\frac{1}{2}} = T^J_{-\frac{1}{2} \mp \frac{1}{2}} = \tfrac{1}{2}(T_{\ell+} \pm T_{(\ell+1)-}), \qquad (1.8)$$

where $\ell\pm$ means $J = \ell \pm \frac{1}{2}$.

The diagonal elements of the matrices $T_{\ell\pm}$ are called $f_{\ell\pm,c}$ (c denotes the channel). They are expressed in terms of the phase shifts $\delta_{\ell\pm}$ and elasticities $\eta_{\ell\pm}$ (supressing the channel index)

$$f_{\ell\pm} = \frac{1}{2iq} (\eta_{\ell\pm}\ e^{2i\delta_{\ell\pm}} - 1), \qquad (1.9)$$

where η varies between 0 and +1. When all channels except one are closed, we have $T_{\ell\pm} = f_{\ell\pm}$ and $\eta_{\ell\pm} = 1$. Also useful are the spin-flip and non-flip amplitudes,

$$-f_2 \sin\theta = g = \sum_{\ell=1}^{\infty} [f_{\ell+} - f_{\ell-}] P'_\ell(\cos\theta)\sin\theta$$

$$\tag{1.10}$$

$$f_1 + f_2 \cos\theta = f = \sum_{\ell=0}^{\infty} [(\ell+1)f_{\ell+} + \ell f_{\ell-}] P_\ell(\cos\theta),$$

for which $d\sigma/d\Omega = |f|^2 + |g|^2$ and the polarization $P(\theta) = 2\text{Im} (f^*g)/d\sigma/d\Omega$. Another useful matrix is $K_{\ell\pm}$, which is defined in terms of $T_{\ell\pm}$ by

$$T_{\ell\pm} - iT_{\ell\pm} QK_{\ell\pm} = K_{\ell\pm} \tag{1.11}$$

By eq. (1.7) and time-reversal invariance, $K_{\ell\pm}$ is real and symmetric.

(iv) <u>Two-baryon channels (NN, or $\Lambda N + \Sigma N$)</u>

$$T^{J\pm}_{\sigma\sigma'} = T^J_{\lambda_a\lambda_b, \lambda_c\lambda_d} \pm T^J_{\lambda_a\lambda_b, -\lambda_c-\lambda_d} \quad (\sigma \equiv 4\lambda_a\lambda_b, \ \sigma' \equiv 4\lambda_c\lambda_d, \ \lambda_a \equiv \tfrac{1}{2})$$

$$\tag{1.12}$$

are matrices of opposite parity and identical J. For T^{J-}, $\sigma = +1$ refers to the spin-singlet states ${}^1S_0$, ${}^1P_1$, ${}^1D_2$ etc, and $\sigma = -1$ to the triplet states with $\ell = J$, viz. $-{}^3P_1$, $-{}^3D_2$ etc. T^{J+} connects the triplet states with $\ell = J\pm1$; here $\sigma = +1$ refers to the combination $(J/(2J+1))^{\frac{1}{2}}|\ell = J - 1\rangle$ $-((J+1)/(2J+1))^{\frac{1}{2}}|\ell = J + 1\rangle$ and $\sigma = -1$ to $((J+1)/(2J+1))^{\frac{1}{2}}$ $|\ell = J - 1\rangle +(J/(2J+1))^{\frac{1}{2}}|\ell = J + 1\rangle$. For $J = 0$ one has $T^{J+}_{1,1} = T({}^3P_0)$ and for $J = 1$

$$T^{1+}_{1,1} = \tfrac{1}{3}T({}^3S_1) + \tfrac{2}{3}T({}^3D_1) - \tfrac{2}{3}\sqrt{2}\ T({}^3S_1 \to {}^3D_1),$$

$$T^{1+}_{-1,-1} = \tfrac{2}{3}T({}^3S_1) + \tfrac{1}{3}T({}^3D_1) + \tfrac{2}{3}\sqrt{2}\ T({}^3S_1 \to {}^3D_1), \tag{1.13}$$

$$T^{1+}_{1,-1} = T^{1+}_{-1,1} = \tfrac{1}{3}\sqrt{2}[T({}^3S_1) - T({}^3D_1)] - \tfrac{1}{3}T({}^3S_1 \to {}^3D_1).$$

For purely elastic NN scattering, the S^{J+} matrix is conveniently parametrized in the (J,ℓ) basis

$$S^{J+} = \begin{bmatrix} n_J e^{2i\delta_J^{J-1}} & i(1-n_J^2)^{\frac{1}{2}}i(\delta_J^{J-1}+\delta_J^{J+1}) \\ i(1-n_J^2)^{\frac{1}{2}}i(\delta_J^{J-1}+\delta_J^{J+1}) & n_J e^{2i\delta_J^{J+1}} \end{bmatrix}, \quad (1.14)$$

with the notation δ_J^ℓ. For NN scattering, T^{J-} is diagonal because of the connection between isospin I, angular momentum J and total spin S:

$$I = 1 \text{ for } J+S \text{ even and } I = 0 \text{ for } J+S \text{ odd} \qquad (1.15)$$

(v) <u>Multipole decomposition for π, η and K-photoproduction</u>

The partial wave helicity amplitudes $T^J_{\lambda_\gamma\lambda\lambda'}$ (where λ_γ, λ and λ' are the helicities of the photon and initial and final nucleon, respectively) are related to the electric and magnetic multipoles $E_{\ell\pm}$ and $M_{\ell\pm}$ by

$$T^J_{1,\frac{1}{2},\pm\frac{1}{2}} = \frac{1}{\sqrt{2}}\left\{\ell M_{\ell+}+(\ell+2)E_{\ell+}\pm(\ell+2)M_{(\ell+1)-}\mp\ell E_{(\ell+1)-}\right\},$$

$$(1.16)$$

$$T^J_{-1,\frac{1}{2},\pm\frac{1}{2}} = \frac{1}{\sqrt{2}}\sqrt{\ell(\ell+2)}\left\{M_{\ell+}-E_{\ell+}\pm M_{(\ell+1)-}\mp E_{(\ell+1)-}\right\},$$

where $\ell\pm$ have the same meaning as in meson-baryon scattering. At threshold the unpolarized differential cross section reduces to

$$\left[\frac{k}{q}\frac{d\sigma}{d\Omega}\right]_{q=0} = |E_{0+}|^2, \qquad (1.17)$$

where k and q are the photon and meson c.m. momenta.

(vi) <u>Threshold behaviour of elastic scattering</u>. The threshold behaviour of a partial wave amplitude depends on its orbital angular momentum ℓ

$$q^{2\ell+1}\cot\delta_\ell = \pm a_\ell^{-1} + \frac{1}{2}q^2 r_\ell, \qquad (1.18)$$

where r_ℓ is the effective range and a_ℓ the scattering
length. (We use the + sign for meson-baryon scattering
and the traditional - sign for baryon-baryon scattering).

(vii) Coulomb corrections to elastic scattering (Mott 65)

Let m denote the reduced mass and e_a and e_b the charges
of particles a and b ($e^2/4\pi = \alpha = 137.036^{-1}$). Define

$$\eta = \frac{m}{q}\frac{e_a e_b}{4\pi}, \quad \sigma_\ell = \arg\Gamma(\ell+1+i\eta) = \sigma_o + \sum_{n=1}^{\ell} \arctan\frac{\eta}{n}, \quad (1.19)$$

$$\sigma_o = -0.5772\eta .$$

The amplitude for pure Coulomb scattering must be added to
the elastic non-flip amplitude f of (1.10). The non relativistic form is

$$f_c(\theta) = -\eta[2q \sin^2\tfrac{1}{2}\theta]^{-1} \exp\{2i\sigma_o - 2i\eta \ln \sin\tfrac{1}{2}\theta\} \quad (1.20)$$

For inclusion of spin and relativistic effects see
(Roper 65). Also multiply each partial wave in (1.10) by
$e^{2i\sigma_\ell}$ for elastic and $e^{i\sigma_\ell}$ for charge exchange scattering.
Further note that the partial wave amplitudes are changed
from their charge independent values so beware (see Oades
70); e.g. for $\pi^+ p \to \pi^+ p$, phase $\delta_{\ell\pm}^{(3/2)}$ becomes $\tau_{\ell\pm}^{(+)}$ which
causes the effective range formula to become (Bethe 49)

$$c^2 q \cot \delta_o + 2q\eta h(\eta) = \pm \frac{1}{a} + \frac{1}{2}r q^2, \quad (1.21)$$

$$c^2 = \left|\Gamma(1+i\eta)e^{-\frac{1}{2}\pi\eta}\right|^2 - \frac{2\pi\eta}{e^{2\pi\eta}-1},$$

$$(1.22)$$

$$h(\eta) = \text{Re}\ \frac{\Gamma'(-i\eta)}{\Gamma(-i\eta)} - \ln\eta = \eta^2 \sum_{n=1}^{\infty} \frac{1}{n(n^2+\eta^2)} - \ln\eta - 0.5772...$$

2.2. Born terms, coupling constants, dispersion relations

(i) Born terms in the s-channel

The amplitudes $T_{\lambda_a \cdots \lambda_d}(s,t)$ are analytic functions of s
and t which may be characterized by their singularities.
Among these are poles, the residues of which are conveniently derived from Born terms, which correspond to the
formation or exchange of a one-particle state. The structure of an s-channel Born term is as follows:

$$T_s^{Born} = \sum_{\lambda_e} V_{\lambda_e, \lambda_c \lambda_d} \frac{1}{m_e^2 - s} V_{\lambda_a \lambda_b, \lambda_e} , \qquad (2.1)$$

where m_e is the mass of the intermediate particle, and λ_e
varies between $+S_e$ and $-S_e$. The vertex functions V depend
on the four-momenta and spins of the three particles that
are coupled at the vertex (see fig. 1(a)). When e is a
resonance (m_e^2 replaced by $m_e^2 - i m_e \Gamma_e$, see sect. 2.5),
$V_{\lambda_a \lambda_b, \lambda_e}$ and $V_{\lambda_e, \lambda_c \lambda_d}$ describe its formation and decay,
respectively. In the Born approximation, V contains the
minimal number of four-momenta which still allows the most
general spin dependence and which is consistent with possible additional symmetry laws (parity, gauge invariance,
chiral symmetry etc. see Pilkuhn 67 p. 287). This determines V up to a few "coupling" constants, normally just
one. For πN, $\bar{K}N$ etc. scattering we have

$$V_{\lambda_b \lambda_e}^{ps} = G_{ab,e} \bar{u}_e(\lambda_e) \gamma_5 u_b(\lambda_b), \quad \gamma_5 \equiv \begin{pmatrix} 0 & \sigma_o \\ \sigma_o & 0 \end{pmatrix}, \quad \bar{u}u = 2m \qquad (2.2)$$

and the Dirac spinor u_e is continued off its mass shell
such that

$$\sum_{\lambda_e = -\frac{1}{2}}^{+\frac{1}{2}} u_e(\lambda_e) \bar{u}_e(\lambda_e) = P_{e\mu} \gamma^\mu + m_e,$$

$$(2.3)$$

$$\gamma^0 = \gamma_o = \begin{pmatrix} \sigma_o & 0 \\ 0 & -\sigma_o \end{pmatrix}, \quad \gamma^i = -\gamma_i = \begin{pmatrix} 0 & \sigma_i \\ -\sigma_i & 0 \end{pmatrix}.$$

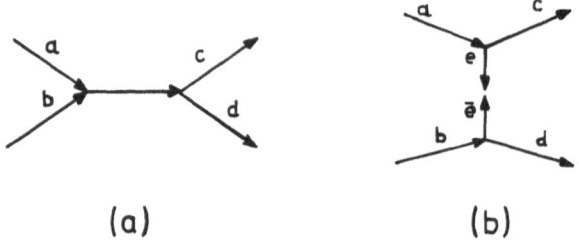

$$(a) \qquad\qquad (b)$$

Fig. 1.(a)s-channel pole, (b)t-channel pole (e and $\bar{e}$ both outgoing).

When the condition $V(P_{a\mu}=0)=0$ is imposed, one takes the pseudovector coupling

$$V^{pv}_{\lambda_b \lambda_e} = \sqrt{4\pi}\ \frac{f}{m_\pi^+}\ \bar{u}_e(\lambda_e)\gamma_5 P_{a\mu}\gamma^\mu\ u_b(\lambda_b), \qquad (2.4)$$

instead of the pseudoscalar coupling (2.2). The product of the spin summation (2.3) and the pole $(m_e^2-s)^{-1}$ is called a propagator.

(ii) <u>Born terms in the t-channel</u>

By crossing symmetry, poles also appear in t and u. A t-channel Born term has the structure (see fig. 1(b))

$$T_t^{Born} = \sum_{\lambda_e} V_{a,ce}\ \frac{1}{m_e^2-t}\ V_{b,d\bar{e}}\ , \qquad (2.5)$$

where the exchanged particle e is treated as an outgoing antiparticle of momentum $-P_e$ at the lower vertex. For π-exchange in NN scattering the V's to be inserted in eq. (2.5) are given by eq. (2.2) or eq. (2.4). For ρ-exchange, the $\rho\pi\pi$ vertex is

$$V_{a,ce}(\lambda_e) = G_{a,ce}(P_a+P_c)_\mu \epsilon^{*\mu}(\lambda_e)\ , \qquad (2.6)$$

where $\epsilon(\lambda_\rho)$ is the normal spin function for a spin-1 particle of mass m_e. It is continued off its mass shell such that

$$\sum_{\lambda_e=-1}^{1} \epsilon_\mu(\lambda_e)\epsilon_\nu^*(\lambda_e) = -g_{\mu\nu} + \frac{e_\mu e_\nu}{m_e^2}, \quad g_{\mu\nu} = \begin{pmatrix} 1 & 0 & 0 & 0 \\ 0 & -1 & 0 & 0 \\ 0 & 0 & -1 & 0 \\ 0 & 0 & 0 & -1 \end{pmatrix},$$

$$e_\mu \equiv (P_a - P_c)_\mu .$$

(2.7)

The vector meson-NN vertex is of the form

$$V_{b,d\bar{e}} = \bar{u}_d(\lambda_d)(G^V_{b,d\bar{e}}\gamma_\mu + G^T_{b,d\bar{e}}\frac{\sigma_{\mu\nu}}{m_b+m_d} e_\nu)u_b(\lambda_b)\epsilon^\mu(\lambda_e), \quad (2.8)$$

with $\sigma^{\mu\nu} = \frac{1}{2}(\gamma^\mu\gamma^\nu-\gamma^\nu\gamma^\mu)$.

(iii) Dispersion relations

Dispersion relations are most conveniently written down for amplitudes that are free from kinematical singularities (see A. Martin 70). For pseudoscalar meson-baryon scattering, these are the amplitudes A and B defined by

$$T(\lambda,\lambda')=\bar{u}_d(\lambda')[A+\tfrac{1}{2}(P_a + P_c)_\mu\gamma^\mu B]u_b(\lambda). \quad (2.9)$$

They are related to f_1 and f_2 of (1.10) by

$$f_1 = \frac{1}{8\pi\sqrt{s}} \sqrt{(E_b+m_b)(E_d+m_d)}[A+B(s^{\frac{1}{2}} - \tfrac{1}{2}m_b - \tfrac{1}{2}m_d)]$$

(2.10)

$$f_2 = \frac{1}{8\pi\sqrt{s}} \sqrt{(E_b-m_b)(E_d-m_d)}[-A+B(s^{\frac{1}{2}}+ \tfrac{1}{2}m_b+ \tfrac{1}{2}m_d)].$$

For certain reactions (including $\pi N, \pi\Lambda, \pi\Sigma$ scattering and $\pi\Lambda\to\pi\Sigma$, but excluding KN and NN scattering), disp. rel. can be derived from axiomatic field theory, at least for a certain range of $t(-18.0\ m_\pi^2 < t < 4m_\pi^2$ for πN scattering. See Sommer 70 for a recent review).

2.3 Isospin and SU(3) properties of coupling constants

The phase conventions of Condon and Shortley are used for Clebsch-Gordan coefficients, and in meson-baryon product states the baryon comes first (compare Particle Data Group 1970). For example, the $|I,I_3\rangle$ pion-nucleon states are

$$|\tfrac{1}{2},\tfrac{1}{2}\rangle = \tfrac{1}{\sqrt{3}}\,|p\pi^0\rangle - \sqrt{\tfrac{2}{3}}\,|n\pi^+\rangle, \quad |\tfrac{1}{2},-\tfrac{1}{2}\rangle = \sqrt{\tfrac{2}{3}}\,|p\pi^-\rangle - \tfrac{1}{\sqrt{3}}\,|n\pi^0\rangle \quad (3.1)$$

(i) <u>Pion-baryon coupling constants</u>

$$G \equiv G(p,p\pi^0) = -G(n,n\pi^0) = -\tfrac{1}{\sqrt{2}}\,G(p,n\pi^+) = \tfrac{1}{\sqrt{2}}\,G(n,p\pi^-) \qquad (3.2)$$

$$G_{\Lambda\Sigma\pi} \equiv G(\Sigma^+,\Lambda\pi^+) = G(\Sigma^0,\Lambda\pi^0) = G(\Sigma^-,\Lambda\pi^-)$$

$$= -G(\Lambda,\Sigma^+\pi^-) = G(\Lambda,\Sigma^0\pi^0) = -G(\Lambda,\Sigma^-\pi^+) \qquad (3.3)$$

The second row of (3.3) follows from the first by replacing a final state pion by an initial state antipion and observing the signs from charge conjugation,

$$C|\pi^\pm\rangle = -|\pi^\mp\rangle, \quad C|\pi^0\rangle = |\pi^0\rangle \qquad (3.4)$$

which follow from the Condon-Shortley phase convention in eq. (3.1)

$$G_{\Sigma\Sigma\pi} \equiv G(\Sigma^+,\Sigma^+\pi^0) = -G(\Sigma^+,\Sigma^0\pi^+) = G(\Sigma^-,\Sigma^0\pi^-) = -G(\Sigma^-,\Sigma^-\pi^0)$$

$$= G(\Sigma^0,\Sigma^+\pi^-) = -G(\Sigma^0,\Sigma^-\pi^+), \qquad G(\Sigma^0,\Sigma^0\pi^0) = 0 \qquad (3.5)$$

$$G_{\Xi\Xi\pi} \equiv G(\Xi^0,\Xi^0\pi^0) = -G(\Xi^-,\Xi^-\pi^0) = -\tfrac{1}{\sqrt{2}}G(\Xi^0,\Xi^-\pi^+) = \tfrac{1}{\sqrt{2}}G(\Xi^-,\Xi^0\pi^-) \qquad (3.6)$$

The SU(3) values of these coupling constants are expressed in terms of G and $\alpha = D/(D+F)$ as follows:

$$\alpha \equiv \sqrt{\tfrac{3}{4}}\,G_{\Lambda\Sigma\pi}/G, \quad G_{\Sigma\Sigma\pi} = 2(1-\alpha)G, \quad G_{\Xi\Xi\pi} = -(2\alpha-1)G \qquad (3.7)$$

(ii) <u>Kaon-baryon coupling constants</u>

$$G_{N\Sigma K} \equiv G(p,\Lambda K^+) = G(n,\Lambda K^0) = -G(\Lambda,pK^-) = G(\Lambda,n\bar{K}^0) \qquad (3.8)$$

$$G_{N\Sigma K} \equiv -G(p,\Sigma^0 K^+) = G(n,\Sigma^0 K^0) = G(\Sigma^0,pK^-) = G(\Sigma^0,n\bar{K}^0) \tag{3.9}$$

$$= \frac{1}{\sqrt{2}}G(p,\Sigma^+K^0) = -\frac{1}{\sqrt{2}}G(n,\Sigma^-K^+) = \frac{1}{\sqrt{2}}G(\Sigma^-,nK^-) = \frac{1}{\sqrt{2}}G(\Sigma^+,p\bar{K}^0)$$

The signs of the $\bar{K}$ couplings in (3.8) and (3.9) follow from

$$C|K^\pm\rangle = -|K^\mp\rangle, \quad C|K^0\rangle = |\bar{K}^0\rangle \tag{3.10}$$

The SU(3) values of $G_{N\Lambda K}$ and $G_{N\Sigma K}$ are

$$G_{N\Lambda K} = -\frac{1}{\sqrt{3}}(3-2\alpha)G, \quad G_{N\Sigma K} = -(2\alpha-1)G = G_{\Xi\Xi\pi} \tag{3.11}$$

(iii) η-baryon coupling constants

Here all Clebsch-Gordan coefficients are +1, and the definitions are obvious. The SU(3)-values are

$$G_{NN\eta} = \frac{1}{\sqrt{3}}(3-4\alpha)G, \quad G_{\Sigma\Sigma\eta} = -G_{\Lambda\Lambda\eta} = G_{\Lambda\Sigma\pi} = \sqrt{\frac{4}{3}}\,\alpha G \tag{3.12}$$

(iv) Vector meson couplings

The BBV coupling constants in (2.8) follow from the BBP ones by replacing $\pi \to \rho$, $K \to K^*$, $\eta \to \phi_8$ = octet component of ω and ϕ:

$$\phi = \phi_8 \cos\theta + \omega_1 \sin\theta, \quad \omega = -\phi_8 \sin\theta + \omega_1 \cos\theta \tag{3.13}$$

Only for the $NN\rho$ and $\Xi\Xi\rho$ couplings, a factor $\frac{1}{2}$ is introduced (Sakurai's universality normalization), e.g.

$$G(p,p\rho^0) \equiv \frac{1}{2}G_{NN\rho} \tag{3.14}$$

The PPV couplings are then obtained by replacing $N \to K$, $\Sigma \to \pi$, $\Lambda \to \eta$. Universality requires (Diu 65)

$$G^V_{NN\rho} = G^V_{\Sigma\Sigma\rho} = G^V_{\Xi\Xi\rho} = G_{\pi\pi\rho} = G_{KK\rho} = 2G_{PPV}(1+\alpha_1), \tag{3.15}$$

where α_1 is SU(3)-breaking ($\alpha_{PPV}=0$ by charge conjugation). Broken SU(3)+universality for the other PPV-couplings:

$$G_{KK\phi_8} = \sqrt{3}\,G_{PPV}(1-\alpha_1), \quad G_{K\pi K^*} = G_{PPV}(1-\alpha_1/2) = -G_{KnK^*}/\sqrt{3} \tag{3.16}$$

2.4 Electromagnetic form factors

The T-matrix for elastic electron scattering on a hadron with initial and final momenta and helicities p, λ and p', λ', respectively, is

$$T(\lambda'\lambda) = e\bar{u}\gamma^\mu u \frac{1}{t} \langle p'\lambda' | J^e{}_\mu{}^m | p\lambda \rangle. \tag{4.1}$$

The electromagnetic vertex functions of the pion and the nucleon are parametrized by form factors as follows

$$\langle p' | J_\mu^{em} | p \rangle = eF_\pi(t)(p+p')_\mu, \quad F_\pi(0)=1 \tag{4.2}$$

$$\langle p'\lambda' | J^{em} | p\lambda \rangle = e\bar{u}_{p'}(\lambda')\{F_1(t)\gamma_\mu + \frac{\kappa}{2M}F_2(t)\sigma_{\mu\nu}(p'-p)^\nu\}u_p(\lambda), \tag{4.3}$$

where M is the nucleon mass and κ is the anomalous magnetic moment (cf. eq. (2.8)). The normalizations are

$$F_{1p}(0) = F_{2p}(0) = F_{2n}(0)=1, \quad F_{1n}(0)=0 \tag{4.4}$$

The relation to the electric and magnetic form factors $G_E(t)$ and $G_M(t)$ is given by

$$G_E(t) = F_1(t) + \frac{\kappa t}{4M^2} F_2(t), \quad G_M(t) = F_1(t) + \kappa F_2(t), \tag{4.5}$$

with the normalizations

$$G_{Ep}(0) = 1, \quad G_{Mp}(0) = \mu_p = 2.7928,$$

$$G_{En}(0) = 0, \quad G_{Mn}(0) = \mu_n = -1.9131. \tag{4.6}$$

2.5 Unstable particles and resonances

(i) Decay rates

The total decay rate Γ is defined by the exponential decrease of itensity, $dI/d\tau = -\Gamma I$, where τ is the proper time. The two-particle decay $(d \rightarrow 1,2)$ of particle d with de-

cay momentum $p = [\lambda(s, m_1^2, m_2^2)/4s]^{\frac{1}{2}}$, with θ_1 the angle between the momentum of particle 1 and spin quantization axis
of d is then given by

$$m\Gamma_{12} = \frac{p}{16\pi\sqrt{s}} \sum_{\lambda_1\lambda_2} \int_{-1}^{1} d\cos\theta_1 \left| T_{\lambda_d\lambda_1\lambda_2}(s, \cos\theta_1) \right|^2. \qquad (5.1)$$

Averaging over λ_d is unnecessary. For identical particles
in the final state ($\pi^0 \to \gamma\gamma$), the integral terminates at
$\cos\theta = 0$.

(ii) Resonance production

Resonance production ab→cd with the subsequent decay
d→1,2 is described by the product of production and decay
amplitudes and a propagator (Pilkuhn 67)
$T(ab \to cd; d \to 1,2)$

$$= \sum_{\lambda_d} T_{\lambda_d\lambda_1\lambda_2}(s_{12}, \cos\theta_1, \phi_1) \frac{1}{m^2 - s_{12} - im\Gamma} T_{\lambda_a\lambda_b\lambda_c\lambda_d}, \qquad (5.2)$$

which leads to the effective mass distribution

$$F(s_{12}) = \frac{\text{const} \cdot m\Gamma_{12}}{(m^2 - s_{12})^2 + m^2\Gamma^2}, \quad s_{12} \equiv (P_1 + P_2)^2, \qquad (5.3)$$

provided $T_{\lambda_a \dots \lambda_d}$ does not vary between $m^2 (=m_d^2)$ and s_{12}.

(iii) Resonance formation

For a resonance in the s-channel (ab→e→cd) the term
"resonance formation" is used. Its contribution to the re-
sonating partial wave amplitude is

$$f_\ell = (m^2 - s - im\Gamma)^{-1} (\frac{m}{q}, \Gamma_{cd} \frac{m}{q} \Gamma_{ab})^{\frac{1}{2}}. \qquad (5.4)$$

In the elastic channel ab→res→ab

$$f_\ell = (m^2 - s - im\Gamma)^{-1} \frac{m}{q} \Gamma_{ab}; \quad \frac{\Gamma_{ab}}{\Gamma} = \text{elasticity x.} \qquad (5.5)$$

(iv) P-wave resonances

The Born approximation for the decay matrix element,

$$T^{Born}_{\ell=1} = Ge_\mu(\lambda_d)(P_2-P_1)^\mu \quad (=2Gp\,\cos\theta_1 \quad \text{for } \lambda_d=0) \tag{5.6}$$

is analogous to eq. (2.6) and leads to the correct threshold behaviour,

$$m\Gamma = \frac{G^2}{4\pi}\,\frac{2}{3}\,\frac{p^3}{\sqrt{s}} \; . \tag{5.7}$$

A more realistic p-dependence is given by the penetration factor (R is the range of the potential)

$$m\Gamma = \frac{G^2}{4\pi}\,\frac{2}{3}\,\frac{p}{\sqrt{s}}\,\frac{p^2}{1+R^2p^2} \; . \tag{5.8}$$

Note: Experimentalists frequently define $m\Gamma$ as the full width at half maximum of the distribution (5.3). The value of $m\Gamma(s=m^2)$ according to eqs. (5.7) and (5.8) is somewhat smaller.

2.6 The ε and f couplings

(i) The ε meson

The Born approximation for the $\varepsilon\pi\pi$ vertex is simply a constant,

$$V_{a,a\varepsilon} = G_{\varepsilon\pi\pi}\cdot m_\varepsilon, \quad a = \pi^+,\pi^0,\pi^- \; . \tag{6.1}$$

$G_{\varepsilon\pi\pi}$ is related to the total $\varepsilon\to\pi\pi$ width by

$$G^2_{\varepsilon\pi\pi}/4\pi = \frac{4}{3}\,\Gamma_{\varepsilon\pi\pi}/p \; . \tag{6.2}$$

The εNN vertex is of the form

$$V_{b,d\varepsilon} = G_{\varepsilon NN}\,\bar{u}_d(\lambda_d)\,u_b(\lambda_b) \tag{6.3}$$

for both p,pε and n,nε .

(ii) The f meson

The Born approximation for the $f\pi\pi$ vertex is

$$V_{a,af}(\lambda_f)= G_{f\pi\pi}(P_a+P_c)_\mu (P_a+P_c)_\nu \, \epsilon^{*\mu\nu}(\lambda_f)/m_f, \qquad (6.4)$$

where $\epsilon^{\mu\nu}(\lambda_f)$ is the spin function for a spin-2 particle of mass m_e ($\epsilon^{\mu\nu}$ is symmetric and traceless, $\epsilon^{\mu\nu}(2)=\epsilon^\mu(1)\epsilon^\nu(1)$). It is continued off its mass shell such that (Pilkuhn 67)

$$\sum_{\lambda_f=-2}^{2} \epsilon^*_{\mu\nu}(\lambda_f)\epsilon_{\rho\sigma}(\lambda_f)= \tfrac{1}{2}(P_{\mu\rho}P_{\nu\sigma}+P_{\mu\sigma}P_{\nu\rho})- \tfrac{1}{3} P_{\mu\nu}P_{\rho\sigma} \qquad (6.5)$$

where $P_{\mu\nu}$ is the spin-1 summation given in (2.7). $G_{f\pi\pi}$ is related to the total $f\to\pi\pi$ width by

$$G^2_{f\pi\pi}/4\pi = \tfrac{5}{8}\, \Gamma_{f\pi\pi}\, m^4_f/p^5. \qquad (6.6)$$

The fNN vertex is of the form

$$V_{b,df}= \bar{u}_d(\lambda_d)\left[G^{(1)}_{fNN}((P_b+P_d)_\mu\gamma_\nu+(P_b+P_d)_\nu\gamma_\mu)/m_p\right.$$

$$\qquad (6.7)$$

$$\left. + G^{(2)}_{fNN}(P_b+P_d)_\mu (P_b+P_d)_\nu/m^2_p\right]u_b(\lambda_b)\epsilon^{\mu\nu}(\lambda_f).$$

We have followed the habit of making the coupling constants dimensionless by introducing some socalled "natural" masses. This habit has led to misunderstandings in connection with SU(2)- and SU(3)-comparisons. For example, for the $f\Lambda\Lambda$ coupling one should also use m_p in (6.7), in order to avoid unmotivated SU(3)-breaking. (The truly natural mass could be m_f or m_{quark}!). Notice in this connection the use of m_{π^+} in (2.4) for all pseudoscalar mesons.

References

Bethe   49 PR 76,38   H. Bethe

Diu   65 NC 35,465   B.Diu, H.R. Rubinstein, J.L.
                          Basdevant

Mott   65 The Theory of Atomic Collisions N.F. Mott,
                          H.S.W. Massey (Oxford Univ.Press)

Roper   65   PR 138,B190 L.D. Roper, R.M. Wright, B.T.
                      Feld

Pilkuhn   67   The Interaction of Hadrons, H. Pilkuhn (North
                      Holland)

A. Martin 70   Elementary Particle Theory, A.D.Martin, T.D.
                      Spearman (North-Holland)

Oades   70   Springer Tracts 55 G.C. Oades

PDG   70   RMP 42,87   Particle Data Group

Sommer   70   FP   G. Sommer, to be publ.

3. STRONG INTERACTIONS

3.1 πN scattering

Units: $\hbar = c = m_\pi^+ = 1$, $M = m_p = 6.722$, $\hbar/m_\pi^+ c = 1.414 \text{fm}$.
Reviews: Hamilton 63; Hamilton 67; Moorhouse 69; Höhler 70

(i) S-wave scattering lengths and πN coupling constants

The S-wave scattering lengths of isospin $I = \frac{1}{2}$ or $\frac{3}{2}$ are

$$a_{2I} \equiv f_{o+}^I (q=0),$$

where f_{o+}^I and q are defined in eqs. (1.9) and (1.2), respectively.

The πN coupling constant is

$$\frac{G^2}{4\pi} \equiv 4M^2 f^2 = 180.770 \, f^2,$$

where G is defined by eqs. (3.2) and (2.2). The nucleon
Born term in the forward dispersion relation for the
isospin-odd amplitude is

$$\frac{T^{(-)}}{4\pi} = \frac{1}{8\pi}\left[T(\pi^- p) - T(\pi^+ p)\right] - \frac{G^2}{4\pi} \frac{2\omega\omega_N}{\omega^2 - \omega_N^2}, \omega_N = \frac{1}{2M} = 0.074, \omega = \frac{s - M^2 - 1}{2M}.$$

a_1-a_3	a_1+2a_3	$f^2 \cdot 10^3$	$G^2/4\pi$	Reference	Method
0.271 ±0.007	-0.002 ±0.008			Hamilton 66	Extrap. of low-energy $\pi^{\pm}p$ cross sections
0.266 ±0.017	0.056[a] ±0.022	for 82.2	14.86	Lovelace 67	Backward d.r., fixed f^2
0.297 0.271		for 81 for 76	14.64 13.74	Höhler 68	Fixed-t d.r., fixed f^2
	0.045			Zovko 69	D.r. for inverse amplitude
0.288 ±0.010	-0.021 ±0.010	80 ± 5	14.46	Höhler 69	Fixed-t d.r.
0.277 ±0.009	-0.026 ±0.008	81.5 ± 1.6	14.73	Samarana. 69	
	-0.010	for 81	14.64	Engels 70a	Backward d.r., fixed f^2
0.288 +0.012 -0.018	0.000 +0.045 -0.035	81 +3 -4	14.64 +0.54 -0.72	Recommended values	

[a] For a critical comment on this number see Engels 70a.

ii) P-wave scattering lengths

$$a_{2I,2J} \equiv \lim_{q^2 \to 0} \frac{f_{1\pm}^{I}}{q^2} .$$

$a_{11}-a_{31}$	$a_{13}-a_{33}$	$a_{11}+2a_{31}$	$a_{13}+2a_{33}$	Reference	Method
-0.045 -0.039	-0.245 -0.237	for $f^2=0.081$ for $f^2=0.077$		Baacke 69	Fixed-t d.r., fixed f^2
-0.051	-0.243	-0.168	0.396	Höhler 69	Fixed-t d.r.
-0.049	-0.250	-0.160	0.431	Collins 69	
-0.045 ±0.006	-0.243 ±0.007	-0.164 ±0.008	0.414 ±0.021	Recommended value	

(iii) The $\Delta(1236)$ resonance

(a) Fit to the $\delta_{1+}^{\frac{3}{2}}$ phase shift

$$M_\Delta \, \Gamma(s) = (M_\Delta^2 - s) \, \tan \delta_{1+}^{\frac{3}{2}}(s) \; .$$

A good fit up to $\sqrt{s} = 1.36$ GeV is possible with eq. (5.8) and

$$\frac{G_\Delta^2}{4\pi} = 47.25, \quad R^2 = 23.7/\text{GeV}^2 \; .$$

(Pilkuhn 69). Eq. (5.7) gives a poorer fit, with $G_\Delta^2/4\pi = 20.7$ (since $1 + R^2 p^2 (M_\Delta^2) = 2.29$). A Born term with relativistic spinors gives a correction to eq. (5.7):

$$M_\Delta \Gamma = \frac{G_{\pi N\Delta}^2}{4\pi} \frac{2}{3} p^3 \frac{(M_\Delta + M)^2 - 1}{4 M_\Delta}, \quad \frac{G_{\pi N\Delta}^2}{4\pi} = 0.33 = 17/\text{GeV}^2,$$

to be compared with 0.34 corresponding to the dimensionless $G_\Delta^2/4\pi = 20.7$ in the non-relativistic formula.—For a further fit see Taguchi 69.

(b) Fit to the region $M-1 \leqslant \sqrt{s} \leqslant M+1$, $|t| < 4$

$$\frac{T_\Delta^{(-)}(t=0)}{4\pi} = - \frac{G_{\pi N\Delta}^2}{4\pi} \frac{2}{9} q_\Delta^2 \frac{(M_\Delta + M)^2 - 1}{M} \frac{\omega}{\omega_\Delta^2 - \omega^2} \; ,$$

$$q_\Delta = 1.65, \quad \omega_\Delta = \frac{M_\Delta^2 - M^2 - 1}{2M} = 2.40.$$

Born term fitted to the Δ-contribution of the fixed-t d.r. evaluated from $\delta_{1+}^{\frac{3}{2}}$ (Höhler 69a)

$$\frac{G_{\pi N\Delta}^2}{4\pi} = 0.26 = 13.5/\text{GeV}^2, \quad f_{\pi N\Delta}^2 = \frac{M}{3M_\Delta} \frac{G_{\pi N\Delta}^2}{4\pi} = 0.066.$$

(To be used at and below threshold. Good to 10% at threshold).

(iv) The $\rho, \varepsilon, \text{S}^*, f$ and g couplings

$G_{\rho\pi\pi}$ and $G_{\rho NN}^{V,T}$ are defined by eqs. (2.6), (2.8) and (3.14).

The coupling constants $G_{\varepsilon\pi\pi}$, $G_{S^*\pi\pi}$, $G_{\varepsilon NN}$, G_{S^*NN}, $G_{f\pi\pi}$ and $G_{fNN}^{(1,2)}$ are defined by eqs. (6.1), (6.3), (6.4) and (6.7). For g exchange we only give the residues $R_g^{(1,2)}$ defined by the Born terms of the isospin-odd amplitudes

$$C_g^{(-)} \equiv A_g^{(-)} + \frac{M}{4P_t^2}\,(s-u)\ B_g^{(-)} = \frac{R_g^{(1)}\ P_3(x)}{m_g^2 - t}$$

$$B_g^{(-)} = \frac{R_g^{(2)}P_3'(x)}{m_g^2 - t}, \quad x = \frac{s-u}{4P_t q_t}, \quad P_t = \left(M^2 - \frac{t}{4}\right)^{1/2}, \quad q_t = \left(\frac{t}{4} - 1\right)^{1/2}.$$

The ε, S^*, f and g particles are denoted by PDG 1970:
$\varepsilon = \eta_{o+}(700)$, $S^* = \eta_{o+}(1060)$, f= f(1260) and g= $\rho_N(1660)$
with $J^P = 3^-$ assumed.

$\dfrac{G_{\rho\pi\pi}G_{\rho NN}^V}{4\pi}$	$\dfrac{G_{\rho NN}^T}{G_{\rho NN}^V}$ a)	$R_g^{(1)}$	$R_g^{(2)}$	Reference	Method
2.8 ±0.1		(m_ρ = 774 MeV)		Sakurai 66	Complete ρ dominance of a_1-a_3, for a_1-a_3= 0.271±0.007
2.85 ±0.3	≡1.85	(m_ρ = 740 MeV)		Hamilton 67	Partial wave d.r.
2.19	3.23	(m_ρ = 765 MeV)		Höhler 68	Cini-Fubini appr.
1.6	3.41 b)	590 c)	+32 c)	Engels 70	Backward d.r.
3.7 c)	1.3 b)c)	340 c)	-65 c)	Schlaile 70	Fixed-u d.r.
1.7 ±0.5	4.7 b) ±1.5	440 ±200	-30 ±15	Strauß 70	Fixed angle d.r.

a) For $G_{\rho\pi\pi}$ we recommend the value $\sqrt{4\pi \cdot 2.13}$ of sect. 6.3(ii).

b) m_ρ = 765 MeV.

c) Due to the large extrap. distance the errors are large.

$\dfrac{G_{\epsilon\pi\pi}G_{\epsilon NN}}{4\pi}$	$\dfrac{G_{S^*\pi\pi}G_{S^*NN}}{4\pi}$	$\dfrac{G_{f\pi\pi}G_{fNN}^{(1)}}{4\pi}$	$\dfrac{G_{fNN}^{(2)}}{G_{fNN}^{(1)}}$	Reference	Method
5.49 ±0.32		3.00 ±0.28	-0.22 ±0.27	Engels 70	Backward d.r.
5.0[a]	4.0[a]	2.0[a]	-1.0[a]	Schlaile 70	Fixed-u d.r.
4.0[a]	2.0[a]	3.20 ±1.00	-1.40[a]	Strauß 70	Fixed angle d.r.

a) Due to the large extrap. distance the errors are large.

References

Hamilton   63   RMP   35   737   J. Hamilton, W.S. Woolcock

Hamilton   66   PL   20   687   J. Hamilton

Sakurai   66   PRL   17 1021   J.J. Sakurai

Hamilton   67   High Energy Physics,Vol.1(ed.E.H.S. Burhop), Academic Press, p.193

Lovelace   67   Pion-Nucleon Scattering(ed.G. Shaw, D.Wong), John Wiley & Sons, p.27

Höhler   68   ZP   214   381   G.Höhler,J.Baacke, F.Steiner

Baacke   69   ZP   221   134   J.Baacke,G.Höhler, F.Steiner

Collins   69   Lund Conf.   B.P.Collins,V.K.Samaranayake, W.S. Woolcock

Höhler   69   ZP   229   217   G.Höhler, H.Schlaile, R.Strauß

Höhler   69a Karlsruhe Preprint   G.Höhler, H.Jakob, R. Strauß

Moorhouse 69   ARNS   19   301   R.G. Moorhouse

Pilkuhn   69   NCL   1   854   H. Pilkuhn, A. Swoboda

Samarana. 69   Lund Conf.   V.K. Samaranayake, W.S.Woolcock

Zovko   69   NP   B11   231   N. Zovko

Taguchi   69   PTP   42 1341   Y. Taguchi, K. Yamamoto

Engels   70   NP   B15   365   J. Engels, G. Höhler, B. Petersson

Engels   70a Kiev Conf.   J. Engels, G. Höhler

Höhler   70   Springer Tracts   G. Höhler, to be publ.

Schlaile   70   Karlsruhe Thesis   H. Schlaile

Strauß   70   Karlsruhe Thesis   R. Strauß

3.2 KN, K̄N, $\pi\Sigma$ and $\pi\Lambda$ scattering

Units: $\hbar = c = 1$ fm, $\hbar/m_{\pi^+}c = 1.414$ fm, $\hbar/m_{K^+}c = 0.400$ fm
Reviews: Bransden 69; below 1 GeV/c: B. Martin 70

(i) KN S-wave scattering

Scattering lengths $a^I_{\ell,2J}$ and effective ranges $r^I_{\ell,2J}$ (see
eq. 1.18)

a^1_s	r^1_s	Reference	Method
-0.29 ± 0.015	0.5 ± 0.15	Goldhaber 62	Fit to K^+p phase shifts, eff. range approx.
-0.28 ± 0.01		Lea 69	Energy-dep. K^+p phase-shift anal.
-0.286 ± 0.004	0.54 ± 0.09		$G^2_y=0$ Energy-dep. K^+p
-0.283 ± 0.004	0.54 ± 0.09	Cutkosky 69	$G^2_y=7$ phase-shift anal. as a function of
-0.292 ± 0.006	0.35 ± 0.09		$G^2_y=14$ G^2_y, a)
-0.31 ± 0.01	0.40 ± 0.14		$K<0.7$GeV/c K^+p partial-
-0.32 ± 0.01	0.22 ± 0.23	A. Martin 69	$K<0.86$GeV/c wave an. with sum rule constraints.
-0.305 ± 0.012	$c^1_s=-0.023$ ±0.004fm^3	B. Martin 70a	Extrap. of K^+p phase shifts using a forward d.r. b)

a) G^2_y is defined in Sec. 3.2(v) below.

b) c^1_s is the S-wave curvature coefficient as defined by
Hamilton 63.

a^0_s	r^0_s	Reference	Method
0.04 ± 0.04		Stenger 64	Impulse approximation an. of K^+d data at 230 MeV/c. a)
$-0.11^{+0.06}_{-0.04}$		Chand 67	

a) Uses K^+p parameters of Goldhaber 62.

(ii) KN p-wave scattering

a_{p1}^{1}	a_{p3}^{1}	Reference	Method		
-0.035 ± 0.006	0.007 ± 0.008	B. Martin 70a	Forward d.r. sum rules.[a]		
-0.009 ± 0.004	0.006 ± 0.002		$G_y^2=0$	Energy-depend. K^+p phase-shift -an.,as func-tion of G_y^2.	
-0.018 ± 0.005	0.011 ± 0.002	Cutkosky 69	$G_y^2=7$		
-0.028 ± 0.003	0.016 ± 0.001		$G_y^2=14$		
$(a_{p1}^{1}+2a_{p3}^{1})= -0.01 \pm 0.01$		A. Martin 69	K^+p partial-wave an. with d.r. sum rule constraints.		

a) Assumes s-wave I=1 phase shifts are of the Goldhaber 62 type.

a_{p1}^{o}	a_{p3}^{o}	Reference	Method
0.11 ± 0.04		B. Martin 70	Scattering length fit to I=0 phase shifts of Stenger 64.

(iii) $\bar{K}N$, $\pi\Sigma$ and $\pi\Lambda$ Interactions, S-waves

For p_K^{lab} <400 MeV, the K-matrix (1.11) may be approximated by a constant matrix (zero-range approximation: ZRA). The effective range approximation (ERA) for K^{-1} is

$$Q^{\ell}(K_{\ell\pm}^{I})^{-1}Q^{\ell}= M_{\ell\pm}^{I} + \frac{1}{2} c_{\ell} (Q^2-Q_o^2)^{1/2} (R_{\ell\pm}^{I})^{1-2\ell} (Q^2-Q_o^2)^{1/2},$$

$c_o = 1$, $c_1 = -3$; Q_o^2 has matrix elements $q_o^2(\bar{K}N) \equiv 0$, $q_o^2(\pi^o\Sigma^o)= 0.0328$ $(GeV/c)^2= 0.843$ fm^{-2}, $q_o^2(\pi^o\Lambda)= 0.0647$ $(GeV/c)^2= 1.661$ fm^{-2}. So far only diagonal matrices have been used for R. In the following tables, the errors are taken from the diagonals of the error matrices.

K^1_{KK}	$K^1_{K\Sigma}$	$K^1_{K\Lambda}$	$K^1_{\Sigma\Sigma}$	$K^1_{\Sigma\Lambda}$	$K^1_{\Lambda\Lambda}$	K^0_{KK}	$K^0_{K\Sigma}$	$K^0_{\Sigma\Sigma}$	reference
.22±.02	.78±.01	.38±.01	.92±.02	-.17±.02	.46±.02	-1.88±.02	.92±.02	-.36±.02	B.Martin 69
.01±.26	-.71±.06	-.38±.06	.34±.30	-.21±.34	.17±.61	-2.40±.28	-1.21±.25	-1.05±.45	A.Martin 70
.56±.02	-.66±.02	.39±.01	1.07±.05	.10±.02	.50±.03	-2.04±.04	-1.28±.04	-.40±.04	Berley 70
-.66±.02	-.99±.01	-.62±.01	1.27±.01	.04±.01	.88±.02	-1.68±.02	-.86±.01	-.09±.03	Thompson 70

M^1_{KK}	$M^1_{K\Sigma}$	$M^1_{K\Lambda}$	$M^1_{\Sigma\Sigma}$	$M^1_{\Sigma\Lambda}$	$M^1_{\Lambda\Lambda}$	M^0_{KK}	$M^0_{K\Sigma}$	$M^0_{\Sigma\Sigma}$	reference
-360±.02	-2.86±.03	2.08±.07	-1.40±.06	1.81±.04	-2.31±.11	.00±.02	-1.11±.04	2.0±.1	Kim 67
-4.34±.09	-4.88±.04	.91±.05	-2.98±.06	1.75±.06	.22±.06	.03±.03	-1.50±.03	2.61±.09	Berley 70

R^1_{KK}	$R^1_{\Sigma\Sigma}$	$R^1_{\Lambda\Lambda}$	R^0_{KK}	$R^0_{\Sigma\Sigma}$	reference
-.13±.07	-.78±.23	-1.22±.45	.54±.08	-.9±.3	Kim 67
-3.3±.2	.9 ±.3	-.6 ±.2	.23±.04	1.3±.4	Berley 70

B. Martin and A. Martin make a pure s-wave analysis of the range 0-300 MeV/c.
It is invariant under sign changes of the triplets ($K^0_{K\Sigma}$, $K^1_{K\Sigma}$, $K^1_{K\Lambda}$) and ($K^0_{K\Sigma}$, $K^1_{K\Sigma}$, $K^1_{\Sigma\Lambda}$)
Kim 67 analyses 0-550 MeV/c, Berley 70 350-430 MeV/c and Thompson 70 0-430 MeV/c.

The s-wave scattering lengths are defined for each elastic channel at its own threshold by

$$f^I_{o+}(q_c=0) = A^I = a^I + ib^I :$$

A_{KN}^1	A_{KN}^0	$A_{\pi\Sigma}^1$	$A_{\pi\Sigma}^0$	$A_{\pi\Lambda}$	Reference	Model
.00±.06 +i(.69±.03)	-1.67±.04 +i(.72±.04)				Kim 65	CSL[a]
-.19±.08 +i(.44±.04)	-1.63±.07 +i(.51±.05)				Sakitt 65	CSL[a]
-.04±.03 +i(.70±.02)	-1.69±.03 +i(.72±.02)				A. Martin 70	CSL[a]
-.13±.02 +i(.51±.03)	-1.65±.04 +i(.73±.02)	.39±.07 +i(.14±.03)	1.09±.23		VonHippel 68	ERA K-matrix
-.09±.03 +i(.54±.02)	-1.66±.02 +i(.69±.02)	.28±.03 +i(.16±.02)	.42±.03	.13±.07	B. Martin 69	ZRA K-matrix
-.05±.04 +i(.63±.06)	-1.74±.04 +i(.70±.01)	-.31±.30 +i(.24±.31)	-.14±.20	-.45±.61	A. Martin 70	ZRA K-matrix

a) CSL ≡ constant scattering length.

(iv) $\bar{K}N$, $\pi\Sigma$ and $\pi\Lambda$ interactions, P-waves

	K^1_{KK}	$K^1_{K\Sigma}$	$K^1_{K\Lambda}$	$K^1_{\Sigma\Sigma}$	$K^1_{\Sigma\Lambda}$	$K^1_{\Lambda\Lambda}$	K^0_{KK}	$K^0_{K\Sigma}$	$K^0_{\Sigma\Sigma}$	Ref.
P$\frac{1}{2}$	.077±.006	-.007±.008	-.065±.005	.01±.20	.02±.05	.19±.02	.03±.01	-.216±.006	.13±.03	Berley 70
P$\frac{3}{2}$	.049±.006	-.029±.006	-.043±.002	-.25±.09	-.05±.02	.05±.01	155±.008	.011±.005	.10±.20	
	-.047±.008 -.071±.006	-.076±.008 -.025±.004	-.045±.005 -.045±.002	-.44±.07 -.06±.05	-.33±.02 -.08±.02	-.07±.01 -.08±.01	.18±.01 .05±.01	-.15±.01 -.02±.01	-.04±.07 -.01±.07	Thompson 70

	M^1_{KK}	$M^1_{K\Sigma}$	$M^1_{K\Lambda}$	$M^1_{\Sigma\Sigma}$	$M^1_{\Sigma\Lambda}$	$M^1_{\Lambda\Lambda}$	M^0_{KK}	$M^0_{K\Sigma}$	$M^0_{\Sigma\Sigma}$	Ref.
P$\frac{1}{2}$	-18.1±.7	7.2±1.1	4.7±.4	-.3±1.1	-6.8±1.0	-3.4±1.3	18±16	-10.6±4.5	-11.4±4.8	Kim 67
P$\frac{3}{2}$	5.2±1.0	-13.8±1.6	-11.9±1.6	-15.8±.9	-13.6±.2	-16±2.0	10.9±.9	-1.4±.2	5.6±.5	
P$\frac{1}{2}$	-14.6+11	6.7±.7	2.9±1.1	2.8±.6	-11.5±1.0	-14.1±29	-4.4±.6	-8.3±0.2	-4.2±0.4	Berley 70
P$\frac{3}{2}$	-21.9+.1	-65±1	-13.0±.4	-1.22+4	-7±1.5	-2.8±.4	8.0±.4	-3.0±.4	16±3	

	R^1_{KK}	$R^1_{\Sigma\Sigma}$	$R^1_{\Lambda\Lambda}$	Reference
	-.7±.2	.27±.05	.3±.1	Kim 67
P$\frac{3}{2}$	-.5±.1	.7±2.1	.40±.04	Berley 70

Kim 67 uses the range $0 < p_K^{lab} < 550$ MeV, Berley 70 use the range $350 < p_K^{lab} < 430$ MeV. Both analyses put $R^0 \equiv 0$. Berley 70 use as starting solution the parameters of B. Martin 69 for the ZRA and of Kim 67 for the ERA. Kim's p-wave solution is not claimed to be unique. Doubt has been expressed about the self-consistency of Kim's solution.

The p-wave scattering lengths are $A^I_{1\pm,(c)} = \lim_{q^2 \to 0} f^I_{1\pm,(c)}/q^2_c$.
A. Martin 70 finds from a CSL analysis below 400 MeV/c:

$$A^1_{1-}(\bar{K}N) = -.067\pm.037+i\ (.0024\pm.0010),$$

$$A^1_{1+}(\bar{K}N) = -.060\pm.037+i\ (.0037\pm.0012).$$

For $\pi\Lambda$ scattering, one can deduce from Ξ decays (using PDG 70) the phase shift difference at $s = m^2_\Xi$:

$$\delta_S - \delta_P = (49^{+14}_{-38})^\circ \text{ from } \Xi^\circ \text{ decay, } -(8\pm18)^\circ \text{ from } \Xi- \text{ decay.}$$

(v) NYK coupling constants

Review: Queen 69, B. Martin 70. $G^2_{N\Lambda K}$ and $G^2_{N\Sigma K}$ are determined from forward dispersion relations. The Born term for $K^\pm p$ scattering in the lab system at $t=0$ is

$$F_\pm(\omega) \equiv \frac{1}{4\pi}\ [A(\omega,0)+ \omega B(\omega,0)]\ = \sum_{Y=\Lambda,\Sigma^\circ} \frac{R_y}{\omega_y \pm \omega}\ ,$$

$$\omega = \frac{s-m^2_p-m^2_K}{2m_p}, \quad \omega_y = \frac{m^2_y-m^2_p-m^2_K}{2m_p}, \quad R_y = \frac{G^2_{NYK}}{4\pi}\ \frac{(m_y-m_p)^2-m^2_K}{4m^2_p}$$

Note: In earlier papers the factor $4m^2_p$ in R is replaced by $4m_p m_y$. Isospin decomposition determines $G^2_{N\Lambda K}$ and $G^2_{N\Sigma K}$ separately. Placing the Σ-pole at the Λ mass yields
$$G^2_y \equiv (G^2_{N\Lambda K} + 0.84\ G^2_{N\Sigma K})/4\pi.$$

From KN forward dispersion relation threshold sum rules:

$G^2_{N\Lambda K}/4\pi$	$G^2_{N\Sigma K}/4\pi$	G^2_y	Reference	$\bar{K}N$ Parametrization
4.1±0.8	≤ 2.5		Lusignoli 66	
4.9±0.9	≤ 3.0		Carter 67	
5.0±1.7	≤ 3.0		Davies 67	
4.0±0.9	≤ 2.4	5.8±0.8	Kim 67a	constant scattering length.
~4.1			A. Martin 67	
5.9±1.3	≤ 1.3		Granovski 68	
		7.8±0.8	A. Martin 69	
3.9±1.6	≤ 3.1		Queen 69a	

$G^2_{N\Lambda K}/4\pi$	$G^2_{N\Sigma K}/4\pi$	G^2_y	Reference	$\bar{K}N$ Parametrizat.
6.2±1.0			Rood 67	
~5.7			A. Martin 67	ZRA
5.0±1.9	≤ 1.0±1.5		B. Martin 69a	K-matrix
4.0±2.4	≤ 0.9±1.5		A. Martin 70	
13.5±2.1	0.2±0.4	13.7±2.5	Kim 67a	
13±3	0±1		Chan 68	ERA
		15.2±2.5	A. Martin 69	K-matrix
11.9±2.7	0.4±0.6		Queen 69a	
	≤ 3	9±3	recommended values	

From fits to $K^{\pm}p$ forward amplitudes using KN forward
dispersion relations:

G^2_y	Reference	$\bar{K}N$ Parametrization
~9	Zovko 66	
5.3±4.5	A. Martin 68	constant scattering
8.0±1.7	Perrin 69	length
8.8±2.0	A. Martin 70	
7.2±2.0	A. Martin 70	ZRA K-matrix
5.8±4.5	A. Martin 68	
10.9±1.7	Perrin 69	ERA K-matrix
10.5±2.0	A. Martin 70	

Other determinations:

$G^2_{N\Lambda K}/4\pi$	$G^2_{N\Sigma K}/4\pi$	Reference	Method
4		Kuo 62	analysis of $\gamma p \rightarrow K^+\Lambda$
5-6		Dufour 64	
2.8±0.6		Dombey 70	
	4.5	Kuo 63	analysis of $\gamma p \rightarrow K^+\Sigma^0$.
3		Banaigs 69	extrap.of Regge residue
14.5±5		A. Martin 70b	
7±16	0±1	Frye 65	KN partial-wave d.r.

$G^2_{N\Lambda K}/4\pi$	$G^2_{N\Sigma K}/4\pi$	Reference	Method
~ 4.4		Rimpault 63	an. of $\pi p \rightarrow K^+ \Lambda$
$(G^2_{N\Lambda K} + 1.1\ G^2_{N\Sigma K})/4\pi \gtrsim 14\pm 4$		Rogers 69	phenomenological bound
$G^2_{N\Lambda K}/4\pi + G^2_{N\Sigma K}/4\pi = 9\pm 3$		Cutkosky 70	$K^+ p$ phase shift an.
$G^2_{N\Sigma K}/G^2_{N\Lambda K} = 0.6\pm 0.25$		Hoogland 70	ratio $(K^- p \rightarrow \Sigma\omega)/(K^- p \rightarrow \Lambda\omega)$

(vi) <u>Pion-hyperon couplings</u>, $G^2_{\Lambda\Sigma\pi}$, $G^2_{\Sigma\Sigma\pi}$

$G^2_{\Lambda\Sigma\pi}/4\pi$	$G^2_{\Sigma\Sigma\pi}/4\pi$	Reference	Method
14.5		Dalitz 63	an. of $\Lambda\Lambda$ hypernucleus
17 ± 6	$\lesssim 2$	de Swart 63	an. of Σp and Σd int.
10.9 ± 0.3		B. Martin 65	partial-wave d.r., $\pi\Lambda$
21.5 ± 7.0	11.4 ± 5.0	Chan 68a[a)]	forward d.r.
11.4 ± 1.2		Pilkuhn 70	Goldberger-Treiman rel. for $\Sigma \rightarrow \Lambda e\bar{\nu}$ decays
$G^2_{\Sigma\Sigma\pi}/4\pi - G^2_{\Lambda\Sigma\pi}/4\pi = 1\pm 0.8$		Everett 69[b)]	Superconv. rel. for $B(\pi^- \Sigma^+ \rightarrow \pi^+ \Sigma^-)$

[a)] Uses ERA K-matrix of Kim 67. Results criticized by Weil 69.

[b)] Updated June 69 by Engels and Pilkuhn.

(vii) <u>The states $\Sigma(1382)$ and $\Lambda(1405)$</u>

The decay widths of the decuplet state $\Sigma(1382)$ are given by eq. (5.7), $G^2/4\pi = 1.5\ m^2 \Gamma(m^2)/p^3$

	$\Sigma^+(1382) \rightarrow \pi\Lambda$	$\Sigma(1382) \rightarrow \pi\Sigma$	comment
Γ_{expt} [MeV]	32.4 ± 3.0	3.6 ± 1.0	from PDG 70
$\Gamma(m^2)$ [MeV]	31.4 ± 3.0	3.2 ± 0.9	comm. after eq.(5.8)
$G^2/4\pi$	10.0 ± 1.0	5.7 ± 1.6	

Let $G^2_{\bar{K}N} = \frac{2}{3}R\ G^2_{\pi\Lambda}$ for the coupling constant $G(\Sigma(1382)\bar{K}N)$.

The K-matrix of Kim 67 gives R<<1. Tripp 68 find R = 1.5 in $K^- p \to \Sigma \pi$, Bunnel 70 find R~1 in $K^- d \to \Lambda \pi^- p$. See also Agarwal 69. SU(3) says R=1.

For $\Lambda(1405) \to \pi \Sigma$, $\quad G^2_{\pi \Sigma}/4\pi = \dfrac{\Gamma(m)}{3p} \dfrac{2m^2}{(m+m_\Sigma)^2 - m_\pi^2} = 0.18 \dfrac{\Gamma}{m_\pi}$

PDG 70 give $\Gamma = (40 \pm 5)$MeV, i.e. $G^2_{\pi \Sigma}/4\pi = 0.052$. Kim 67 (ERA) gives 50 ± 5MeV, B. Martin 69, A. Martin 70 give 20-30 MeV.

The coupling to the $\bar{K}N$ channel is given by

$\dfrac{1}{2\pi} \displaystyle\int \text{Im } f^o_s(\bar{K}N) d\omega = \dfrac{G^2_{\bar{K}N}}{4\pi} \dfrac{(m_N + m)^2 - m_K^2}{4m^2}$, integration extending over the resonance. Kim 69 gives $G^2_{\bar{K}N}/4\pi = 0.32 \pm 0.04$, using Kim 67. The solution of B. Martin 69 gives 0.19 ± 0.03 . Tripp 68 get ~0.28. SU(3) without mixing says $G^2_{\bar{K}N} = \frac{2}{3} G^2_{\pi \Sigma}$.

References

Goldhaber 62   PR   9,135   S.Goldhaber, W.Chinowska, G.Goldhaber, W.Lee, T.O.Hallloran, T.F.Stubbs, G.M.Pjierrou, D.H. Stork, H.K. Ticho.

Kuo   62   PR 129,2264   T.K. Kuo

Dalitz   63   PL   5,53   R.H. Dalitz

Hamilton   63 RMP   35,737   J. Hamilton, W.S. Woolcock

Kuo   63   PR 130,1537   T.K. Kuo

Rimpault   63   NC   31,56   M. Rimpault

de Swart   63   PR 130,319   J.J. de Swart, C.K. Iddings

Dufour   64   NC   34,645   J.D. Dufour

Stenger   64   PR 134B,1111   V.J. Stenger, W.E. Slater, D.H. Stork, H.K. Ticho, G. Goldhaber, S. Goldhaber

Frye   65   PR 138B,947   G. Frye, R.L. Warnock

Kim   65 PRL   14,29   J.K. Kim

B. Martin 65   PR 138B,1136 B.R. Martin

Sakitt   65   PR 139B,719   M. Sakitt, T.B. Day, R.G. Glasser, N. Seeman, J. Friedman, W.E. Humphrey, R.R. Ross

Lusignoli 66   PL   21,229   M. Lusignoli, M. Restignoli, G.A. Snow, G. Violini

Zovko   66   ZP 192,346   N. Zovko

Carter   67 PRL   18,801 A.A. Carter

Chand   67 ANP   42,81 R. Chand

Davies   67 NP   B3,616 G.H. Davies, N.M. Queen,
                            M. Lusignoli, M. Restignoli,
                            G. Violini

Kim   67 PRL   19,1074 J.K. Kim

Kim   67a PRL 19,1079 J.K. Kim

A. Martin   67 PL 25B,343 A.D. Martin, F. Poole

Rood   67 NC 50A,493 H.P.C. Rood

Chan   68 PRL   20,568 C.H. Chan, F.T. Meiere

Chan   68a PL 28B,125 C.H. Chan, F.T. Meiere

Granovskii 68 SJNP 6,444 Ya. Granovskii, V.N. Starikov
A. Martin 68 PL 26B,527 A.D. Martin, G.G. Ross
Tripp   68 PRL 21,1721 R.D. Tripp, R.O. Bangerter,
                            A. Barbaro-Galtieri, T.S. Mast

VonHippel 68 PRL   20,1303 J.K. Kim, F. Von Hippel

Agarwal   69 NCL   2,758 B.K. Agarwal, M.C. Sharma,V.P. Seth

Banaigs   69 NP   B9,640 J. Banaigs, J. Berger, C. Bonnel,
                            J. Duflo, L. Goldzahl, F. Pilouin,
                            W.F. Baker, P.J. Carlson,
                            V. Chabaud, A. Lundby

Bransden   69 High Energy Physics, Vol. III (ed. E.H.S.
                Burhop, Academic Press), p.1.

Dombey   69 PL 30B,646 N. Dombey

Everett   69 PR 177,2561 A.E. Everett

Kim   69 PR 184,1961 J.K. Kim, F. Von Hippel

Lea   69 Lund Conf. A.T. Lea, B.R. Martin, G.C. Oades

A. Martin 69 NP B10,125 A.D. Martin, R. Perrin

B. Martin 69 PR 183,1345 B.R. Martin, M. Sakitt

B. Martin 69a PR 183,1352 B.R. Martin, M. Sakitt

Perrin   69 NP B12,26 R. Perrin, W.S. Woolcock

Queen   69 FP 17,467 N.M. Queen, M. Restignoli,
                            G. Violini

Queen   69a NP B11,115 N.M. Queen, S. Leeman,
                            F.E. Yeomans

Rogers   69 PR 178,2478 T.W. Rogers

Weil   69 PL 29B,501 C. Weil

Berley   70 PR   D1,1996 D. Berley, P. Yamin, R. Kofler,
                            A. Mann, G. Meisner, S.Yamanoto
                            J. Thompson, W. Willis

Bunnell   70 NCL   3,224   K.O. Bunnell, D. Cline,
                              R. Laumann, J. Mapp, J.L. Uretsky

Cutkosky 70 PR   D1,2547 R.E. Cutkosky, B.B. Deo

Hoogland 70 NP   B21,381 W. Hoogland, J.C. Kluyver,
                              G.G.G. Massaro, A.G. Tenner,
                              A. Minguzzi-Ranzi, S. Focardi,
                              D. Merrill, G. Lamidey,
                              V. Karshon, G. Yekutieli

A.Martin 70 NP   B16,479 A.D. Martin, G.G. Ross

A.Martin 70a NP   B20,287 A.D. Martin, R. Perrin

A.Martin 70b PL   32B,297 A.D. Martin, C. Michael

B.Martin 70 Springer Tracts Vol. 55 B.R. Martin

B.Martin 70a NP   B.R. Martin, G.D. Thompson

PDG   70 RMP   42,87 Particle Data Group

Pilkuhn 70 Springer Tracts 55 H. Pilkuhn

Thompson 70 Brookhaven preprint J. Thompson .

3.3 NN Scattering

(i) Scattering lengths and effective ranges

Units: $\hbar=c=fm=1$. Reviews: Breit 67, Henley 69, Kramer 70.

For np and nn scattering, the S-wave scattering lengths
a and effective ranges r are defined by eq. (1.18) for $\ell=0$.
For pp scattering, eq. (1.21) applies. The pp and nn parameters refer to the $^1S_0$ state, np parameters with subscripts
s and t refer to the $^1S_0$ and $^3S_1$ states, respectively.

	a_s	a_t	r_s	r_t	Reference	comment
pp	$-7.786\pm.008$		$2.840\pm.009$		Slobodri.68	vacuum pol. included
nn	-16.1 ± 1.0		$3.2\ \pm1.6$		Baumgart.68	
	$-13.1\ ^{+3.4}_{-2.4}$				Butler 68	
	$-16.4\ ^{+2.6}_{-2.9}$				Zeitnitz 70	
np	-23.714 $\pm.013$	5.425 $\pm.004$	2.704 $\pm.087$	1.763 $\pm.005$	Houk 68	Wilson 68 quotes $r(^3S1)=1.749\pm.008$

The pp scattering length in the absence of the Coulomb potential, a_N, is approximately related to a and r by

$$\frac{1}{a_N} = \frac{1}{a} - 2q\eta(\ln 2q\eta r + 0.330) = \frac{1}{a} + 0.0690,$$

i.e. $a_N = -16.8$, which agrees with $a_s(nn)$ but not with $a_s(np)$.

(ii) <u>Meson-nucleon coupling constants</u>

Definitions: $G_{\pi NN}$ in 3.2, $G_{NN\eta}$ in 3.12, $G_{NN\sigma}$ and $G_{NN\epsilon}$ in 6.1, $G_{\rho NN}$ and $G_{\omega NN}$ in 2.8 and 3.14.

Separate determinations of the pion pole:

$G^2_{\pi NN}/4\pi$	References	Method
14.72±.83	Mac Gregor 68	phase-shift anal.
14.1 +1.4-0.8 13.9 +1.6-1.7	Cutkosky 68	350 MeV optimized polynomial 400 MeV expansion of np diff. cross sections.

this section continues next page.

$\boxed{\text{To section 3.4ii}}$ $\Sigma^\pm p$ scattering lengths. Notation $a_{J,I}$. For I= 3/2 <u>a</u> is real, for I= 1/2 one has instead A= a+ib (analogous to K^-p but with reversed sign convention). Analysis of Alexander 70, including the data of Eisele 70 (priv. comm. by Alexander) obtain $a_{1,3/2}\sim 0$, $b_{0,1/2}\sim 0$ and (for constructive Coulomb interference)

$a_{0,3/2}$	$a_{1,1/2}$	$b_{1,1/2}$	$a_{0,1/2}$	comment
2.23±.18	-2.21±.67	3.77±.94	.38±.24	solution 1
2.2 ±.18	0.0±0.8	1.64±.24	-5.46± .8	solutions 5,6

(ii) Meson-Nucleon Coupling Constants

m_π	$\dfrac{G_{\pi NN}^2}{4\pi}$	m_η	$\dfrac{G_{\eta NN}^2}{4\pi}$	m_σ	$\dfrac{G_{\sigma NN}^2}{4\pi}$	m_ϵ	$\dfrac{G_{\epsilon NN}^2}{4\pi}$	m_ρ	$\dfrac{(G_{\rho NN}^V)^2}{4\pi}$	$\dfrac{G_{\rho NN}^T}{G_{\rho NN}^V}$	m_ω	$\dfrac{(G_{\omega NN}^V)^2}{4\pi}$	Ref.
140	14.4	550	10.2 ± 5.9	400	1.6 ± 0.1	700	2.5 ± 1.3	765	4.8 ± 0.8	3.66	782	4.2 ± 2.0	Köpp 66
136.5	14.52	–	–	468	2.79	–	–	763	4.9	4.0	783	0.94	Arndt 66
136.5	14.7	550	0.2 ± 2.3	–	–	613	14.1 ± 1.5	750	7.52 ± 4.0	2.5 ± 1.0	750	13.5 ± 1.7	Bugg 68
138.7	14.7	548.7	2.0	416	2.35	782	14.7	763	2.6	3.74	782.8	23.0	Green 67
138.7	14.02	548.7	1.205	416	1.92	1070	7.9	763	3.12	2.39	782.8	7.97	Green 69
140	14.4	550	8.0	400	1.4	700	6.8	711	2.4	4.78	781	9.05	Schierh 69,70

Meson Masses in MeV. The underlined numbers are input to the fits. Köpp 66: combined analysis of forward d.r. and partial wave amplitudes in Born approximation. Arndt 66: Born approximation with unitarity correction based on the Mandelstam representation. Bugg 68: Analysis of forward d.r.. An additional particle $P(I(J^{PG})=1(0^{--}))$ with mass m_P=550 MeV and $G_{PNN}^2/4\pi=6.6\pm1.7$ is included. Green 67: Nonrelativistic Schrödinger equation with regularized OBE potential. An additional particle $\sigma_1(I(J^{PG})=1(0^{+-}))$ with mass m_{σ_1}=763 MeV and $G_{\sigma_1 NN}^2/4\pi=0.65$ is included. Green 69: Generalized potentials. Schierh. 70: Relativistic three-dimensional integral equation with generalized OBE potential. Schierholz uses $G_{\omega NN}^T/G_{\omega NN}^V = -0.1$, the other determinations have $G_{\omega NN}^T=0$. A more complete review on the various methods is given in Kramer 70.

273

References

Arndt   66 PL 21,314   R.A. Arndt, R.A. Bryan,
                            M.H. MacGregor

Baumgart.66 PRL 16,105 E. Baumgartner, H.E. Conzett,
                        E. Shield, R.J. Slobodrian

Köpp   66 PL 23,494   G. Köpp, P. Söding

Breit   67 High Energy Physics Vol.I (ed. E.H.S. Burhop,
                Academic Press), G. Breit, R.D. Haracz

Green   67 NP B2,267   A.E.S. Green, T. Sawada

Bugg   68 NP B5,29   D.V. Bugg

Butler   68 PRL 21,470   P.G. Butler, N. Cohen, A.N. James,
                            J.P. Nicholson

Cutkosky 68 PRL 20,1272 R.E. Cutkosky, B.B. Deo

Houk   68 RMP 40,672E T.L. Houk, R. Wilson

MacGregor68 PR 169,1128 M.H. MacGregor, R.A. Arndt,
                         R.M. Wright

Slobodri.68 PRL 21,438   R.J. Slobodrian

Wilson   68 CNPP 2,141   R. Wilson

Schierh. 68 NP B7,483   G. Schierholz

Henley   69 Isospin in Nuclear Physics (ed.D.H.Wilkinson,
                North Holland

Green   69 NP B10,289 T. Ueda, A.E.S. Green

Zeitnizt 70 NP A149,449 B. Zeitnitz, R. Maschuw, P. Suhr

Kramer   70 Springer Tracts 55 G. Kramer

Schierh. 70 to be publ. G. Schierholz

3.4 YN Scattering

Units: $\hbar=c=fm=1$ Reviews: Argonne 69, Deloff 69, Gal 70.

(i) Λp Scattering

a_s	r_s	a_t	r_t	Reference	Method
-2.25	+3.29	-2.08	+3.40	Herndon 70	HBE+ES
-1.8	+2.8	-1.8	+3.3	Alexand.68	ES[a]
-15to0.0	0.0to+15	-3.2to-0.6	+25to+15	Sechi-Zo.68	ES[a]
-1.7±0.5	+2.5 $^{+1.0}_{-0.5}$	-1.5±0.05	+2.0±0.05	Fast 68	ES[b]
		-2.0±0.5	+3.0±1.0	Tan 69	FSI

HBE= from hypernuclear binding energies; ES= from elastic scattering cross section; FSI= from final-state interaction in K⁻d→π⁻pΛ at rest.

a) The errors are large and strongly correlated.

b) Fast 68 assume a Λp resonance at 2126 MeV. The resonance interpretation is questioned by Alexander 69.

(ii) Σ^+p Scattering

Fast 69 give a_s= -6±1, r_s= 2.1±.3, a_t= -.2±.05, r_t= -40
Eisele 70 find $\sigma_{el}(\Sigma^+p)$=(85±8)mb for 140-170 MeV/c and indication of constructive Coulomb interference
The subsequent solutions of Alexander 70 are given in
References   section 3.3ii.

Herndon   67 PR 159,853 R.C. Herndon, Y.C. Tang

Alexander 68 PR 173,1452 G. Alexander, U. Karshon,
                          A. Shapira, G. Yekutieli, R. Engelmann,
                          H. Filthut, W. Lughofer

Sechi-Zo. 68 PR 175,1735 B. Sechi-Zorn, B. Kehoe, J.Twitty,
                          R.A. Burnstein

Alexander 69 PRL 22,483 G. Alexander, B.H. Hall, N. Jew,
                          G. Kalmus

Argonne   69 Proc. of the Int. Conf. on Hypernuclear
                Physics, Argonne Nat. Lab. (ed8 A.R. Bodmer,
                L.G. Hyman).

Deloff   69 FP 17,129 A. Deloff

Fast   69 PRL 22,1453 G. Fast, J.C. Helder, J.J.deSwart

Tan   69 PRL 23,395 T.H. Tan
Alexander 70 NP B22,583   G.Alexander, Y.Gell, I. Stumer
Eisele   70 Kiev Conf. and ZP F. Eisele, H. Filthuth,
                          W. Föhlisch, V. Hepp, E. Leitner, G. Zech

Gal   70 High-En. Physics and Nuclear Structure,
                S. Devons ed.(Plenum Press),p.516 A. Gal .

3.5 Meson-meson Scattering

(i) Pion-pion scattering lengths

Convention $\delta_\ell^I(q)\ q^{-(2\ell+1)} \to a_\ell^I$ for q→0, units: $\hbar=c=m_{\pi^+}=1$.

a_0^0	a_0^2	a_1^1	References	Method
0.3-0.4 0.5-1.0 0.6-1.2	↦down-up ↦up-up ↦up-down		Malamud 69	Our linear fits to(q/ω) cotg δ from β-solutions of Malamud 69 in the region 400-800 MeV.
	-0.052		Baton 67	Chew-Low extrap. in the ρ-region and extrap. to threshold.
0.20	-0.06	0.038	Weinberg 66	Current algebra, linear extrap.
0.2	-0.06		Khuri 67	Current algebra
0.345	-0.09		Wagner 69	One-channel unitarization of Veneziano model
0.288	-0.063		Lovelace 69	Two-channel ($\pi\pi$, $K\bar{K}$) unitarization of Veneziano model
0.16	-0.05	0.035	Morgan 68,70	D.r. for forward scattering
		0.037	Olsson 67	D.r. for forward scattering
		0.045-0.10 (a) 0.035 0.077(b)	Pišút 68	Extrap. to threshold of physical region fits with (b) and without (a) CDD pole in δ_1^1
0.15-0.32		0.040	Schareng. 70	Extrap. of forward-backward asymetry
$a_0^0 > -0.5$			Goebel 68	Rigorous bound with mild phenom. ingredients
$a_2^0 - a_0^0 = -0.42$			Blair 70	Data $\pi^- p \to \pi^- \pi^+ n$ fitted near threshold by Gribov formula

Due to the lack of a direct experimental evidence, pion-pion scattering lengths are obtained by extrapolations, from models, by disentangling genuine three-body inter-actions, or by combinations thereof. No reliable methods of estimating true errors are available and we therefore do not quote the errors. The present opinion about pion-pion scattering lengths is mostly due to the influential paper by Weinberg 66. His values of scattering lengths are at present consistent with the bulk of experimental data (see Morgan70a). Still, changes are not excluded. For a review of methods to extract $\pi\pi$ phase shifts from $\pi N \to \pi\pi N$, see Gutay 69.

References

Weinberg 66   PRL 17,616   S. Weinberg

Baton   67   NP B3,349   J.P. Baton, G. Laurens, J. Reignier

Khuri   67   PR 153,1477 N.N. Khuri

Olsson   67   PR 162,1338 M.G. Olsson

Goebel   68   PL 27,291   C.J. Goebel, G. Shaw

Morgan   68   NP B10,1387 D. Morgan, G. Shaw

Pišút   68   NP B6,325   J. Pišút, M. Roos

Gutay   69   NP B12,31   L.V. Gutay, F.T. Meiere,
                              D.D. Carmony, F.J. Loeffler,
                              P.L. Csonka

Lovelace 69   CERN TH 1041, PAC p.562[*)]   C. Lovelace

Malamud 69   PAC p.93[*)]   E. Malamud, P.E. Schlein

Wagner   69   NC 64A,189 F. Wagner

Blair   70   PL 32B,528 I.M. Blair, H. Müller, G. Torelli,
                          E. Zavattini, G. Mandrioli

Morgan   70   PR D 2,520   D. Morgan, G. Shaw

Morgan   70a Springer Tracts 55   D. Morgan, J. Pišút

Pišút   70   Springer Tracts 55   J. Pišút

Schareng.70   NP   J.H. Scharenguivel, L.J. Gutay,
                            D.H. Miller, F.T. Meiere,
                            S. Marateck

[*)]PAC: Proceedings of the Argonne Conference on $\pi\pi$ and $K\pi$ scattering, May 1969.

4.1 Photon Couplings

(i) Vector-meson-photon couplings (α^{-1}=137.036).

Reviews: Lohrman 69, Silverman 69, Braccini 70, Gourdin 70, Marshall 70, Perez-Y.J. 70.

Vector Meson Dominance (VMD)

$$J_\mu^{e.m.}(x) = -\{\frac{m_\rho^2}{f_\rho}\rho_\mu(x) + \frac{m_\omega^2}{f_\omega}\omega_\mu(x) + \frac{m_\phi^2}{f_\phi}\phi_\mu(x)\}$$

Leptonic decay width

$$\Gamma(V \to e^+ e^-) = \frac{1}{3} \alpha^2 m_V a_V \left(\frac{f_V^2}{4\pi}\right)^{-1} , \quad V = \rho, \omega, \phi.$$

a_V is a model dependent finite width correction (Gounaris 69, Renard 70).

Coupling Constants from leptonic decays (vector meson on mass shell)

$\dfrac{f_\rho^2}{4\pi}$	$\dfrac{f_\omega^2}{4\pi}$	$\dfrac{f_\phi^2}{4\pi}$	Reference
1.99 ± 0.19	14.0 ± 2.8	11.0 ± 0.9 ($a_V = 1.07, 1.00, 0.85$ $V = \rho, \omega, \phi$ resp.)	Bizot 70
1.85 ± 0.13	13.9 ± 2.5	12.85 ± 1.10 ($a_V = 1.00$, all V)	Gourdin 70

Novosobirsk results for ρ and ϕ leptonic decay widths (Auslander 69, Sidorov 69) are compared with Orsay data by Perez-Y.J. 70. Using a Gounaris-Sakurai fit, values of the former are $\Gamma(\rho \to e^+ e^-) = 6.05 \pm .50$ KeV, $\Gamma(\phi \to e^+ e^-) = 1.42 \pm .15$ KeV.

Coupling Constants from Photoproduction on nuclei (photon on mass shell)

V	$\dfrac{f_V^2}{4\pi}$	$\sigma_{Total}(VN)$ (mb)	$\dfrac{Re\ FVN}{IM\ FVN}$	Reference
ρ	4.4 ± 0.8	30^{+6}_{-4}	0.0 [a]	Bulos 69
ρ	2.28 ± 0.48	26.7 ± 2	-0.2	Alvensleb. 70
ρ	2.48 ± 0.48	26.8 ± 2.4	-0.2	Behrend 70
ρ	2.68 ± 0.16 [b] 27.7 ± 1.2 2.52 ± 0.16 [c] 27.1 ± 1.1		-0.24 to -0.27	McClellan 70
ω	29.2 ± 4.0	27 ± 2 [d]	-0.2 [d]	Behrend 70a
ω	20 ± 4.4	$27^{+6}_{-5.5}$	0.0 [e]	Draccini 70a
ϕ	13.6	12	-0.35 [f]	McClellan 69

a) For a discussion see Behrend 70.

b) Nuclear density parameters from optical model analysis of nucleon-nucleus total cross sections.

c) Nuclear density parameters from electron scattering.

d) Taken equal to corresponding parameters for ρ as predicted by quark model.

e) Coupling constant and total cross section for other values of this parameter also reported.

f) Chosen to produce agreement with results from $\phi \to e^+ e^-$ decay.

Lohrmann 69 and Sakurai 69 have shown that the prediction for $\sigma_{Total}(\gamma p)$ from a Vector Meson dominated Compton amplitude is in agreement with experiment for the values of f_ρ^2 obtained from leptonic decays.

(ii) Pion form factor

a) From π^+ electroproduction using the dispersion theory of Zagury (Mistretta 68)

$$F_\pi(t) = \left[1 - \frac{t}{(0.56 \pm 0.08)^2 \text{GeV}^2} \right]^{-1}, \quad (0 \leqslant -t \leqslant 0.4 \text{ GeV}^2).$$

b) Analysis of $e^+ e^- \to \pi^+ \pi^-$ colliding beam experiments (Roos 69 and private comm.)

$$F_\pi(t) = \exp \left\{ \frac{t}{\pi} \int_{4\mu^2}^{\infty} \frac{\phi(t')dt}{t'(t'-t)} \right\},$$

where $\phi(t) = \text{arccot}(m_\rho^2 - t)/(m_\rho \Gamma(t))$ for $4\mu^2 < t < (1.5 \text{ GeV})^2$ and $\phi(t) = \pi$ for $t > (1.5 \text{ GeV})^2$, and the t-dependence of Γ is given by eq. (5.7)

$$\Gamma(t) = \Gamma_\rho \frac{m_\rho}{\sqrt{t}} \left(\frac{p}{p_\rho} \right)^3, \quad m_\rho = 770 \pm 4 \text{ MeV}, \quad \Gamma_\rho \equiv (m_\rho^2) = 122 \pm 6 \text{ MeV}.$$

c) Vector-meson dominance with finite width for the ρ-meson (Gounaris 68)

$$F_\pi(t) = \frac{m_\rho^2 \left[1 + d(\Gamma_\rho/m_\rho) \right]}{m_\rho^2 - t - i m_\rho \Gamma_\rho (p/p_\rho)^3 (m_\rho/\sqrt{t})} \quad (t \text{ near } m_\rho^2),$$

$$d = \frac{3}{\pi} \frac{m_\rho^2}{p_\rho^2} \ln \left(\frac{m_\rho + 2p_\rho}{2m_\pi} \right) + \frac{m_\rho}{2\pi p_\rho} - \frac{m_\pi^2 m_\rho}{\pi p_\rho^3},$$

$$d = 0.48 \quad \text{for} \quad m_\rho = 770 \text{ MeV}.$$

(iii) <u>Nucleon form factors</u>

Reviews: Rutherglen 69, Wilson 70.

The following gross properties are valid within 8-10%:

(a) Scaling law

$$G_{Ep}(t) = \frac{G_{Mp}(t)}{\mu_p} = \frac{G_{Mn}(t)}{\mu_n} = G_D(t),$$

where μ_p and μ_n are the proton and neutron total magnetic moments.

(b) Dipole fit G_D

$$G_D(t) = \left(1 - \frac{t}{M_D^2}\right)^{-2}, \quad M_D^2 = 0.71 \text{ GeV}^2.$$

(c) $G_{En}(t)$ different from zero, but accuracy of data is much lower than for the other form factors; the only re-liable number is the slope at t=0, measured in scattering of thermal neutrons from atoms

$$\frac{dG_{En}(t)}{d(-t)}\bigg|_{t=0} = 0.496 \pm 0.010 \text{ GeV}^{-2} \qquad (\text{Krohn } 66).$$

Recent measurements, however, show deviations from dipole fit and from scaling law:

(a) $G_{Ep}/G_D < 1$ for $0.4 \leqslant -t \leqslant 3 (\text{GeV/c})^2$; $G_{Mp}/\mu_p G_D < 1$ for $-t < 0.8 \ (\text{GeV/c})^2$ and $G_{Mp}/\mu_p G_D > 1$ for $-t > 0.8 \ (\text{GeV/c})^2$. Refs.: Albrecht 66, Bartel 67, Berger 68, Coward 68, Berger 69, Bartel 70, Goitein 70.
G_{Mn} is not in disagreement with the dipole fit, but data are accurate only to $\pm$ 5% (Bartel 69).

(b) Scaling law $\mu_p \, G_{Ep}/G_{Mp} = 1$ valid for $-t \leqslant 1 \ (\text{GeV/c})^2$ within the experimental errors of $\pm$ 5%, but experiments deviate from unity by two standard deviations for $-t > 1.2 \ (\text{GeV/c})^2$ (Berger 68, Berger 69, Bartel 70, Litt 70).

(c) $G_{En}(t)$ can be obtained from elastic e-d scattering; a fit for $-t \leqslant 1 \ (\text{GeV/c})^2$ gives

$$G_{En}(t) = \frac{-\mu_n \tau}{1 + p\tau} G_{Ep}, \quad p = 10.7, \quad \tau \equiv -\frac{t}{4M^2},$$

the value of p obtained depends on the model used for the
analysis (Galster 70; see also Bartel 69).

(iv) <u>Two-photon decays of π° and η°.</u>

$$T_{\gamma\gamma}^{Born}=g_{\gamma\gamma}\alpha\varepsilon_{\alpha\beta\gamma\delta}P_1^\alpha\varepsilon_1^{*\beta}P_2^\gamma\varepsilon_2^{*\delta}, \quad \Gamma_{\gamma\gamma}=\frac{1}{16}\alpha^2 m^3 \frac{g_{\gamma\gamma}^2}{4\pi}.$$

$\Gamma_{\gamma\gamma}$		Reference	$g_{\gamma\gamma}^2/4\pi \left[GeV^{-2}\right]$
π°	11.7 ± 1.2 eV	Bellettini 70	1.43 ± 0.14
η°	1.01 ± 0.23 keV[*)]	PDG 70a	1.84 ± 0.42

[*)]possible corrections discussed by Benfatto 70.

4.2 <u>π and η photoproduction</u>   (units $\hbar=c=m_\pi=1$)

(i) <u>Multipoles at threshold and Panofsky ratio</u>

Table of $100E_{o+}$ (E_{o+} given in 1.16, 1.17).

π^+	π^-	π°	Reference	comment
$2.86\pm.02$	$-3.15\pm.06$		Adamovich 68,69	comb. of various measurem.
$2.78\pm.05$	$-3.27\pm.07$		Adamovich 69a	from σ_{tot}.
		$-.18\pm.09$	Govorkov 67	from $d\sigma/d\Omega$ at 90°
		$-.24\pm.04$	Lebedev 67	
		$-.17$	Müllensief.68 from extrap. in fig. 1	data of Hitzeroth 69 and Govorkov 68

See Pfeil 70, Gehlen 70 for theor. values of E_{o+}.

Panofsky ratio $\sigma_{tot}(\pi^-p\to\pi^\circ n)/\sigma_{tot}(\pi^-p\to\gamma n)=1.533\pm.021$
Cocconi 61. For analyses of η photoproduction, see
Donnachie 69, Pfeil 70, for K photoprod. see Orito 69,
Meyer 70.

(ii) <u>NN[*]γ magnetic dipole moment</u>

Reviews: Rollnik 69, Pfeil 70
Relation between magnetic dipole moment $\mu^*(\tilde{\mu}^*)$ of $\Delta(1236)$

and $I = \frac{3}{2}$ multipoles $M_{1+}^{3/2}$, $E_{1+}^{3/2}$ (Breit-Wigner resonance form assumed)

$$\mu^* = \sqrt{\frac{8\pi q_\Delta M_\Delta \Gamma}{3m_p k_\Delta^2}}\, |M_{1+}^{3/2}(M_\Delta)|\,, \qquad \tilde{\mu}^* = -i\sqrt{\frac{8\pi q_\Delta M_\Delta \Gamma}{3m_p\, k_\Delta^2}}\,(M_{1+}^{3/2}+E_{1+}^{3/2})\sqrt{\frac{\Gamma}{\Gamma_{\pi N}}}$$

k and q are the initial and final momenta at resonance energy M_Δ, $\Gamma/\Gamma_{\pi N} = 1.006$.

μ^*/μ_p	Reference	Method
1.21±0.02	Dalitz 66	
1.24±0.02	Pfeil 70	updating Dalitz 66
1.17±0.04	Donnachie 67	
1.16±0.03	Grilli 67	
1.07±0.01	Ash 67	

Predictions from different models (Pfeil 70)

$\mathrm{Im}\,M_{1+}^{3/2}$ $\times 10^2$	$\dfrac{\mathrm{Im}\,E_{1+}^{3/2}}{\mathrm{Im}\,M_{1+}^{3/2}}$	μ^*/μ_p	$\tilde{\mu}^*/\mu_p$	Reference	Method
3.53	-0.045	1.23	1.18	Engels 68	Disperison-relations
3.41	-0.016	1.18	1.17	Adler 68	"
3.43	-0.045	1.20	1.15	Gehlen 70	"
3.64	-0.01	1.27	1.26	Schwela 69	"
3.38	-0.035	1.18	1.14	Berends 69	"
3.55	-0.059	1.24	1.18	Pfeil 68	Isobar mod.
3.37	-0.045	1.18	1.13	Walker 69	Multipole-analysis
3.51	-0.01	1.23	1.22	Noelle 70	

References

Cocconi 61 NC 22,494   V.T. Cocconi, T. Fazzini, G. Fidecaro, M. Legros, W. Merrison

Albrecht 66 PRL 17,1192   W. Albrecht, H.J. Behrend,F.W. Brasse,W. Flanger, H. Hultschig, K.G. Steffen

Dalitz   66   PR 146,1180   R.H. Dalitz, D.G. Sutherland

Krohn   66   PR 148,1303   V.E. Krohn, G.R. Ringo

Ash   67   PL 24B,165   W.W. Ash, K. Berkelman,
                               C.A. Lichtenstein, A. Ramanauskas,R.H.Sieman

Bartel   67   PL 25B,236   W. Bartel, B. Dudelzak,
                               H. Krehbiel, J.M. McElroy, U.Meyer-Berkhout,
                               R.J. Morrison, H. Nguyen Ngoc, W. Schmidt,
                               G. Weber

Donnachie 67   NP 87,556   A. Donnachie, G. Shaw

Govorkov   67 SJNP 4,265   B.B. Govorkov, S.P. Denisov,
                               E. Minarik

Grilli   67   NC 49,326   M. Grilli, M. Nigro,E.Schiavuta

Lebedev   67   Lebedev and Tamm, Proc. Lebedev Institute 34
                 (1967)1

Adamovich 68 SJNP 7,643   M.I. Adamovich, V.G. Larionova,
                               R.A. Latypova, S.P. Kharlamov,F.R. Yagudina

Adler   68 ANP 50,189   S.L. Adler

Berger   68   PL 28B,276   Ch. Berger, E. Gersing, G. Knop,
                               B. Langenbeck, K. Rith, F. Schumacher

Coward   68 PRL 20,292   D.H. Coward, H.De Staebler,
                               R.A. Early, J. Litt, A. Minten, L.W. Mo,
                               W.K.H. Panofsky, R.E. Taylor, M. Breidenbach,
                               J.I. Friedman, H.W. Kendall, P.N. Kirk,
                               B.C. Barish, J. Mar, J. Pine

Engels   68   PR 175,1951 J. Engels, A. Müllensiefen,
                               W. Schmidt

Gounaris   68 PRL 21,244   G. Gounaris, J.J. Sakurai

Govorkov   68 SJNP 6,370   B.B. Govorkov, S.P. Denisov,
                               E.V. Minarik

Mistretta 68 PRL 20,1523   C. Mistretta, D. Imrie,J.A.Appel,
                               R. Budnitz, L. Carrol, M. Goitein, K. Hanson,
                               R. Wilson

Müllensi. 68   ZP 211,360   A. Müllensiefen

Pfeil   68   Thesis, Bonn W. Pfeil
                 Diplomarbeit, Bonn Kim Yong Taik

Adamovich 69 SJNP 8,685   M.I. Adamovich, V.G. Larionova,
                               S.P. Kharlamov, F.R. Yagudina

Adamovich 69a SJNP 9,496   M.I. Adamovich, V.G. Larionova,
                               S.P. Kharlamov, F.R. Yagudina

Auslander 69 SJNP 9,144   V.L. Auslander, G.I. Budker,
                               E.V. Pakhtusova, Ju.N. Pestov, V.A. Sidorov,
                               A.N. Skrinsky, A.G. Khabakhpashev

Bartel   69 PL 30B,285 W. Bartel, F.W. Büsser, W.R. Dix,
                R. Felst, D. Harms, H. Krehbiel,P.E.Kuhlmann,
                W. Schmidt, W. Walter, G. Weber

Berends   69 NP B4,1   F. Berends, A. Donnachie, D.Weaver

Berger   69 Univ. Bonn preprint Ch. Berger, V. Burkert,
                G. Knop, B. Langenbeck, K. Rith

Bulos   69 PRL 22,490 F. Bulos, W. Busza, R. Giese,
                R.R. Larsen, D.W.G.S. Leith, B. Richter,
                V. Perez-Mendez, A. Stetz, S.H. Williams,
                M. Beniston, J. Rettberg

Donnachie 69 NP B12,   A. Donnachie, Bajpai

Gounaris 69 PR 181,2066 G.J. Gounaris

Hitzeroth 69 NC 60A,467 W. Hitzeroth

Lohrmann 69 Lund Conf.   E. Lohrmann

McClellan 69 Cornell Univ. preprint CLNS-70 G. McClellan,
                N. Mistry, P. Mostek, H. Ogren, A. Osborne,
                J. Swartz, R. Talman, G. Diambrini-Palazzi

Orito   69 Univ. Tokyo preprint S. Orito

Rollnik   69 Methods in subnuclear physics, Vol. III, ed.
                Nicolic (Gordon and Breach, New York, 1969)

Roos   69 NP B10,563 M. Roos, J. Pišút

Ruthergl.69 Liverpool Conf.   J.G. Rutherglen

Sakurai   69 Daresbury Conf.   J.J. Sakurai

Schwela   69 ZP 221,71   D. Schwela, R. Weizel
                ZP 221,158 D. Schwela

Sidorov   69 Daresbury Conf.   V.A. Sidorov

Silverman 69 Daresbury Conf.   A. Silverman

Walker   69 PR 182,1729 R.L. Walker

Alvensleb.70 PRL 24,786 H. Alvensleben, U. Becker,
                W.K. Bertram, M. Chen, K.J. Cohen, T.M.Knasel,
                R. Marshall, D.J. Quinn, M. Rohde,G.H.Sanders,
                H. Schubel, S.C.C. Ting

Bartel   70 Kiev Conf.   W. Bartel, F.W. Büsser, W.R. Dix,
                R.Felst, D. Harms, H. Krehbiel,P.E.Kuhlmann,
                J. McElroy, G. Weber

Behrend   70 PRL 24,336 H.J. Behrend, F. Lobkowicz,
                E.H. Thorndike, A.A. Wehmann

Behrend   70aPRL 24,1246 H.J. Behrend, F. Lobkowicz,
                E.H. Thorndike, A.A. Wehmann, M.E.Nordberg Jr.

Bellettini70 NC 66A,243 G. Bellettini, C. Bemporad,
                P.L. Braccini, C. Bradaschia, L. Foà,
                K. Lübelsmeyer, D. Schmitz

Benfatto 70   NC 69A,102   G. Benfatto

Bizot   70   PL 32B,416   J. C. Bizot, J. Buon, Y. Chatelus,
                J. Jeanjean, D. Lalanne, H. Nguyen Ngoc,
                J.P. Perez-Y.Jorba, P. Petroff, F. Richard,
                F. Rumpf, D. Treille

Braccini 70   Vth Rencontre de Moriond, Meribel P.L.Braccini

Braccini 70a DESY preprint 70/33 P.L. Braccini,
                C. Bradaschia, R. Castaldi, L. Foà,
                K. Lübelsmeyer, D. Schmitz

Galster 70   Kiev Conf. S. Galster, H. Klein, J. Moritz,
                K.H. Schmidt, D. Wegener, J. Bleckmenn

Gehlen   70   NP B20,102   G. von Gehlen

Gehlen   70a Kiev Conf. G. von Gehlen

Goitein 70   PR 1D,2449 M. Goitein, R.J. Budnitz,L.Carroll
                J.R. Chen, J.R. Dunning, K. Hanson, D.C.Imrie,
                C. Mistretta, R. Wilson

Gourdin 70   Ruhestein Conf. M. Gourdin

Litt   70   PL 31B,40   J. Litt, G. Buschhorn, D.H.Coward,
                H. De Staebler, L.W. Mo, R.E. Taylor,
                B.C. Barish, S.C. Loken, J. Pine, J.I.
                Friedman, G.C. Hartmann, H.W. Kendall

Marshall 70   DESY preprint 70/32   R. Marshall

McClellan 70   Cornell Univ. preprint CLNS-123 G.McClellan,
                N. Mistry, P. Mostek, H. Ogren, A.Silverman,
                J. Swartz, R. Talman

Meyer   70   Kiev Conf. V. Meyer zu Hörste, W. Pfeil

Noelle   70   Kiev Conf. P. Noelle, W. Pfeil, D. Schwela

PDG   70a PL 33B,1   Particle Data Group

Perez-Y J.70 Orsay preprint LAL 1234   J. Perez-Y-Jorba

Pfeil   70   Springer Tracts 55 W. Pfeil, D. Schwela

Renard   70   NP B15,267 F.M. Renard

Wilson   70   Kiev Conf. R. Wilson

5. Weak Interactions

Units: $\hbar$=c= 6.582183×10^{-22} MeV sec (see Taylor 69)

Reviews: Marshak 69, Brene 68.

5.1 μ-decay

$\Gamma_\mu = (2.9942 \pm 0.0011)10^{-10}$ eV,

$g_\mu = (192 \ \pi^3 \Gamma_\mu \ m_\mu^{-5})^{\frac{1}{2}} = (1.1634 \pm .002)10^{-5}/\text{GeV}^2$

$\text{GeV}^{-2} = 1.231081×10^{-44}$ erg cm^3

Inclusion of radiative corrections:

$$g_\mu, \text{corr} = g_\mu (1 + \frac{\alpha}{4\pi}(\pi^2 - \frac{25}{4})) = (1.1659 \pm .0002)10^{-5}/\text{GeV}^2$$

5.2 Beta decay

The matrix element of neutron decay is approximately

$$T(n \to pe\bar{\nu}) = 2^{-\frac{1}{2}} \bar{u}_p \gamma_\mu [g_V - g_A \gamma_5] u_n \bar{u}_e \gamma^\mu (1 - \gamma_5) u_\nu$$

$g_V [10^{-5}/\text{GeV}^2]$	$g_A [g_V]$	$g_A [10^{-5}/\text{GeV}^2]$	Reference
$1.1494 \pm .0009$	$1.226 \pm .011$	$1.409 \pm .013$	Blin-St. 70

5.3 Pion decay

partial width $\Gamma_{\pi\mu} = (2.5293 \pm .0023)10^{-8}$ eV

$|g_\pi| = (4\pi \Gamma_{\pi\mu} m_\pi)^{1/2}(m_\pi/m_\mu)/(m_\pi^2 - m_\mu^2) = (1.0578 \pm .0005)10^{-6}/\text{GeV}$,

$f_\pi \equiv \sqrt{2} g_\pi/g_V = 130.15$ MeV $= .932$ m_{π^+} .

5.4 Goldberger-Treiman relation

$$-g_\pi(t=0) \ G_{NN\pi}(t=0) = \frac{1}{2}(m_p + m_n)g_A$$

$g_\pi(t=0) \ G_{NN\pi}(t=0) = (.922 \pm .030)g_\pi G_{NN\pi}$ for $G_{NN\pi} = \sqrt{4\pi \cdot 14.64}$

5.5 Cabibbo angles for leptonic hyperon decays

The vector and axial vector Cabibbo angles θ_V and θ_A are defined through the weak hadronic current

$$J_\mu = \cos\theta_V \ V_\mu^{\Delta S=0} + \sin\theta_V \ V_\mu^{\Delta S=1} + \cos\theta_A \ A_\mu^{\Delta S=0} + \sin\theta_A \ A_\mu^{\Delta S=1}$$

By CVC, V_μ has no D-type coupling, while $A_\mu = (1 - \alpha_{\text{weak}})F_\mu + \alpha_{\text{weak}} D_\mu$

θ_V	θ_A	α_{weak}	Reference	comment
	$.246 \pm .004$	$.624 \pm .011$	Brene 68, updated Sept. 70	1-angle fit
$.253 \pm .009$	$.238 \pm .011$	$.615 \pm .016$		2-angle fit

5.6 Pionic hyperon decays

Matrix element for the decay $Y^a \to \pi^b + baryon^c$:

$$T(Y^a_b) = \bar{u}_c(\lambda_c)(A - B\gamma_5)u_a(\lambda_a)$$

Relation to S-wave and P-wave amplitudes:

$$S = A \quad \text{(parity violating)} \qquad k^2 = \frac{(M-M')^2 - m_\pi^2}{(M+M')^2 - m_\pi^2},$$

$$P = kB \quad \text{(parity conserving)}$$

$$\Gamma = C(|S|^2 + |P|^2) = C(|A|^2 + k^2|B|^2), \qquad C = p\frac{(M+M')^2 - m_\pi^2}{8\pi M^2}$$

Y^a_b	Λ_-	Σ^-_-	Σ^+_+	$\Sigma^+_0, \gamma > 0$	$\Sigma^+_0, \gamma < 0$
$A \cdot 10^7$	$3.30 \pm .04$	$4.06 \pm .07$	$.04 \pm .09$	$3.32 \pm .30$	$2.50 \pm .39$
$B \cdot 10^7$	$22.67 \pm .71$	$-1.19 \pm .93$	$41.43 \pm .76$	-25.02 ± 4.0	-33.35 ± 3.04

Ξ^-_-	Ξ^0_0	Reference: Filthuth 69.
4.05 ± 0.07 3.32 ± 0.11		$\gamma = \dfrac{|S|^2 - |P|^2}{|S|^2 + |P|^2}$
-16.11 ± 1.41 -9.86 ± 2.21		

Our A and B are dimensionless. For dimension $10^5/\sec^{\frac{1}{2}}$, Γ contains an additional $1/m_{\pi^+}$. The conversion factor to dimensionless A and B is $2.1715 \cdot 10^{-7}$.

References

Brene   68 NP B6,255 N. Brene, M. Roos, A. Sirlin

Filthuth 69 Top. Conf. Weak Interactions CERN report 69-7, p. 131 H. Filthuth

Marshak 69 Theory of Weak Interactions in Particle Physics (Wiley-Intersc., N.Y.)R.E. Marshak, Riazuddin, C.P. Ryan

Taylor   69 RMP 41,375 B.N. Taylor, W.H. Parker, D.N. Langenberg

Blin-Sto.70 preprint   R.J. Blin-Stoyle, J.M. Freeman

6. SU(3)-Comparison

6.1 Pseudoscalar mesons-baryon octet (BBP)

SU(3)-invariance for pv-couplings f_{ABi} with $\alpha = 0.62$

$$\frac{G^2}{4\pi} = \frac{1}{m_{\pi^+}^2} (m_A + m_B)^2 f_{ABi}^2 \quad \text{(see eqs. 2.2 and 2.4)}$$

	NNπ	$\Lambda\Sigma\pi$	$\Sigma\Sigma\pi$	$\Xi\Xi\pi$	NΛK	NΣK	NNη	$\Lambda\Lambda\eta$	$\Sigma\Sigma\eta$
f^2	.0815	.042	.048	.004	.045	.002	.007	.042	.042
$G^2/4\pi$	14.7	11.4	13.8	1.4	9.3	0.5	1.3	10.7	12.2

Note: The recommended value $G_Y^2 \equiv (G_{N\Lambda K}^2 + 0.84\ G_{N\Sigma K}^2)/4\pi = 9 \pm 3$ of sect. 3.2v is smaller than the SU(3)-prediction even for larger values of α. Therefore α has been determined from $f_{\Lambda\Sigma\pi}^2/f_{NN\pi}^2$ and the resulting kaon couplings have been reduced by a factor 0.5 (Pilkuhn 70). The same factor could be applied to η couplings. Values for $G_{NN\eta}^2/4\pi$ are given in section 3.3ii. Chase 69 find $G_{NN\eta}^2/G_{NN\pi}^2 = 0.18 \pm 0.06$ from comparison of backward peaks in $(\pi^- p \to \eta n)/(\pi^- p \to \pi^\circ n)$.

6.2 Pseudoscalar mesons-baryon decuplet (PBB*)

	$\Delta(1238)$	$\Sigma(1382) \to \pi\Lambda$,	$\to \pi\Sigma$	$\Xi(1530)$
$\Gamma(m^2)$ [MeV]	113	31.8 ± 3.0	3.2 ± 1.0	7.3 ± 1.7
$G^2/4\pi$	20.7	10.0 ± 1.0	5.7 ± 1.6	8.2 ± 1.9
$G^2/4\pi$(SU3)	20.0	10.0	6.67	10.0

References: $G^2/4\pi$ in eq. 5.7, Δ in sect. 3.1iii, $\Sigma(1382)$ in sect. 3.2vii, $\Gamma(\Xi^*)$ from PDG 70. Further SU(3)-predictions

	$\Delta \to \Sigma K$	$\Sigma^* \to N\bar{K}$	$\Sigma^* \to \eta\Sigma$	$\Xi^* \to \Lambda\bar{K}$	$\Xi^* \to \Sigma\bar{K}$	$\Omega \to \Xi\bar{K}$
$G^2/4\pi$	20.0	8.87	10.0	10.0	10.0	20.0

Exper. evidence for $G^2(\Sigma^* \to N\bar{K})$ in sect. 3.2ii.
Decay rates of all baryon multiplets have been compared with SU(3) by Plane 70. See also Astier 70 for meson spectroscopy.

6.3 Vector mesons

(i) ϕ-ω mixing

The Gell-Mann-Okubo mass formula $4m_{K^*\circ}^2 = m_{\rho\circ}^2 + 3m_{\phi_8}^2$

yields $\underline{\theta = (39.7\pm0.5)^\circ}$ for the mixing angle (3.13).

(ii) VPP-couplings

	ρ°	K^{*+}	ϕ	ω
m	772 ± 6	892.1 ± 0.4	1019.5 ± 0.6	783.7 ± 0.4
$\Gamma(m^2)$	111 ± 6	50.1 ± 0.8	$4.09\pm0.29^{a)}$	12.7 ± 1.2
$G^2/4\pi$	2.13 ± 0.15	0.832 ± 0.007	$1.55\pm0.18^{b)}$	
SU(3):$^{c)}$	2.13	0.824	1.66	1.14

a) From Bizot 70. The following branching ratios have been used: K^+K^-(48.8±2.1)%, $K^\circ\bar{K}^\circ$(30.6±1.3)%, 3π(20.6±3.5)%.
b) From Gourdin 70 .
c) From eqs. (3.15) and (3.16), with $G_{PPV}^2/4\pi=0.72$, $\alpha_1=-0.140$

(iii) V-γ couplings

The ω-ϕ mixing angle is related to f_V of sect. 3.5(i):
$$m_\phi^{2+\alpha}/f_\phi = \cos\theta \, m_\rho^{2+\alpha}/f_\rho \sqrt{3} \quad \text{(a)}$$

$$m_\omega^{2+\alpha}/f_\omega = -\sin\theta \, m_\rho^{2+\alpha}/f_\rho \sqrt{3} \quad \text{(b)}$$

with $\alpha = -1$ (current mixing) or-2 (mass mixing)

α	$\cos\theta$(a)	$\cos\theta$(b)
-1	13.4±15.6	41.5±5.7
-2	42.6± 3.9	40.8±5.5

The ϕ data exclude the current mixing model. Finite width corrections (Gourding 70) raise $\cos\theta$(b) by 3%, reduce $\cos\theta$(a) for mass mixing by about 10% and render $\cos\theta$(a) imaginary for current mixing.

References

Chase   69 PL 30B,659   R.C. Chase, E. Coleman, H.W.
                Courant, E. Marquit, E.W. Petraske, H.F. Romer,
                K. Ruddick

Astier   70 Kiev Conf.   Astier

Bizot   70 Univ. Paris preprint L.A.L. 1233   J.C. Bizot,
                J. Buon, Y. Chatelus, J. Jeanjean, D.Lalanne,
                H. Nguyen Ngoc, J.P. Perez-Y-Jorba, P. Petroff,
                F. Richard, F. Rumpf, D. Treille

Gourdin 70 Springer Tracts 55   M. Gourding

PDG   70 PL 33B,1   Particle Data Group

Pilkuhn 70 Springer Tracts 55   H. Pilkuhn

Plane   70 NP B   D.E. Plane, P. Baillon, C. Brincman
                M. Ferro-Luzzi, J. Meyer, E. Pagiola, N.Schmitz
                E. Burkhardt, H. Filthuth, E. Kluge, H.Oberlack
                R. Barloutaud, P, Granet, J.P. Porte, J.Prevost